IEE TELECOMMUNICATIONS SERIES 21

Series Editors: Professor J. E. Flood
Professor C. J. Hughes
Professor J. D. Parsons

SPC DIGITAL TELEPHONE EXCHANGES

Other volumes in this series:

before

SPC DIGITAL TELEPHONE EXCHANGES

F. J. Redmill and A. R. Valdar

Peter Peregrinus Ltd. on behalf of the Institution of Electrical Engineers

Published by: Peter Peregrinus Ltd., London, United Kingdom

© 1990: Peter Peregrinus Ltd.

British Library Cataloguing in Publication Data

Redmill, Felix, 1944-
    SPC digital telephone exchanges.
    1. Telecommunication systems. Stored program controlled networks
    I. Title II. Valdar, A. R. III. Series
    621.38
ISBN 0 86341 147 9

Printed in England by Eastern Press, Reading

To C K Reddi
New Delhi, India

Contents

Acknowledgments

The idea for this book originated during a four-week course presented to the Bangladesh PT&T in March 1985 under the sponsorship of the UN's International Telecommunications Union (ITU). The ITU Project Manager at that time was Mr C. K. Reddi, of the Indian PT&T, to whom this book is dedicated.

In addition to Brian Down of British Telecom (BT), who, with us, prepared and ran the course, and who assisted in developing the initial structure of the book, we would like to mention Tony Lavender and Rick Manterfield, also of BT, for their assistance in its preparation. Thanks too are extended to Phillip Knight and Martin Pigott for their help in the preparation of Chapter 20. We are also indebted to the many colleagues who have been kind enough to provide useful comments on draft material, particularly Marian Petrie for her considerable help in editing. Above all, we are grateful to Professor Flood for his guidance and help in editing the book.

We would like to thank BT for its support during the writing of this book. Acknowledgement is also made to the editor of the African Technical Review for permission to use extracts from articles published by the authors in that Journal.

Finally, sincere acknowledgment must go to our friends and families for their enduring support and patience.

FJR & ARV
July 1989

Preface

In recent years, telecommunications has seen analogue switching and transmission give way to digital, and relay-logic control to SPC. Now, too, there is a programme of replacement of the numerous incompatible signalling systems with a single, unified, message-based system designed for inter-processor communication.

This book explains the new technology for all who are interested in telephony. It is suitable as an introductory text: Part I presents the fundamentals of telephony in simple terms, and the succeeding details are developed on this foundation. At the same time, the breadth and depth of the coverage of digital switching and stored-program control in Parts II and III make it both a textbook for serious students and a reference book for experts.

In Part IV the authors step back and view the network as a whole. The trend towards the use of information systems to control networks and support administrations is explained; common-channel signalling and the CCITT No. 7 system are described; planning in a digital environment is discussed; finally, the introduction of new services and, in particular, the integrated services digital network, is considered. This part of the book will be of interest and value to network planners everywhere.

Part I
Fundamentals of Telephony and Introduction to Digital Exchanges

Chapter 1

Introduction to telecommunications networks

Telecommunications networks exist to convey communication signals from one point to another. The principal components of a network are nodes, or switching centres, and transmission links.

The complexity of a network is a function of the volume of telecommunications traffic to be conveyed, the number of nodes and the number of links. Networks, therefore, vary from being exceedingly simple to extremely complex. The most important example of a telecommunications network is provided by telephony and, since this book is concerned with telephone switching centres, or exchanges, a discussion of telephone networks will be used as an introduction to telecommunications networks in general.

A telephone network provides the facilities for voice communication. Such communication began, a hundred years ago, with small local networks.[1] It progressed with the growth and interconnection of these to form integrated national networks. Now an international network allows a connection to be made between almost any two telephones in the world. Subscribers have come to take this for granted and, indeed, to expect a consistently high standard of service.

1.1 Network Structure

1.1.1 Subscribers' connections

The simplest conceptual telephone network consists of two basic telephones with a simple transmission link between them, i.e. the 'two tin cans connected by a wire' model. As this simple network expands to include further subscribers, each subscriber acquires a direct link to every other, as well as a separate telephone for each link.

The first refinement to this replaces the numberous telephones at a subscriber's premises with a single one. Some form of switching mechanism is then required to connect the telephone to the desired link. This system (see Fig. 1.1) is satisfactory if the subscribers remain few and close together; otherwise the

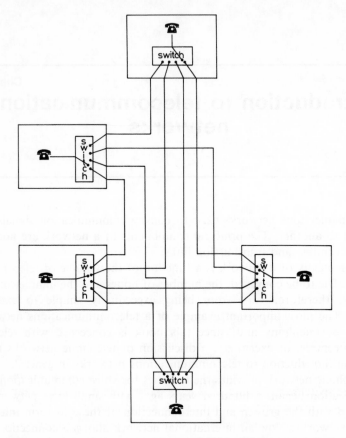

Fig. 1.1 *A fully interconnected customer network consisting of five terminals and ten links. Each terminal consists of a telephone and a switch*

number of links becomes prodigious and the cost of the fully-interconnected network prohibitive. For N subscribers, each needing to be connected to $N-1$ others, but with each connection functioning in both directions, a total of $N(N-1)/2$ connections is required. Thus, 100 subscribers would require 5000 links and 10 000 subscribers would require 50 000 000.

A network of practical proportions can be created by concentrating all the switches in a central location — a switching centre, or 'exchange' — and providing each subscriber with a single link to that exchange (see Fig 1.2). This reduces the number of transmission connections required to $N,$ and achieves a considerable saving.

Centralised switching also offers equipment flexibility. For a network of subscribers, each with his own switch, the number of outlets of each switch must be equal to or greater than the number of subscribers (excluding the

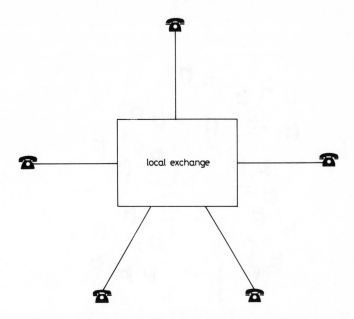

Fig. 1.2 *Customers are connected to the local exchange, where all switching is carried out*

owner of the switch). As the number of subscribers grows, the requirement for outlets exceeds the number of outlets of each switch and, therefore, each switch must be replaced by a larger one. On the other hand, expansion in a centralised switching system can be achieved by addition rather than replacement, so that investment in equipment is safeguarded.

Further economy accrues because centralised switching implies centralised control of the network and centralised maintenance. Switch repair, which would otherwise be carried out at subscribers' premises, is performed within the exchange, travelling time and abortive call-outs are avoided, and staff numbers are reduced.

Efficiency in providing and maintaining the transmission connections between subscribers and the exchange is also attained. Instead of laying each connection, or circuit, individually, a number which are all destined for the same area are pacelled together as a cable. This is provided, as an entity, between defined termination points, within the exchange at one end and a point of distribution to the local subscribers at the other (see Fig. 1.3). The circuit to a subscriber traditionally consists of a pair of wires which form an electrical circuit or 'loop' between the exchange and the subscriber. Now, other forms of connection, such as an optical fibre or microwave link, are being used between subscriber and exchange.

In order to allow flexibility of connection between the pairs in a cable and the exchange switching equipment, a main distrubtion frame (MDF) is introduced.[2] This consists of a frame, on either side of which is a rack of

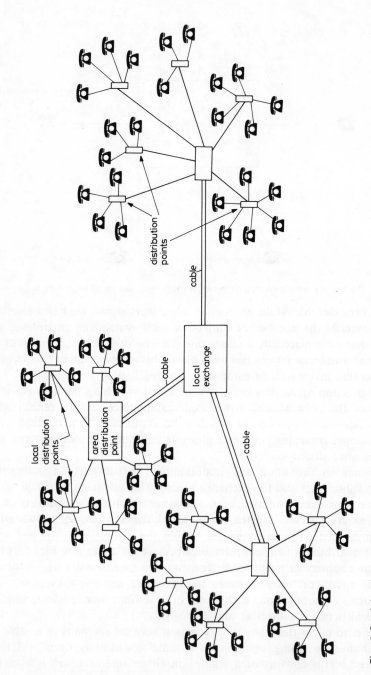

Fig. 1.3 *A local network*

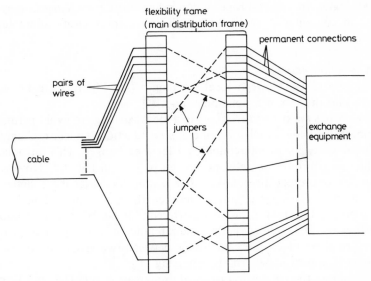

Fig. 1.4 *Main distribution frame provides flexibility of connection between transmission circuits and switching equipment*

terminations (see Fig 1.4). One set of terminations provides permanent connections for the individual pairs of wire leaving the exchange in a cable; the other terminates the connections to the subscribers' line-terminating units (SLTUs) within the exchange. Between the two racks are 'jumper wires' or 'jumpers' which can be changed easily and therefore enable any circuit to be connected to any SLTU. In addition, the MDF provides a point at which instruments may be connected to test a circuit in either direction — into or out of the exchange. It also is a point at which fuses and voltage-protection devices (e.g. against lightning) can be installed. In some cases, greater flexibility is achieved by inserting an intermediate distribution frame[3] (IDF) between the MDF and the exchange equipment.

Both logical and physical components in the exchange and the local network are identified by numbers. Each SLTU within the exchange is allocated a permanent equipment number (EN). Also, each customer's circuit, or pair of wires, within a cable is given a unique number. Thus, by virtue of the flexibility of the MDF (as shown in Fig. 1.4), any EN may be connected to any pair. On the other hand, a subscriber is allocated a directory number (DN) which is independent of either equipment of pair numbers. A means of translation between DN and EN must therefore be provided within the exchange, for a number of reasons. For example, the exchange receives address information, from a calling subscriber, in the form of the directory number of the called subscriber, but it needs to determine which equipment to activate. In electromechanical exchanges, the relationships between equipment and directory numbers are stored electromechanically in the form of wired logic. In stored-program-controlled (SPC) exchanges, they are recorded in computer storage and are therefore more easily alterable and accessible.

1.1.2 A junction network

The interconnections between local exchanges (see Fig 1.5) are known as junctions, and they form a junction network.[4-6]

Just as savings are derived from using a local exchange to avoid permanent interconnections between all subscribers, so a junction tandem exchange (T) removes the necessity to interconnect all local exchanges. This introduces a hierarchial structure to the network, as can be seen in Fig. 1.5. However, it is economical to provide direct links between some local exchanges, depending on the volume of telephone traffic between them.[7] A direct link is often provided on the basis that if the traffic offered to it exceeds a certain amount, the overflow traffic is routed via the junction tandem exchange. This provides one or more alternative routes for traffic between two given exchanges, and the process is known as 'alternative routeing'.

Fig. 1.5 shows the connection of local exchanges A,B,C,D,E and F to the junction tandem exchange T; some are also directly connected to each other. Thus, traffic between A and D is routed via T. However, traffic between A and F is routed over a direct link, with overflow via T. It is notable that, although exchange C is geographically close to D, E and B, it possesses direct routes to none of these; its direct link to A is justified by a higher level of traffic, and the geography loses its relevance.

A local exchange may be used as a tandem point when this is advantageous.

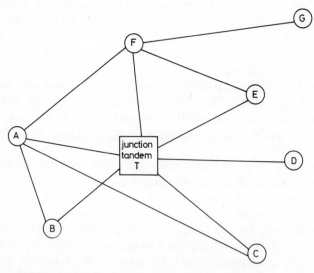

Fig. 1.5 *Local exchanges interconnected by a junction network*

For example, owing to the particular traffic volumes between exchanges A, G and F, it may be arranged for traffic between A and G to be routed via F instead of via T. It would also be possible to route calls from A to E via F, while sending those from E to A via T. Each exchange contains routeing algorithms for calls originating in it or transiting it. These algorithms are parts of a routeing plan which is continually being revised and refined in the light of regular measurement of telephone traffic on each route and the introduction of new routes. In electromechanical exchanges, the algorithms are built in the form of wiring, so changing them is laborious and error-prone. In SPC exchanges, they are kept in computer storage, which allows greater speed and flexibility in making changes.

1.1.3 A national network

It is possible to conceive of a network in which long-distance calls are set up by means only of local and junction tandem exchanges. The disadvantages, however, are prohibitive. Call set-up would take a long time; it would be prone to failure due to unavailablity of plant; it would be inefficient in its use of exchange equipment because it would occupy switching equipment, for the duration of the call, in each of its many transit exchanges. Long-distance routes, occupying fewer intermediate switching points, or exchanges, overcome these disadvantages.

A national network consists of a hierarchy of networks, interconnected so as to offer an achievable routeing strategy for calls between any two subscribers. In most countries, therefore, there is a long-distance or 'trunk' network[4,8,9] which is an entity distinct and accessible from the junction networks, as shown in Fig 1.6. 'Primary centres' form the interfaces between the trunk and junction networks. Each local exchange is connected to a primary centre, either directly or via a junction tandem exchange.

A primary centre thus forms the first tier of a trunk network, with other tiers used according to the size of the country and the routeing strategy employed. Fig 1.6 shows a trunk network consisting of two tiers, with each local exchange connected to its primary centre and each primary exchange connected to a secondary centre. The highest tier of the trunk network, in this case the second, is fully interconnected, with each of its component switching centres connected to the country's international exchange.

1.2 Tools of Network Planning

1.2.1 Numbering plan

In the early days of telephony, when only small local telephone networks existed, each exchange was independent, with its subscribers having no access to those on other exchanges. No exchange identification was required in a subscriber's directory number, which, therefore, directly reflected the capacity of

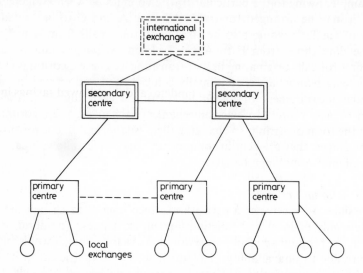

Fig. 1.6 *A national network with trunk exchanges in two tiers*

the exchange. For example, a directory number in the range of 000–999 showed that the exchange could accommodate up to a thousand terminations.

As soon as one exchange was connected to another by junctions, a subscriber had to inform the exchange if the call being made was to another local subscriber (an 'own exchange call') or to a subscriber on another exchange. In early manually-operated exchanges, the subscriber could tell the operator. In automatic exchanges, the information was conveyed by an expanded numbering scheme. The total number consisted of two fields, one defining the called subscriber's local exchange and the other being the individual subscriber's number. This carried disadvantages; a greater onus was placed on the calling subscriber to dial correctly; and the complexity of the control equipment within the exchange was increased. However, the price of gaining direct access to more telephone users was having to employ more digits to address them uniquely.

In the past (and it is still the case in some places) it was normal for one exchange to be accessed by different exchange codes, depending on where the call originated. This was because many exchanges did not contain equipment to translate the dialled number into a routing number, and the number as seen by the originating exchange had to designate an outgoing route rather than a distant exchange. The limit to the addressing power of a small number of decimal digits meant that the same number was used in different exchanges to designate different routes. This problem was accepted in rural junction networks and was overcome in the trunk network by placing the calls via operators. Eventually, it was overcome in urban areas by providing local exchanges with the necessary control equipment to receive and store the dialled digits (registers) and to translate them into routeing codes (translators). A numbering plan[10,11] which dissociates routeing from numbering is known as a linked numbering scheme. The world numbering plan is a linked numbering scheme, as are most national numbering plans.

Once the dialled number contained an 'exchange field', it became possible to address exchanges not directly connected to the calling exchange. An intermediate exchange could be used as a tandem, and this allowed savings in transmission plant (cables and terminating equipment). As discussed earlier, direct connections did not have to be provided to link exchanges between which there was less than a defined minimum volume of traffic.

The creation of a trunk network meant that the 'area' of a local exchange now had to be defined. The dialled number had to contain another code (the 'area code') and, therefore, more digits. This created the 'national number' of a subscriber, which is longer than the local directory number.

Dialling strategy can demand that all calls use the complete national number. However, with a large proportion of calls being local-area calls, it is advantageous to omit the area code from these: the subscriber dials fewer digits and the administration provides less control equipment. The control equipment must then have a method for distinguishing between local and trunk calls. This would be possible by ensuring that digit combinations used for the two types of code did not overlap. This method, however, reduces the numbering range available for the designation of areas and of local exchanges within an area. It therefore imposes a restriction on expansion of the exchange. Another method, and that chosen by most countries, is to distinguish trunk calls by one or more prefix digits ('0' is the internationally agreed standard[12]).

International calls too are identified by one or more prefix digits. '010' is used in the UK, but '00' is now the internationally agreed standard.[13] With international telephony comes the need for international standards, and the body which regulates these is the International Consultative Committee for Telegraphs and Telephones (CCITT), which is the standardisation arm of the International Telecommunications Union (ITU). Both are based in Geneva. To create a world number plan, the world has been divided into zones, each designated by a single digit (see Table 1.1), and the countries within these zones are identified by a 'country code' which begins with the zone digit. For example, the zone of Africa is designated by '2' and the country codes for Nigeria and Kenya are 234 and 254, respectively.

Accepted convention limits international numbers to a maximum of 12 dialled digits, excluding the international prefix. This assists subscribers by minimising the chance of incorrect dialling, while allowing each country an adequately-large number range. A call from the UK to Nigeria, for example, would consist of the following sequence of digits: 010 234 A–B P–Q X–Y. In this, '010' is the UK international-access code; '234' is Nigeria's country code, of which the '2' identifies the Africa zone; A–B is the code of the trunk

Table 1.1　The world numbering zones

INTERNATIONAL DIGIT	WORLD ZONE
0	not allocated
1	North America
2	Africa
3	Europe
4	Europe
5	South America
6	Australasia
7	USSR
8	Far East
9	Middle East and Asia

exchange, or primary centre, in Nigeria, to which the call must be routed; P−Q identifies the terminating local exchange; and X−Y is the called subscriber's number. The final three fields depend for their size on the country's own numbering plan, but, in total, they should not exceed 12 − N digits, where N is the number of digits in the country code. It is arranged that countries with large populations, or large numbers of likely telephone subscribers, are allocated fewer country digits to allow them the flexibility of having more digits in the national number. Usually a national number, including the trunk prefix, is permitted a total of ten digits.

It should be noted that a disadvantage of the use of prefix digits is a reduction of the available numbering range on every local exchange. For instance, if the trunk prefix digit is '1', no number commencing with '1' can be used as a subscriber's number. This has also been the case with digits used for access to service or emergency numbers. As will be seen, however, these limitations may be overcome with SPC.

National numbering plans may be (though many are not) geographically based. A country is divided into areas which are identified by a particular first digit. Each area may then be subdivided, so that the second digit addresses a given part of the area, and so on. This is the basis of the world's zone numbering plan, and it allows the control equipment in an exchange to commence routeing a call after receiving only one routeing digit, following the trunk prefix digit.

1.2.2 Transmission plan

The actual structure of any network depends on a number of factors, one of the most important being transmission standards. Any transmitted signal incurs an attenuation which, without repeaters for amplification, is proportional to the length of the transmission path. The process of switching in an exchange also attenuates signals. In order that all calls encounter acceptable, if not uniform, performance, so that speech is always intelligible to the listener, a transmission plan[14,15] for the network is required. A transmission plan lays down maximum allowable losses for all types of transmission path[16] — and minimum losses, too, since the stridency of unattenuated speech is

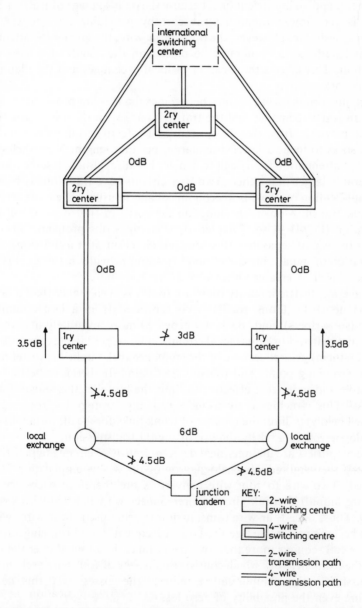

Fig. 1.7 *Example of a transmission plan (losses based on UK transmission plan 1960)*

unacceptable. Fig. 1.7 illustrates an example of a transmission plan in which the figures show nominal losses in decibels (dB). Such losses are achieved in different ways in the different networks.

In local networks, subscribers' connections consist of pairs of wires (usually copper) which must be provided per subscriber. They represent a substantial investment, and an inefficient one, because the daily average of traffic per subscriber is very low. Minimising the cost by providing low-gauge wire is, therefore, desirable. However, the thinner the wire, the greater the attenuation per unit length, so a limit must be placed on the length of a subscriber's connection. This affects the location of local exchanges and the planning of local network areas.

In the junction network, routes between exchanges are provided in accordance with traffic demand, and the traffic intensity on them is higher than in the local network. It is, therefore, more cost effective to use a heavier gauge of wire so as to incur a lower attenuation per unit length. Nevertheless, uniformity of attenuation is difficult to achieve because, although some routes are considerably larger than others, two-wire circuits have traditionally been used and amplification on these is uneconomical and unusual. The more recent use of digital transmission in the junction network, in the form of pulse-code modulation (PCM) — see Chapter 5), overcomes this problem. Inherent in PCM is the use of separate paths for each direction of transmission, and the regeneration of signals, instead of amplification, provides better-quality transmission as well as greater uniformity of performance.

The average traffic intensity on trunk routes is even higher than it is in the junction network. Trunk traffic is concentrated from a larger number of sources (subscribers), and routes are provided more accurately in accordance with actual demand (whereas local circuits must be provided regardless of the traffic on them). Moreover, use of the trunk network implies a greater number of both switching points and transmission paths. It therefore becomes both imperative and more cost effective to limit the losses at this stage of a call's progress. This is achieved by using a routeing strategy (or routeing plan, discussed below) to limit the number of trunk links for a call, by amplification on analogue routes, and by the use of digital transmission.

Since amplifiers are undirectional devices, four-wire circuits (a pair for each direction) are required on analogue routes on which amplication is to be provided. Two-wire to four-wire convertors are therefore necessary where four-wire amplified trunk circuits interconnect with the two-wire switching centres. Thus, once four-wire transmission is being used, four-wire switching centres become preferable (see Fig 1.7). A frequently-used routeing strategy is that, if a call requires more than two trunk links, it is routed over the highest tier of the trunk network which consists exclusively of four-wire exchanges and transmission paths. Amplification reduces the losses over this network, providing even the possibility of zero loss.

The attenuation problem is considerably reduced in networks in which digital transmission and switching predominate. The nature of digital transmission makes it possible to achieve a uniform transmission performance, owing to the complete regeneration of digital signals by repeaters, rather than the amplification of an existing signal along with its noise, as in analogue transmission. Indeed, in a wholly-digital network, attenuation, rather than amplification, is introduced artifically in order to make transmitted speech agreeable to the ears of subscribers. Thus, in a digital environment, all connections are good, instead of there being a wide range of performance, from very good to very bad. Moreover, digital switching is now cheaper than analogue. All modern systems are based on both digital switching and digital transmission, and the future therefore promises better network performance than has been achieved in the past. Indeed, in most countries, optical fibres are rapidly replacing all other transmission media.

Clearly, an international call, which uses a number of transmission links in at least two countries, requires amplification or regeneration. All international calls are therefore provided with four-wire transmission paths as well as four-wire switching at international exchanges. Submarine coaxial cables and microwave paths provided by communications satellites form the basis of intercontinental communications, and microwave line-of-sight links are used extensively in continental networks. Tropospheric scatter,[17,18] in which microwave signals are reflected off a perturbed layer of the atmosphere known as the troposphere, is used occasionally for communication beyond the horizon — for example, between a country's mainland and off-shore islands or oil rigs. All new international transmission paths, via both satellite and submarine cable, are digital; new-technology plant, such as optical-fibre cable, is bringing improved quality to international communications.

1.2.3 Routeing plan

The third plan (numbering and transmission plans have already been discussed) which is essential to an administration, if its network is to operate efficiently, is a routeing plan.[19,20] This defines the criteria for the routeing of calls through the network under all circumstances. It has already been shown that in a junction network a call may be routed between two exchanges either over a direct link or via one or more transit points. The direct link is provided according to certain criteria — such as if there is more than a given amount of telephone traffic between the two exchanges — and the rules that embody these criteria are a part of the routeing plan.

Similarly, for the trunk network, the routeing plan consists of the rules which determine the need for trunk exchanges, how they should be interconnected, and whether they should be structured hierarchically (as the two-tier trunk network in Fig 1.6) or all on a par with each other. In an analogue network, the routeing plan is affected by the transmission plan which defines the maximum number of unamplified links allowable on a call, and thus the maximum number of junction links, since all trunk links are amplified, and also the maximum number of four-wire amplified links, when two-wire switching is employed. This is because each link must have a finite loss (typically 3dB) to guarantee stability. In a digital network, there are other planning considerations, which will be discussed in Chapter 21, after the principles of digital switching and control have been explained. Signalling, discussed in the next chapter, also influences the routeing plan.

There are other aspects of the routeing plan. For instance, whether the circuits on any route are 'undirectional' or 'both-way'; that is, whether they may be seized, to carry a call, at one end only or at either end. The routeing plan must contain rules for deciding, on economic and technical grounds, if both-way circuits are justified on any route.

Another consideration is if alternative routeing[21-23] should be used. Alternative routeing is the process of offering a second-choice route to calls which have encountered congestion on their first-choice route. For example, in Fig 1.8 there is a direct route between exchanges A and B, and traffic between the two exchanges would normally be offered to that route. If, however, there are no free circuits on the direct route, any new calls will be lost unless there is an alternative. In this Figure, such a second choice is shown via exchange C. The alternative-routeing facility not only provides an alternative route in the overall service, but is often designed to ensure efficient use of both first-choice and alternative routes. The first choice may be designated a 'high-usage' route,[23] in which case it is dimensioned with too few circuits for the total expected traffic. The overflow traffic then shares the alternative, 'fully

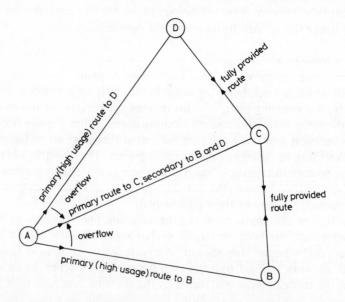

Fig. 1.8 *Automatic alternative routeing*

provided', route with other traffic. Both routes are therefore always efficiently used. Routes AB and AD are high-usage, and route AC is fully provided to carry overflow from AB and AD as well as direct traffic from A to C.

With electro-mechanical control equipment, routeing instructions are built-in with complex wiring. It is therefore difficult and time-consuming to change them. Modern SPC exchanges are more flexible; routeing instructions exist as software in computer storage and are easily and quickly changed. Therefore, even dynamic alternative routeings can be offered to allow immediate re-routeing of traffic on a temporary basis, in times of serious congestion or failure of network components. Dynamic rerouteing of traffic comes within the scope of network management,[24-26] the aim of which is to optimise the use of the network under all conditions.

1.2.4 Telephone traffic

The number of calls which a circuit, or group of circuits, can carry in a given time depends on the holding times and arrival pattern of the calls. For example, if the holding time of all calls is 3 mins, and calls arrive at 3 min intervals, so that one's arrival is immediately after the clear-down of its predecessor, a single circuit can theoretically carry 20 calls per hour. The occupancy of that circuit during that hour will be almost exaclty 60 mins, or 100% of the time. If a 21st call arrives during that hour, it will encounter congestion, and fail.

If, on the other hand, the holding time of each call were 2 min, the circuit would be able to carry a theoretical maximum of 30 calls. In practice, however, calls have different holding times, and their arrival rate is not regular. If, indeed, 20 calls arrived during a period of 1 hr, the chances are that they would overlap, so that, even if their average holding time were 3 min or less, a number of them would be lost. Then, in spite of these losses, the circuit's occupancy would be less than 100%.

While it may be held that occupancy is related to the number of calls carried, it is not simply related to the number of calls offered. Occupancy is a measurable entity and is referred to as the 'traffic'[27-30] carried. The total duration of the calls, divided by the duration of the monitoring period (with consistent units), is the traffic intensity, (the unit of which is the 'erlang' (E)). i.e.

$$\text{Occupancy} = \Sigma \ \frac{\text{Call durations}}{\text{time}} = \text{traffic in E, for 1 circuit}$$

In the above examples, where a circuit was engaged for 60 min during a period of 1hr, the traffic intensity was 1 erlang.

Similarly, the traffic intensity can be calculated for a group of circuits. Fig. 1.9 shows a group of five circuits each carrying a number of calls (hatched areas) over a period of 2hrs. Offered calls which are lost due to congestion (as might be the case at time = 20 mins) are not considered.

For this group:

Total call duration   = 349 min

Traffic intensity (A)   = $\dfrac{349}{2 \times 60}$ = 2.9 erlangs (E)

Traffic intensity per circuit = $\dfrac{2.9}{5}$ = 0.58E

Traffic intensity can also be calculated by instantaneous measurements, in which case it is equal to the average number of calls in progress, i.e.

$$A = \dfrac{Ch}{T}$$

where C = number of calls in progress at a given time

h = average call holding time

T = total time.

In order to determine if exchanges and routes are accurately 'dimensioned' (i.e. provided with the optimum quantities of equipment and circuits, respectively), and to forecast future traffic intensities so as to revise network plans, it is essential to measure the traffic at various points in the network. While it is desirable to obtain perfectly-accurate results, the permanent attachment of a traffic meter to every circuit terminated on an exchange has not been economically feasible, and sampling methods have been used. (In SPC exchanges, recording is carried out in software, and it is possible to perform total monitoring of an exchange. However, post processing of the data can be a massive and expensive task.)

A traditional means of sampling has been to test circuits for occupancy at regular intervals. The sum of the occasions on which occupancy is detected is divided by the number of tests to obtain the average occupancy. For example, if tests of the group of circuits in Fig. 1.9 were performed every 10 mins, as shown by the vertical dotted lines, the occurrences of busy circuits would total 36 over the 2hr period. Since there were 12 samples, the average traffic carried by the group is calculated to be $36 \div 12 = 3.0$E. This is very close to the actual average of $349 \div 120 = 2.9$E, obtained by dividing the total of the actual busy times by the duration of the monitoring period (both in minutes).

Typically, traffic varies according to the time of day, the day of the week, the season and the place. While individual subscribers may be assumed to make calls randomly, each exchange and each route experiences peak periods of use every day. For exchanges in predominantly business areas, these usually occur in the mornings. In residential districts these may occur in the evenings, particularly when cheap rates are made available,exhibiting how tariffs can be used to alter or regulate traffic profiles. In business areas, traffic is negligible

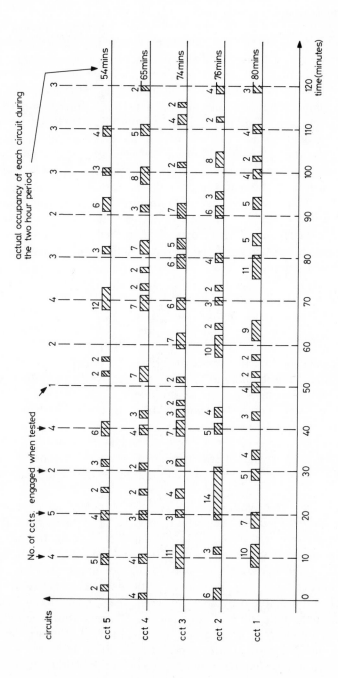

Fig. 1.9 *Diagrammatic representation of the occupancy by calls of a group of five circuits, which are monitored every 10 min*

on Sundays, often rises to a peak at midweek and then decreases again. On the other hand, domestic traffic frequently shows a peak at weekends when families are at home and charge rates are low. Seaside resorts show increased traffic in summer when they are full, skiing resorts in winter, etc.

In the same way, traffic from individual subscribers is peaky. Over a period of a day, several calls, representing spikes on the subscriber's traffic profile, may amount only to an intensity of about 0.03 erlang. However, because the traffic from many subscribers converges on an exchange, there is a much more predictable average level of traffic there at any given time. As the traffic progresses through the stages of equipment in the exchange, it becomes more and more 'smoothed', i.e. it exhibits less-exaggerated peaks. On junction routes it is relatively smooth, and on trunk routes more so. These routes are therefore dimensioned to carry the predicted rather than the maximum possible traffic. Likewise, an exchange is dimensioned with equipment to carry the forecast traffic rather than on the assumption that all subscribers will simultaneously initiate calls. This recognises that there will always be some probability that a call being offered to the exchange will be lost due to the unavailability of equipment.

To dimension for the highest possible traffic load would incur the administration in huge expenditure on seldom-used equipment, so an acceptable level of loss is chosen and dimensioning is carried out to obtain a given probability of this loss during the 'busy hour',[31] i.e. the busiest hour of the day. For example, if it is considered that the loss of one call in a hundred is acceptable during the busiest period, the 'grade of service'[32, 33] (GOS) is set at 1 in 100. This means that the probability of loss is 0.01. Clearly, the probability of loss of a call at less busy times is more favourable to the subscribers.

The mathematical study of telephone traffic,[27] or teletraffic theory, is used to ensure that dimensioning is such that the probability of loss of a call is maintained at a level acceptable to subscribers and, at the same time, economic to the administration. It should be remembered, however, that local lines must be provided on a per-subscriber basis, and that these are the ultimate sources of all traffic.

The correct amount of equipment, or circuits, to be provided is calculated from tables derived by using teletraffic theory. As in all mathematical applications, various assumptions need to be made concerning the conditions within the exchange, the randomness of traffic, and the use of statistical distributions to approximate to the behaviour of traffic quantities, such as arrivals.

To allow adequate planning, therefore, traffic measurements have traditionally been taken during the busy hour. In recent years, owing to increased use of the telephone, both domestically and in business, busy periods have extended over a number of hours and a consistent period of measurement has not always coincided exactly with the peak traffic. Sometimes the result is inadequate dimensioning, but more complete recording in digital SPC exchanges can alleviate this problem.

1.3 Network planning

The demands on the various networks are not static. New businesses introduce new traffic. New customers require terminations on local exchanges. New ideas, such as radio phone-ins, can generate large traffic peaks which may require special arrangements if they are not to cause congestion in the network. New services which use the telephone network, such as electronic mail, facsimile and data transmission, may have different traffic characteristics from traditional telephone traffic.

In order that the various networks can continue to meet the demands of changing traffic patterns, they must frequently be adjusted. They must be reviewed in the light of such questions as the following: (i) When is it necessary to provide a direct route between two exchanges or to increase the number of circuits on an existing route? (ii) When is it necessary to install a new exchange? (iii) Where should it be located?

The taking of such decisions constitutes the discipline of network planning.[34] This requires evidence, rules of operation and a framework within which to function. The evidence comes from regular traffic measurement at all exchanges. The rules of operation comprise the guidelines laid down by teletraffic theory, the economics of the various options, and the capabilities of exchange and transmission equipment. The rules are designed to optimise the cost and traffic-handling capacity of the network. The framework must define the boundaries within which the rules apply, and it is provided by the plans already discussed (numbering, transmission and routeing) as well as signalling constraints (discussed in the following chapter) and a charging plan.[35] The latter defines how the costs of providing, maintaining and administering the network should be recovered from subscribers. It addresses such matters as how tariffs should be set. For example, it states what the proportion of fixed charges (rentals etc.) to call-related charges should be, and whether calls should be charged for on a per-call basis or on a time basis.

These plans clearly affect each other and, in their interrelation, they combine technical criteria (e.g. transmission limits) with political and economic considerations (e.g. charging). In the end, however, all decisions must be cost effective. Thus, not only is it necessary to know that a new local exchange is required in a certain area, but it is also essential to determine exactly where the new exchange should be located. The sum of the costs of switching equipment, transmission plant, and accommodation should, ideally, be a minimum. The balance of the equation depends on current costs, which, in turn, depend on the technologies in use. Thus, new technologies are introduced, not for their own sakes, but when they become cost effective.

It is not within the scope of this book to describe network planning in detail. Such a discourse could fill a book of its own. However, recent leaps in technology have resulted in the commencement of a new era in telecommunications. The new digital methods described in this book — switching and, to some extent, transmission in Part II, control in Part III, and signalling in Part IV — have overturned the traditional rules and equations of network planning.[36] Moreover, network planners in many administrations have the unprecedented job not merely of expanding and improving the old networks but, indeed, of designing and introducing wholly new ones based on digital technology.[37] These networks will cater for other services as well as telephony[38] and, since many services of the future are not yet imagined, flexibility is vital in modern network planning. In order, therefore, to show how the digitalisation of the individual telephony functions affects the network as a whole, Chapter 21 offers an insight into network planning in the wholly-digital era.

1.4 References

1 ROBERTSON, J.H. (1947): *The Story of The Telephone* (Pitman).
2 ATKINSON, J. (1950): *Telephony,* Vol. 1 (Pitman), Chap. XIV
3 ATKINSON, J.: *Telephony,* Vol. 1. Ibid.
4 POLLARD, J.R. (1975): 'Introduction' in *Telecommunications Networks,* Ed. FLOOD, J. E., (Peter Peregrinus Ltd), Chap. 1
5 KYME R.C. (1976): *The Junction Network, Post Office Electrical Engineers' Journal* **69,** p.175
6 ELLIOTT, J.L.C. (1972): *The Design and Planning of the Junction Network,* Ibid., **65,** p.12
7 ATKINSON, J.: *Telephony,* Vol. 1, Chap. XII, Op. cit.
8 WARD, K.E. (1981): *The Evolution of the Inland Telecommunications Network,* Post Office Electrical Engineers' Journal, **74,** p.162
9 BOAG, J.F. and SEWTER, J.B. (1971): *The Design and Planning of the Main Transmission Network,* Ibid. **64,** p.16
10 FLOOD, J.E.: 'Numbering, routing and call charging' in *Telecommunications Networks,* Ed. FLOOD, J.E., Chap 6, Op. cit.
11 CCITT (1981): *Economic and Technical Aspects of the Choice of Telephone Switching Systems,* GAS 6, Geneva, Chap IV
12 CCITT (1985): *Red Book Fascicle II.2,* Recommendation E 163
13 CCITT: Ibid.
14 FRY, R.A.: 'Transmission standards and planning' in *Telecommunications Networks,* Ed. FLOOD, J.E., Chap. 7, Op. cit.
15 CCITT, GAS 6, Chapter IV, Op. cit.
16 TOBIN, W.J.E. and STRATTON, J. (1960): *A New Switching and Transmission Plan for the Inland Trunk Network,* Post Office Electrical Engineers' Journal, **53,** p.75
17 HILL, S.J. (1982): *British Telecom Transhorizon Radio Services to Offshore Oil/Gas Production Platforms, Part 1 — Service Requirements and Propagation Considerations,* British Telecommunications Engineering, **1,** p.42
18 HILL, S.J. (1982): *British Telecom Transhorizon Radio Services to Offshore Oil/Gas Production Platforms, Part 2 — Radio Techniques and Networking Arrangements,* Ibid. **1,** p.70
19 FLOOD, J.E.: 'Numbering, routing and call charging', in *Telecommunications Networks,* Ed. FLOOD, J.E., Chap. 6, Op. cit.

20 CCITT, GAS 6: Chap. IV, Op. cit.

21 CCITT: Red Book, Vol. II, Fascicle II.3 (1985), Recommendation E600

22 CLOS C. (1954): *Automatic Alternative Routing of Telephone Traffic*, Bell Lab. Record, **32**, p.51.

23 WILKINSON R.I. (1972): *Theories for Toll Traffic Engineering in the USA*, Bell Systems Technical Journal, **35**, p.421

24 GIMPELSON, L.A. (1974): *Network Management: Design and Control of Communications Networks*, Electr. Commun., **49**, p.4

25 JENKINS, C.H. and TAR Z.J. (1984): *The Need for Network Management*, Telecommunication Journal, **51**, II, p.78

26 STRATHAM P.E.: *The History of the Study of Network Management in the CCITT*, p.82, Op. cit.

27 BEAR D. (1976): *Principles of Telecommunication-Traffic Engineering*, (Peter Peregrinus Ltd)

28 POVEY, J.A.: 'Teletraffic engineering', in *Telecommunications Networks*, Ed. FLOOD J.E., Chap. 8, Op. cit.

29 REDMILL F.J. (1986): *Teletraffic Engineering*, Telecommunication Journal, **53**, II, p.83

30 ATKINSON, J. (1950): *Telephony*, Vol. 2 (Pitman), Chap. 2

31 POVEY, J.A. (1967): *A Study of Traffic Variation and a Comparison of Post-selected and Time-consistent Measurement of Traffic*, 5th Teletraffic Congress

32 CCITT (1985): Red Book, Vol.II, Fascicle II.3, Section 5

33 Le GALL, P (1964): *Variations in Traffic and the Definition of the Grade of Service*, 4th Teletraffic Congress

34 BACK, R.E.G.: 'Network planning' in *Telecommunications Networks*, Ed. FLOOD, J.E., Chap. 14, Op. cit.

35 CCITT. GAS 6, Chap. IV, Op. cit.

36 CROOKS, K.R. (1985): *Local Network Strategy — Today's Plans for Tomorrow's Network*, British Telecommunications Engineering, **3**, p.297

37 GARBUTT, B.N.: *Digital Restructuring of the British Telecom Network*, Ibid., p.300

38 BROWN, D.W. and LIDBETTER, E.J: *The Future Network*, Ibid. p.318

Chapter 2

Introduction to switching systems

2.1 Analysis of a call

For the purpose of introducing switching systems, the progress of a telephone call is arbitrarily considered to consist of ten stages. These are described below and summarised in two diagrams, to which the descriptions refer. Fig. 2.1 illustrates the progress of a local call as seen by the calling subscriber and the local exchange. Fig. 2.2 shows the stages of a call involving a second exchange, from the first exchange's point of view.

2.1.1 Off-hook signal

A subscriber wishes to make a call and lifts a telephone handset. This necessarily produces the 'off-hook' or 'seize' signal, which informs the exchange that it must prepare to handle a call. The lifting of the handset releases a switch which completes an electric circuit (or 'loop') between the exchange and the telephone.[1] When this circuit is made, a device in the exchange is activated, and a train of signals to the appropriate parts of the exchange is initiated. Depressing the switch, to its position when the handset is at rest, provides a simple means of signalling to the exchange that a call has ceased, and allows the loop to be disconnected when a call is not in progress, thus saving power. Power to the subscriber's line is provided by batteries in the exchange, because a direct-current (DC) voltage is required. The batteries are charged by the mains supply, via rectifiers, and are capable of maintaining the exchange's electrical supply for a specified time in the event of a mains failure.

2.1.2 Identification of calling subscriber

The call is detected at the calling subscriber's line-terminating unit (SLTU) in the exchange, which has a designated equipment number[2] (EN). This needs to be translated into the subscriber's directory number (DN). Translation tables are therefore required. In an electromechanical exchange, these are held in wired logic. In SPC, they are held in computer storage, and the details of how they are used are presented in Chapter 15.

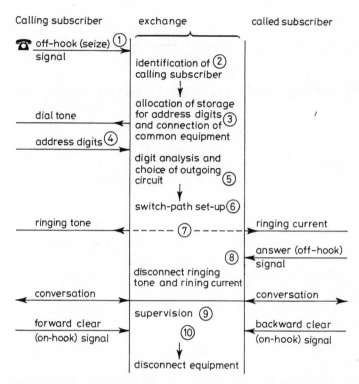

Fig. 2.1 *The life cycle of a local call*

The control system also needs to identify the calling subscriber for two other reasons. First, the subscriber needs to be charged for the call. Second, a check must be made to discover if the subscriber is permitted to make an outgoing call. The necessary information is stored in the subscriber's class-of-service record. There is a class-of-service record for each termination on the exchange for the purpose of storing various types of information about the termination. The nature and the use of these are described in Chapter 14.

2.1.3 Allocation of storage and connection of common equipment

A principal function within an exchange is control. Some logic must interpret events during the course of a call, make decisions on what actions are necessary, and initiate these actions. When an 'off-hook' signal is received by the exchange, the control system must allocate 'common equipment' to the call and provide a path to it from the calling line. This takes the form of 'long-holding-time' equipment, which is needed throughout the call, and 'short-holding-time' equipment, which is required only during the call set-up stage. An example of the former, in analogue exchanges, is the transmission bridge,[3] which separates the speech path, on which AC is used, from the DC

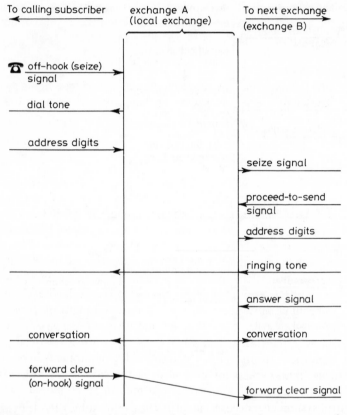

Fig. 2.2 *Progress of a call which is routed via a second exchange, as seen from the first (local) exchange*

path through the exchange. An example in SPC exchanges is the call record, which is an area of memory, held for the duration of a call, in which all details of the call are stored (see Chapter 14). Short-holding-time equipment includes a receiver and storage for the digits which constitute the 'number address' of the called subscriber. These digits not only identify the called subscriber, but also provide the information necessary for the routeing of the call through the network.

In an electromechanical exchange, digits are stored in an item of equipment known as a 'register'[4] and, in an SPC exchange, in an area of memory. When storage has been allocated, a signal (dial tone) is sent to the calling subscriber to indicate that the exchange is ready to receive the address digits. Because the exchange is dimensioned with storage devices on the basis of forecast call-arrival rates, rather than on the maximum possible number of simultaneous call arrivals, there is a finite (though small) probability of unavailability.

Traditionally, the subscriber has been aware of this because of the temporary lack of dial tone. In SPC exchanges, this probability can be reduced by increasing the amount of storage available, although this is only of advantage if the processing power is capable of handling the increased call-arrival rate.

2.1.4 Address digits
After receiving the dial tone, the subscriber inputs the address digits by dialling or keying. The digits are sent as signals to the exchange and are stored there. Signalling is an important aspect of telephony and its principles are presented in Sections 2.2 and 2.3 of this chapter.

2.1.5 Digit analysis
The control system must analyse the address digits[5] to determine the routeing of the call out of the exchange. If the call is to an ordinary subscriber terminated on the exchange, there is only one circuit on which it can be routed: the called subscriber's line. If this is engaged, the call cannot proceed, and the 'busy' tone is sent to the caller. If, on the other hand, the call is to a subscriber on a distant exchange, it may be allocated to any one of the circuits on the appropriate route out of the originating exchange, with, perhaps, the further option of an alternative route. If all circuits are engaged, the 'busy' tone signal is sent to the calling subscriber and the call is abandoned. If an appropriate circuit is found to be free, it is seized, so that it cannot be used for any other call. In electromechancial exchanges, seizing takes the form of applying an electrical condition to the circuit's terminating equipment, and is often referred to as 'marking'.[6] This may also be the case in an SPC exchange. However, it is more usual for circuit information to be stored in tables in software, in which case a designated code in a given data field indicates the status of a circuit.

2.1.6 Switch path set-up
At this point, the control system knows the identities of both the input and the output circuits. Its next task is to choose a path between them[7] through the exchange's switches. Within the control systems there are algorithms for selecting appropriate switch paths. Each switch point on a chosen path must be checked to ensure that it is not already in use, and seized if it is not. Again, this is carried out in electromechanical exchanges by testing for and applying electrical conditions, and in SPC exchanges by interrogating and inserting entries in sorted tables. In electromechanical exchanges, the register used for the receipt and storage of digits must be disconnected when the path has been set up.

2.1.7 Ringing current and ringing tone
A signal must be dispatched to the 'distant end' to enable the call to progress. If the called subscriber is local, this entails sending ringing current, which

activates the telephone bell. Otherwise, a seize signal must be sent to the next exchange, as shown in Fig. 2.2, to activate it to initiate its own actions. These actions are similar to those described in Sections 2.1.2 to 2.1.6 above, and involve signals being sent back to the originating exchange. When all connections have been made to allow the call to be set up, either locally or via the junction or trunk network, ringing current is sent to the called subscriber and ringing tone to the calling subscriber.

2.1.8 Answer signal

An answer signal, received from the called subscriber (in which case it is the 'off-hook' signal), or another exchange, is recognised by the local exchange's control system. Transmission must now be permitted[8] on the selected switch path through the exchange. The ringing current and ringing tone must be removed from the called and calling subscriber's lines, respectively. The two parties are thus connected, and charging of the call to the calling subscriber is initiated.

2.1.9 Supervision

While the call is in progress, monitoring, referred to as 'supervision',[9] takes place, so that charges can be determined and the 'clear' signal detected. (Supervision also refers to scanning all line terminations on an exchange to detect the seize signal of a new call.)

2.1.10 Clear signal

When the clear signal ('forward clear' generated by the calling subscriber, and 'backward clear' by the called) is received, the exchange equipment, or storage, used in the connection of the call must be freed and made available for use in setting up other calls.

For the network to be managed and maintained effectively, the administration requires data to be collected on each call. When calls fail because of faulty equipment or because of the unavailability or circuits or exchange equipment, data are required for maintenance, network management and planning. Performance monitoring[10] is necessary for management statistics, as well as for maintenance and network management. Data on successful calls are required for billing and accounting. Thus there is a significant administrative function in telephony. In electromechanical exchanges, this is effected via wired connections between individual items of equipment and monitoring points, from which the data are collected. In SPC, because control is implemented by computers, administrative dataT are collected and stored in software. Processing may then be carried out by the control processors or by postprocessing computers external to the exchange.

2.2 Channel-associated signalling

The description of the life cycle of a call, in the preceding Section, showed that the set-up, control and clear-down of a call depends on signalling,[11] both between the subscriber and the local exchange and between exchanges. Until recently, telephony signalling was performed on a call-by-call basis and the signals were transmitted on the same path as the call with which they were associated. Signalling was, thus, channel-associated;[12] and signalling systems were dependent on the switching and transmission systems with which they functioned. Consequently, in order to achieve efficiency and economy, new signalling systems were designed to match every change in switching and transmission technology. With the advent of SPC, it was clear that fast, cheap signalling could be carried out directly between processors, because interprocessor communication was already widely used in commercial computing. This Section offers a description of channel-associated signalling, while Section 2.3 introduces the principles of interprocessor 'common-channel' signalling.

There are many criteria of signalling. One is that signalling must take place in both the 'forward' and 'backward' directions. Forward signals are signals sent by the calling subscriber or by an exchange which is seen, from the point of view of the progress of the call, to be the initiating party. For example, in Fig. 2.1, signals from the calling subscriber to the exchange are forward signals. Signals from the exchange to the calling subscriber are therefore 'backward' signals. Similarly, in Fig. 2.2, signals from exchange A to exchange B are forward signals, while those from exchange B to exchange A are backward signals.

Another criterion of signalling is that there are two main categories of signalling information. These are 'line' signalling and 'selection' signalling. The CCITT[13] defines line signalling as 'a signalling method in which signals are transmitted between equipments which terminate and continuously monitor part or all of the traffic circuit'. Examples of line signals are circuit 'seize' and 'clear'. Selection signalling conveys the information relating to the routeing of the call. This includes the called subscriber's number (i.e. the 'address') as well as call classification (e.g. class-of-service) information. A specification for a signalling system must, therefore, define a range of line signals, as well as selection signals, and how they should be produced, received and identified. Its implementation must consist of the means of producing, receiving and identifying these signals.

Various factors influence how signalling is carried out, but among the most important is the need for a signalling system to be compatible with both the switching and the transmission equipment. Further, signalling equipment has traditionally been expensive, so designs which are economical quickly gain popularity. For these reasons, new signalling systems have been introduced with each new design of exchange. Consequently, numerous systems exist.

However, these apply mainly to interexchange communication, whereas there has been remarkable consistency in subscriber signalling.

2.2.1 Subscriber signalling

Since the facility for signalling must be provided in every telephone, an economy in subscribers' signalling equipment can result in enormous savings. Simplicity has therefore been important. As described earlier, the traditional way of providing the subscriber's seize signal has been to make use of the DC circuit (loop) which is created when the handset is lifted. Address digits are coded as numbers of pulses, consisting of breaks in the loop, and the method is known as 'loop-disconnect' signalling.[14] The process is a simple mechanical one: the telephone dial breaks the DC loop the appropriate number of times on its way back to the rest position, creating a train of pules whose number corresponds to the digit being transmitted. Recognition of the pulses within the exchange is achieved by timing, as loop breaks are made by the dial at a nominal rate of ten per second. Recognition of digits is by counting the pulses in a digit-pulse train. The break-make ratio varies according to the equipment in use, but is typically about 60 ms/40 ms. The duration of the interdigital pause depends on the subscriber's speed of dialling, but is always recognisably longer than either the break or make times. Individual digits are, therefore, identified by the pause between digit-pulse trains. The clear signal takes the form of a permanent break in the loop.

Although loop-disconnect signalling is still widespread, the advent of common-control exchanges (see Section 2.5.2 of this chapter) allowed the introduction of multifrequency (MF) signalling in the local network. In this, the line signals, such as seize and clear, are still based on the DC loop, but each selection digit is coded, not as a sequence of pulses, but as a combination of two AC frequencies (or tones). In telephones using MF signalling, the mechanical dial is replaced with push buttons, with each button representing a digit. When a button is depressed, the two frequencies describing that digit are generated and sent, simultaneously, to line. To avoid the inadvertent imitation of tones by speech, an adequate separation of the two tones is required. In the CCITT-recommended system,[15] whose frequencies are shown in Fig. 2.3, the two frequencies consist of one from each of two separated blocks. Interpretation in the exchange is achieved by passing signals through a system of filters to detect which frequencies are present. This method of signalling is much faster than loop-disconnect methods.

2.2.2 Interexchange signalling

DC signalling systems[16] have been and still are used between exchanges in junction networks. However, the fact that they cannot be amplified makes them unsuitable for long distances. (DC regenerators have been used, but these are expensive and cumbersome.) Being slow, DC systems introduce unacceptable delays if used for setting up calls which require a number of

	1209 Hz	1336 Hz	1477 Hz	1633 Hz
697 Hz	1	2	3	13 (spare)
770 Hz	4	5	6	14 (spare)
852 Hz	7	8	9	15 (spare)
941 Hz (spare)	11	0 (spare)	12 (spare)	16

Fig. 2.3 *The matrix of CCITT-recommended subscriber access signalling*

interexchange links. They require a metallic path, and so can only function on wire circuits. They cannot be used, for example, with multiplexed line systems, or optical-fibre or radio-transmission systems. In general, therefore, there is still a large residue of DC signalling over audio circuits in junction networks, but AC systems predominate in trunk networks.

In AC systems, selection signalling (historically referred to as 'interregister signalling', because digits are stored in 'registers' at electromechanical exchanges) is achieved in the form of MF signals. Line signals are coded in various ways and use various protocols. Thus, the selection-signalling and line-signalling functions are separated, with the result that modes of each which are most appropriate to an application may be combined in the design of a signalling system. AC systems using frequencies in the range up to 4 kHz are referred to as voice frequency (VF) systems, because this is the range of an audio telephony channel. However, the 'voice band' occupies the frequency range 300 Hz to 3400 Hz, and signals within this range are said to be 'in-band'.[17] The majority of line-signalling systems fall within this category, but there are some which make use of the range 3400 Hz to 4 kHz. For example, the CCITT R2 system[18] uses the frequency 3825 Hz. Such systems are referred to as 'out-band',[19] because they function outside the defined band of speech frequencies.

In out-band systems, filters transparent to the 300–3400 Hz speech band are used to detect and remove the line signals, so signalling and speech may occur simultaneously without annoyance to subscribers. In the line signalling of the CCITT R2 system, mentioned above, the 3825 Hz tone is maintained on all idle circuits. A seize signal thus takes the form of the tone being switched off in the forward direction, and the answer signal is the removal of the tone in the backward direction. Address digits are sent as MF signals.

When a call needs to be routed through a number of exchanges in tandem, each exchange must send forward line signals to its successor in the chain, and backward line signals to its predecessor, in order to set up and monitor a circuit for the call. The selection signals, however, may be sent in one of two ways. In 'end-to-end' signalling, just enough selection digits are passed from one exchange to its successor to allow the latter to determine how the call should be routed. Then, when the complete circuit has been set up, the digits of the called subscriber's number are sent directly and very rapidly from the originating to the terminating exchange. In 'link-by-link' signalling,[20] all the address digits are sent by each exchange to its successor as the circuit is being set up, incurring a great deal more time and holding common equipment in the various exchanges for longer.

In order that signals are correctly recognised when received, they must be dispatched according to certain protocols. This is particularly important for VF signals, because they might otherwise be imitated by speech. Therefore, standards for signal duration and power are laid down for the various systems. Indeed, in some systems, signals are transmitted not merely for a defined duration, but until an acknowledgment of a defined type has been received. This is known as 'compelled signalling' and has the advantage of minimising the occurrences of timing discrepancies between sending and receiving equipment. The CCITT R2 line signalling, discussed above, is compelled.

However, compelled signalling takes longer than noncompelled signalling, and there are instances when time is significant. For example, on satellite links there is a delay of a quarter of a second because of the time taken by the signals to travel the 70 000 km to and from the satellite. If the digits of an international number were transmitted using compelled signalling, call set-up times would be unacceptably long. Systems have traditionally been tailored to meet their applications. For example, the international CCITT No.5 system uses compelled line signalling and noncompelled selection signalling.[21]

The CCITT signalling system No.5 line signals are each composed of one or both of two frequencies, 2400 Hz and 2600 Hz. The principal signals are shown in Fig. 2.4. Because the interpretation of a signal depends on its direction (forward or backward) and its place in the sequence of events different signals may be allocated the same frequency or combination of

SIGNAL	DIRECTION	FREQUENCIES
seize	forward	2400 Hz
proceed to send	backward	2600 Hz
(digits are sent at this point)	forward	
answer	backward	2400 Hz
answer acknowledgment	forward	2400 Hz
clear forward	forward	2400 Hz + 2600 Hz
release guard	backward	2400 Hz + 2600 Hz

Fig. 2.4 *CCITT signalling system No. 5 principal line signals*

frequencies. Again, considering the table, the extra 'acknowledgment' signals are necessary because the signals are compelled. The 'seize' signal is maintained in the forward direction until a 'proceed-to-send' signal is received to indicate that the 'seize' has been received and recognised, a register has been reserved for the digits, and the digits may now be sent. Likewise, the 'proceed-to-send' signal is maintained until the first digit has been received. The 'release guard' is a backward signal to indicate that the 'clear forward' signal has been received and that the equipment and switch path associated with the call will be cleared down.

The MF address signalling is similar in principle to that described above for subscriber signalling. Once the 'proceed-to-send' signal has been received, it is known that the distant exchange is awaiting the address digits, so they are dispatched rapidly in noncompelled mode.

2.2.3 Pulse-code modulation channel-associated signalling

Since the 1960s, pulse-codes modulation (PCM — see Chapter 5) has been used as a means of converting analogue speech signals into digital form and multiplexing them in preparation for transmission. The multiplexed speech channels are each allocated to a 'time-slot' in a continuous train of digital pulses. Each speech channel has continuous use of its allocated time slot. The signals, too, are converted into digital form. MF selection signals are encoded as speech and transmitted in the time slot allocated to the speech channel. They therefore need to be detected by filters when the speech is reconverted into analogue form at the receiving exchange. All line signals and loop-disconnect selection signals are separated from the speech path. Then, the signals for all channels in the PCM system are made to share a single time slot, taking turns to occupy it. In the 30-channel PCM system, this is time slot 16 (TS16), which simultaneously contains the signals for two speech channels. It is arranged that each speech channel uses TS 16 on its every 16th appearance. A speech channel's appropriate share in TS16 is permanently reserved for that speech channel. Thus, when a call is in the conversational stage, or the channel is idle, and signals are not being dispatched, its information-carrying capacity is unused. It should be noted that, although the signalling for all 30 channels is carried in a separate channel (TS16), the signalling is still on a channel-associated basis. In the 24-channel PCM system, signals are carried by a method known as 'bit stealing', which is described in Chapter 5.

PCM channel-associated signalling does not offer an independent signalling system, but a means of transmitting, over a PCM system, signals which already exist, for example, as loop-disconnect signals arriving at a local exchange from a subscriber. A full description is given in Chapter 5.

2.2.4 Limitations of channel-associated signalling systems

By their nature, channel-associated systems are subject to a number of limitations, which include the following:

(i)   Compared with signalling in commercial computer networks, they are slow.

(ii)   Their signal repertoirs are small.

(iii)   Expansion is only possibly in certain cases, and then it is difficult, expensive and limited.

(iv)   The signals apply to telephony only, and cannot be made to cover new services.

(v)   There is no facility for transferring non-call-related information, for example, to aid the management of the network.

(vi)   Signals can only be sent at certain times during the life-cycle of a call.

(vii)   Signal tones are audible to subscribers, and, therefore, annoying.

(viii)   They all depend on great accuracy in timing, frequency and, in some cases, power; a high proportion of telephony faults in electromechanical systems are due to timing differences between sending and receiving signalling equipment.

(ix)   The signalling capacity allocated to a circuit is always reserved for that circuit, whether or not the circuit is in use and whether or not signals are being sent; its information-carrying capacity is, therefore, used inefficiently.

2.3 Common-channel signalling

The equivalent of interregister signalling in SPC exchanges is interprocessor signalling, i.e., direct communication between exchange-control systems. It was, therefore, a natural step for modern signalling systems to be based on the established principles of commercial intercomputer communication over digital links.

In telephony, calls have traditionally been circuit-switched, i.e. a circuit is dedicated to a call for the duration of the call. Although this has been inevitable, it has not used the circuit efficiently, because a call typically consists of more silence than speech. An important consideration in the development of the principles of digital intercomputer networks was, therefore, to employ circuits more efficiently. This led to packet-switched networks, in which circuits were not allocated to individual calls, and, later, to local-area networks, which were also based on the transmission of discrete 'messages' or 'packets'.

2.3.1 Principles of message-based communication

In message-based communication, the data to be sent from one computer to another are divided into modules of fixed size which fit into the data fields of 'messages' of defined format. Each message (see Fig. 2.5) must consist of a minimum of five fields:

(i) The address of the destination computer. This is analysed by any computer which receives the message and compared with that computer's own address. Mismatch results in the message being forwarded to another computer.

(ii) The address of the originating computer. This not only lets the destination computer know the source of the arriving data, but it also allows a request for retransmission if the original message is corrupted in transit or fails to reach its destination.

(iii) The message number. This allows all the messages of a data transmission to be assembled in the correct sequence, to be checked for continuity and completeness, and to have their data extracted in the correct order.

(iv) The data field. This contains the data being transmitted.

(v) An error-check field, which allows the data be checked for correctness after transmission.

Because circuits are not dedicated to individual calls, a sequence of messages on any given circuit may be associated with different calls and directed to different destinations. Similarly, all the messages of a call are not necessarily routed in the same way. Typically, they are transmitted according to a routeing algorithm based on the destination and the availability and loading of circuits.

When a message has been assembled, it is transmitted towards some chosen node (computer) in the network. When it arrives there, its destination address is interrogated and it undergoes an error check. If it has not reached its destination and it is found to be correct, it is again dispatched. The error check ensures that corrupted messages do not congest the network and, if the message is found to be corrupted, a request for retransmission is sent to the originating computer. When a sound message is recognised as having reached its ultimate destination, its data are extracted and stored in accordance with its message number.

It is acceptable for messages to arrive out of sequence because of the different routeings by which they may be sent through the network, but, if an expected message does not arrive within a given time (in the form of a message) for retransmission is sent to the originating computer.

2.3.2 Message-based signalling

Common-channel signalling (CCS) uses message-based data communication. Signals are sent, as messages, between the control systems of SPC exchanges. Because the information in the data field of a message can be used not only to define the signal, but also to identify the call to which it refers, common-

address of destination	address of origin	packet number	data field	error check field

Fig. 2.5 *The minimum contents of a message*

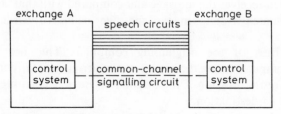

Fig. 2.6 *Separate speech and signalling circuits*

channel signalling can be remote from the speech path (hence its name). Thus, between two exchanges, the signalling for a number of speech circuits can take place over a single signalling circuit, as in Fig. 2.6. Further, because the handling of such signalling is at the speed of computer processing, the signalling circuit does not have to be a direct connection between the two exchanges. Indeed, once signalling is detached from speech transmission, there is the opportunity for an independent signalling network to exist. For example, Fig. 2.7 shows seven exchanges interconnected by separate speech and signalling networks. Dimensioning a signalling network is dependent not only on carrying capacity and economics, but also security, and the principles are described in Chapter 21.

Common-channel signalling overcomes all the limitations of channel-associated signalling mentioned in Section 2.2.4 above. Considering these limitations, the advantages of common-channel signalling include:

(i)   Being digital transmission directly between processors, it is extremely fast.
(ii)   The signal repertoire, being dependent only on the size of the data field, is potentially vast.
(iii)   Given (ii), above, expansion will not be limited by technical constraints, but only by international agreement.

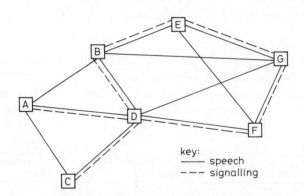

Fig. 2.7 *Independent speech and signalling networks*

(iv)   The large repertoire will provide signals for all services and not only for telephony.

(v)   Network-management signals will be facilitated by the fact that they can be made available from the repertoire and by the speed and flexibility of the signalling network.

(vi)   Because signalling links are independent of speech paths, signals may be sent at any time.

(vii)   Because signals are not transmitted over the speech path, they are inaudible to subscribers.

(viii)   Message transmission is dependent on accuracy in timing, but problems are immediately detected and automatically overcome, not only by the error checks built into individual messages, but also by a number of self-checking mechanisms in the network itself.

(ix)   Because in common-channel signalling a single message of a few hundred bits contains all the information necessary to set up a call, a single signalling channel can accomodate the messages relating to a large number of speech channels. Signalling channels are, therefore, efficiently used; unlike in channel-associated signalling, there is no permanent allocation of the signalling-channel capacity to circuits. Signalling-channel dimensioning is thus done on a traffic basis.

In addition, because common-channel signalling takes places directly between processors, signals are handled by software and do not require the large amounts of expensive and space-consuming equipment necessary in channel-associated systems.

The development of CCS is the greatest quantum leap in the history of telephony signalling. Being interprocessor, it is independent of switch or transmission technology; so new signalling systems do not need to be introduced in parallel with every other development in telephony. Standards can be laid down which are constant for all applications. Thus, for the first time, there can be a system which may be used in both national and international networks. Indeed, the CCITT signalling system No.7, which is described in Chapter 20, is already in use both nationally and internationally.[22] It is also constantly being expanded to provide signalling for new services and network management. Common-channel signalling is also employed between subscribers and their local exchanges, using the ISDN architecture (see Chapter 22).

2.4 Switching

At each exchange there are inlets and outlets, which comprise the terminations of subscribers' lines, junctions, trunks or international circuits. Whereas an exchange may, in the most general sense, be referred to as a switch, it in fact

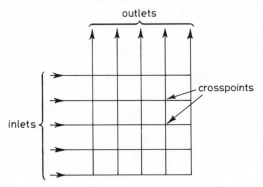

Fig. 2.8 *A switch, comprising a matrix of crosspoints (switch controls are not shown)*

consists of a large number of individual switches, or crosspoints. These may be arranged in various ways in order to achieve efficiency and economy.

Each crosspoint is an electrical connection which may be made or broken, so that it can, when made, form part of the 'switch path' of a call through the exchange. One way of providing the crosspoints in an exchange is as a 'switch block' consisting of a matrix, as in Fig. 2.8. Electromechanical crosspoints have traditionally been expensive (the Figure does not show that each connection consists, not of a single wire, but of two or four wires, as well as a number of control wires, which the control system uses for operating the chosen crosspoints). Thus, while the single matrix is the simplest conceptual way of providing a switch block, economising on crosspoints is an advantage. Simply reducing the number of crosspoints in the matrix merely reduces the size of the exchange, so savings are achieved, instead, by having a number of 'switching stages' instead of a single matrix.

2.4.1 Staged switching

A simple example of staged switching is shown in Fig. 2.9, and a more detailed discussion of the subject is given in References 23 and 24. In the Figure, the circuits incoming to the exchange are connected, in groups of 100, to A-stage switches, each of which has only ten outlets. There is therefore an immediate concentration within the exchange, and a consequential saving on crosspoints. This, however, is at the expense of introducing the possibility that an arriving call cannot be connected, due to the unavailability of a switchpath (in this case, an outlet from the A-stage switch). This condition is known as 'blocking'. Because all inlets do not give rise to calls at the same time, this probability can, in practice, be kept low by allocating an acceptable grade of service and provisioning the exchange to meet it, as described in the previous chapter.

In Fig. 2.9 the A-stage switches simply provide the concentration which

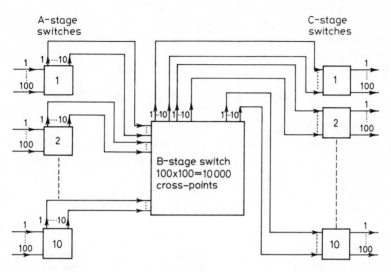

Fig. 2.9 *The principle of staged switching*

allows an economy of crosspoints. Similarly, the C-stage switches provide expansion, so that the required number of outgoing circuits can be provided on the exchange. The C-stage switches also provide routeing, i.e. to particular outgoing circuits. The B-stage provides routeing through the exchange. When a call arrives at an A-stage switch, the only necessary action is to find an outlet to the B-stage. The B-stage must then close the appropriate crosspoint for the call to be routed to the correct C-stage switch. The total number of crosspoints, for 1000 incoming circuits and 1000 outgoing circuits (summing the A,B and C stages, respectively) is:

$$10 \ (100 \times 10) + 100 \times 100 + 10 \ (10 \times 100) \ = \ 30 \ 000$$

This number compares favourably with the figure of 1 000 000 which would be necessary if a single matrix (as in Fig. 2.8) were used.

A more advanced form of staged switching is shown in Fig. 2.10. Here, all the switching stages perform an element of routeing. Each B-stage switch is seen to possess only one outlet to each C-stage switch. If the A switch simply allocated an arriving call to an arbitarily chosen outlet, there would be an unacceptably high probability of finding the required outlet from the chosen B switch engaged. The system in the Figure requires co-ordination in the choices of outlets across the whole switching system. Whereas the control of the switching stages in the system of Fig. 2.9 could be handled in a step-by-step manner, the control of this system must be co-ordinated so that an outlet from an A-stage switch to a B-stage switch is chosen only if it is known that the outlet from that B-stage switch to the required C-stage switch is free. (The 'common control' necessary for this is discussed in Section 2.5 below.) In this way, switch paths are never partially connected; if there is no complete path, switching equipment is not held unnecessarily. This 'link' or 'conditional' switching has been the basis of all modern exchange systems.

The number of crosspoints in the switching system of Fig. 2.10 (summing the A,B and C switches, respectively) is

$$10 \ (100 \times 10) + 10 \ (10 \times 10) + 10 \ (10 \times 100) \ = \ 21 \ 000$$

2.4.2 Switching Technology

In analogue exchanges, which are only now being superseded by wholly digital exchanges, switching has been space-divided: an exclusive switch path is set up for the benefit of a single call and exists for the duration of the call. With digital switching, however, the switching of calls is achieved by the opening and closing of logic gates at regular intervals, to allow electrical signals, in the

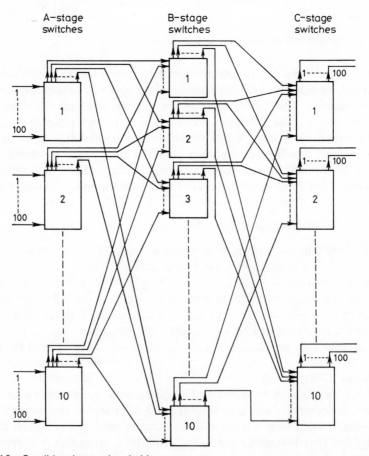

Fig. 2.10 *Conditional staged switching*

form of binary digits, on to the physical switch paths. In this way, a number of calls time-share the same switch path; their signals are not transmitted continously, but in selected 'time slots' in a composite pulse train.

It is worth defining here the distinctions between the modes of switching. Three parameters are essential to the precise definition of a switch (or switching stage, since any switching system may consist of a mixture of types): transfer characteristic, path provision and type of switching:

(i) Transfer characteristic
The transfer characteristic, also known as the 'switching mode', of a switching stage may be either analogue or digital.

An analogue switch can pass any input signal voltage within the working range. Normally, such switches pass signals that are electrical analogues of sound pressure variations originating from speech or music. However, an analogue switch can also pass digital signals.

A digital swtich is one which passes signals having voltages only at n defined levels. For a binary digital switch, $n = 2$.

(ii) Path provision
A switch, consisting of a number of crosspoints, is capable of providing a number of simultaneous connections, and two methods are employed.

In space division (SD), each call or channel is allocated an exclusive physical path through the switch for the duration of the call. Paths through the switch are identified by their position.

In time-division multiplexing (TDM) a switch is time-shared by a number of channels. Each channel is periodically allocated a short time slot during which it has exclusive access to a common path through the switch. Before audio-telephony channels are passed through a TDM digital-switching network, they are converted into digital form.

(iii) Type of switching
The type of switching describes the particular function of the switch. There are two types: space and time. In a space switch, connections are made between different physical positions (i.e. between one link and another) without the introduction of delay to the transmitted speech signal. In a time switch, connections are made at different points in time. The information in a given time slot on the inlet to the switch is transferred to a selected time slot on the outlet. This necessarily involves storing the speech signal for a defined period, and so introduces a delay (which, however, is imperceptible to subscribers).

The different switch types can be distinguished precisely, using these three parameters. Thus, a switch may be: '(analogue or digital)/(SD or TDM)/(space or time)'. For example, an electromechanical switch block might be described as 'analogue/SD/space'. It is interesting to note that space switches may be either analogue or digital, whereas, for practical reasons, time switches must be digital.

2.5 Exchange control

The control system is the brain of an exchange. It contains the logic capable of deciding what actions need to be carried out and of dispatching the electrical signals necessary for initiating them. For example, when a seize signal is received by the exchange, the control system searches for a free storage facility for the ensuing digits and, having found one, initiates a proceed-to-send signal (dial tone, if the seize signal is on a local line). When digits have been received, the control system interprets them, determines which outgoing circuit the call should use, and selects an appropriate switch path through the exchange. When the clear signal arrives, the control system clears down the switch path, thus making its components available for another call. Control also involves supervision of the exchange, including the collection of data for billing, maintenance, planning, etc.

2.5.1. Implementation in manual exchanges

In manual exchanges, control and switching are both performed by an operator. While the number of circuits remains small, this is satisfactory, for it provides maximum flexibility. As long as there is a free connecting cord, the operator can connect any calling subscriber's line or incoming junction to any free subscriber's line or outgoing junction. The operator can choose routes known to offer the best chance of success and, when the network is impenetrable (due either to congestion or a fault), can avoid repeat attempts by explaining to subscribers why they should not call again within a given time. In using local knowledge of the network to achieve the switching function, the operator implicitly provides centralised control and an effective form of network management.

This basic form of control collocates all the control functions in the operator's head and hands. The operator receives the routeing information verbally from the calling subscriber in the form of the name of the called party, interprets this to determine which outgoing line is required, tests the line, sets up a connection across the switchboard by means of a cord, monitors the transmission path to ensure that the parties are in communication with each other, records the call for billing purposes, monitors the call for clear down, and, ultimately, disconnects the circuit by removing the connecting cord.

Whereas this provides an effective means of control, it is extravagant in its use of resources. A call demands the exclusive attention of an operator during the set-up period, which means that all other calls arriving during that time are forced to queue. Inconvenience, however, can be minimised by employing an appropriate number of operators.

In a small exchange, where the call-arrival rate only justifies a single operator, there is fully centralised control. The distinguishing characteristic of this is that all the control functions for all the lines are provided by a single unit, in this case, the operator. One penalty for entrusting all the control functions to a single unit is the loss of the entire exchange when that unit is not functioning, e.g. when, for one reason or another, the operator is not in position. This is overcome by providing a relief operator, who may then share the load when the principal operator is present. A further penalty is the finite probability of a call having to queue, or being lost, because the control unit is busy. This can be adjusted to an acceptable value by optimising the number of control units, having regard to the calling rate on the lines and the cost of provision.

2.5.2 Common control

The sharing of control resources between calls is 'common control'. This may either be centralised, as in the manual case, or distributed. During the evolution of control, both have been employed. In the 'marker-based' exchange systems[25] in use prior to the arrival of SPC, a combination of centralised and distributed control was used. While a 'marker', containing the algorithms (in the form of relay logic) for routeing calls through and out of the exchange, was central, other items of control equipment (such as registers for storing address digits) were dispersed.

In the first SPC exchanges,[26] the control functions were centralised in a single computer which was duplicated for security. Now, with the reduced cost and increased power of microprocessors, control has, in the latest generation of SPC exchanges, once more been distributed significantly. Functions such as signalling control, data handling, digit receipt and switch control are all entrusted (in varying degrees, according to the design) to microprocessors dispersed around the exchange, with a central processor providing co-ordination. Modern control is thus provided almost exclusively in software, and this is described in Part III.

2.6 References

1 ATKINSON, J. (1948): *Telephony,* Vol. 1 (Pitman), Chap. VIII
2 TIPPLER, J. (1975): 'Switching Systems', in *Telecommunication Networks,* Ed. FLOOD, J.E., (Peter Peregrinus Ltd), Chap. 4
3 ATKINSON J. (1950): *Telephony,* Vol. 2 (Pitman), Chap. VIII
4 FLOWERS, T.H. (1976): *Introduction to Exchange Systems* (John Wiley)
5 PEARCE, J. GORDON (1981): *Telecommunications Switching* (Plenum Press, New York), Chap. 3
6 TIPPLER, J.: 'Switching Systems', Op. cit.
7 HILLS, M.T. (1979): *Telecommunications Switching Principles* (Allen & Unwin)
8 HILLS, M.T. and EVANS, B.G. (1973): *Telecommunication Systems* (Allen & Unwin)
9 PEARCE, J. GORDON: Op. cit. Chap. 12
10 REDMILL, F.J. (1985): *Performance Monitoring — A Telecommunications Maintenance Tool Using Live Traffic,* Reliability '85, Birmingham, UK, July 1985
11 WELCH, S. (1979): *Signalling in Telecommunications Networks* (Peter Peregrinus Ltd)

12 WELCH S: op. cit. Chap. 11
13 CCITT (1984): Red Book, Vol. VI, Fascicle VI.1. Recommendation Q9
14 WELCH S: Op. cit., Chap. 2
15 CCITT (1973): Green Book, Vol.6, Part 1. Recommendation Q16
16 WELCH, S.: Op. cit., Chap. 3
17 WELCH S.: Op. cit., Chap. 8
18 CCITT (1984): Red Book, Vol.VI, Fascicle VI.2. Recommendations Q140–164
19 WELCH, S.: Op. cit., Chap. 9
20 WELCH, S.: Op. cit., Chap. 10
21 CCITT (1984): Red Book, Vol.VI, Fascicle VI.2. Recommendations Q140–164
22 CCITT (1984): Red Book, Vol. VI, Fascicle VI.7, Recommendation Q700
23 HILLS, M.T.: Op. cit., Chap. 7
24 BEAR, D. (1976): *Principles of Telecommunication Traffic Engineering* (Peter Peregrinus Ltd), Chap. 1
25 HILLS, M.T.: Op. cit., Chap. 9
26 No. 1 ESS, Bell System Technical Journal, **43**, September 1964, Special issue in 2 parts

Chapter 3

Introduction to SPC digital telephone exchanges

3.1 Introduction

The previous chapter introduced the concept of SPC and the difference between analogue and digital switching. Modern SPC exchanges, which use digital-switching technology, are making a major impact on the telecommunications networks of the world. Whether introduced as parts of integrated digital transmission and switching networks, or as straight replacements for analogue switching units, such exchanges offer many advantages. The telecommunication administrations benefit from operational savings and features available from such systems, while the subscribers enjoy an improved quality of service and a range of new services and facilities. In setting the scene for the rest of the book, this chapter gives an introduction to SPC digital telephone exchanges, their advantages, and the services and facilities that they offer.

3.2 The development of SPC digital exchanges

The current fully electronic SPC digital exchanges represent the successful coupling of electronic and computer technology with telephony. The first signs of success of such a marriage came in the early 1960s. There then followed two decades of development, in which successive generations of exchange systems contained increased quantities of electronics.[1] This development was motivated by the desire to improve upon the cost, quality, maintainability and flexibility of electromechanical telephone exchanges by exploiting the proven advantages of the rapidly evolving electronic and computer technologies.

The first application of electronic devices to telephone exchanges was in the control area: stored-program control. The first public SPC exchange, the No. 1ESS, developed by AT&T Bell Laboratories, was introduced at Succasunna, New Jersey, USA, in May 1965.[2,3] This historic event initiated worldwide interest in SPC, which resulted in the introduction during the 1970s of a range of new exchange systems incorporating various degrees of computer-control technology. However, these early systems all used electromechanical switching devices (e.g. crossbar[4] and reed relay[5]) because of the problems in developing suitable semiconductor switching arrays for public telephony applications. (Such hybrid-exchange systems were thus 'semi-electronic', although, for reasons of prestige, they were often referred to as 'electronic' when first introduced.)

There were two obstacles hindering the use of semiconductor switches for telephony. The first was the difficulty of producing large semiconductor switch matrices with adequately low crosstalk characteristics. The switches forming such matrices require very high off-resistances if interference between circuits is to be eliminated. Working in the analogue mode, semiconductor switches were unable to compete with the transmission linearity and the near-infinite off-resistances of existing electromechanical switches. The second obstacle was the inability of the available semiconductor devices to handle the high voltages and ringing currents required by the conventional telephone.

It is interesting to note that some small PABXs were successfully developed with analogue electronic switches (using PAM/TDM techniques, as described in Chapter 4). Their small size, typically terminating a maximum of 200 extensions, enabled crosstalk to be kept adequately low, while the high-voltage obstacle was minimised by the use of special telephones.[6] Clearly, such conditions did not exist in public telephone switched networks (PSTNs). The application of semiconductor devices to switching in public telephony had to await the use of digital technology. The shift towards digital technology, and the overcoming of the two obstacles, was influenced by the introduction of digital transmission into the PSTNs and developments in semiconductor integrated-circuit (IC) devices, as described in the remainder of this Section.

Many countries began to introduce digital transmission, in the form of pulse-code modulation (PCM), into their networks during the late 1960s. The PCM systems were used originally in the junction networks to expand the capacity of existing audio-pair cables by virtue of their 24- or 30-channel multiplexing (see Chapter 5), known as 'pair gain'.[7] The application of digital transmission to long-distance routes did not start until the late 1970s, when higher-capacity systems, multiplexed from 24- or 30-channel groups, were used on coaxial cable. Now, microwave-radio and optical-fibre digital transmission systems are also being deployed.

The first application of digital technology to exchange systems was in the role of tandem switching between digital (PCM) junction routes. This overcame the problem of crosstalk because digital signals were sufficiently insusceptible to it. Thus, large semiconductor switching matrices could be used. Clearly, tandem and toll exchanges were not affected by the second obstacle because, with no subscriber lines involved, there were no high voltages or ringing currents to deal with. Hence, it was possible for an experimental digital junction-tandem exchange to be installed in London by the British Post Office in 1968. This carried live traffic successfully for a number of years.[8] CIT-Alcatel led the world with the first public digital tandem system (E10) in 1970 in Lannion, France.[9] In the USA, Bell introduced full public all-electronic digital toll and tandem exchanges, using the 4ESS system, from January 1976 onwards.[10]

A key advantage of digital switching is the elimination of the multiplexing equipment normally associated with the PCM digital-transmission systems terminating at the exchange. Thus, one of the major incentives for the introduction of digital switching into PSTNs was the potential for the elimination of analogue-to-digital conversion equipment from the trunk (toll) and junction networks. The planning aspects of co-ordinating the introduction of digital switching and transmission into a PSTN, to form an 'integrated digital network' (IDN), are discussed in Chapter 21.

However, successful application of digital semiconductor technology to local exchanges depended on a solution to the second obstacle, namely the handling of the high voltages and currents associated with subscriber lines. The solution adopted universally was to handle all the high-voltage and DC-path requirements of subscriber lines in interface units at the periphery of the exchange,[6] as described later in this chapter. This enabled the electronic switches to be developed unhindered by the demanding requirements of the subscriber lines.

Thus, successful application of semiconductor technology to subscriber-line switching required economical designs of subscriber-line interfaces. The main cost component was the analogue-to-digital conversion equipment. Until the early 1980s, the cost of subscriber-line interfaces made digital switches unattractive compared to the standard analogue switches then available (e.g. crossbar and reed relay). Therefore, the first generation of digital local exchanges (e.g. E10,[9] System X,[11] and AXE10[12]) each comprised two forms of switching systems. One was an analogue reed-relay unit which terminated the subscriber lines and concentrated their traffic on to internal highly loaded trunks that could economically be connected to analogue-to-digital convertors. The second form of switch was a digital system which interconnected the internal digital trunks with external trunk and junction digital routes. This hybrid analogue-digital architecture had the advantage of avoiding the provision of expensive PCM-encoding equipment for each subscriber's line, and it exploited the inherent DC metallic path through reed relays to perform the subscriber-line support functions (see Chapter 7). The exception to this approach was Northern Telecom, who, in 1980, produced the world's first all-digital local exchange system (the DMS 100), which exploited the analogue-to-digital convertors developed for their digital PABX system (SLI).[13,14]

The advent, in the early 1980s, of generally available integrated-circuit devices, that provided cheap analogue-to-digital conversion, enabled the cost of subscriber-line interfaces to decrease sufficiently for all-digital switching systems to be competitive with the analogue-digital hybrids. Thus, the current generation of SPC exchanges (local, toll, trunk and international) comprise stored-program control and electronic digital switching. With the exception of some of the components within the subscriber-line interfaces, these exchanges use only digital technology.

However, the rapid rate of progress in semiconductor technology is continuing to broaden the options for SPC digital-exchange designers. For example, the AT&T 5ESS system uses specially developed analogue semiconductor devices for some of the subscriber-line interface functions.[15]

3.3 Advantages of SPC digital exchanges

SPC digital exchanges offer many advantages to the administration and its subscribers (see the following Sections). However, it is fair to say that some of these result from the virtues of SPC, so analogue SPC exchanges would also offer them. In addition, the full range of advantages does not accrue until the SPC digital exchanges are incorporated into a digital transmission environment.

3.3.1 Advantages of stored-program control

3.3.1.1 Flexibility. Chapter 2 described how, in an SPC exchange, the hardware is controlled by programs and data held in electrically alterable storage. This control process offers a high degree of flexibility in the way that the exchange hardware is made to operate. Flexibility has long- and short-term aspects.

The long term will be considered first. At the switching-system-development stage, a range of programs may be produced to enable a basic exchange system to provide a variety of capabilities and facilities to suit the requirements of an administration. This software tailoring covers generic characteristics of local exchanges in the network, e.g. numbering, charging and routeing rules, types of call offered, administrative and subscriber facilities.

An important aspect of the long-term flexibility provided by SPC systems is the ability for established exchanges to be upgraded without disruption to the service. This enables new capabilities and facilities, not known or specified earlier, to be incorporated during the life of an exchange system. Some of these enhancements may be achieved merely by the incorporation of new software, e.g. the introduction of a closed user-group facility for a certain class of subscriber. Other enhancements, such as the introduction of digital data switching (see Chapter 22), also require the addition of new hardware.

SPC offers short-term flexibility due to its ability to alter the status of exchange equipment simply by changes to data. Thus, the operation of an exchange can be made to react rapidly to network conditions. For example, routeing algorithms may be changed so that calls are re-routed to avoid congestion in the network. The short-term flexibility of SPC enables a wide range of administration and subscriber facilities to be provided economically and with ease of operation, as described below.

3.3.1.2 Subscriber facilities. SPC exchanges enable a wide range of subscriber facilities to be provided more cheaply and easily than in non-SPC exchanges. The facilities are allocated by the administration, as appropriate. Thereafter, many may be invoked by subscribers on a call-by-call basis. The following are examples of possible facilities:

(i) Short-code dialling: Subscriber-selected telephone numbers are called by dialling pre-input short codes (by convention, the term 'dialling' is used to cover telephone input operations using a dial or a keypad).

(ii) Call transfer: Incoming calls to one telephone number are automatically redirected to another.

(iii) Ring back when free: The exchange is requested to set up a connection to an engaged telephone as soon as it becomes free.

(iv) Automatic alarm call: The exchange rings at a predetermined time on a single or daily basis.

(v) Outgoing- or incoming-call barring: This enables owners of telephones used by others to restrict the type of calls made or received.

(vi) Itemised billing: The provision of bills with detailed listings of calls made and the charges raised.

(vii) Malicious-call tracing: The subscribers, or the appropriate authorities, are notified of the original of a malicious call.

(viii) Centrex: The provision, by the local exchange, of extention-to-extension communication within a business subscriber's private system. In addition to PABX-type facilities, centrex subscribers have normal telephone service for calls leaving or entering their private system.

Many of the above facilities require an additional subscriber-signalling capability, as provided by multifrequency (MF) push-button telephones (see Chapter 2). Similarly, the extention of these facilities beyond the exchange area requires an appropriate interexchange signalling capability, such as the CCITT No.7 signalling system (see Chapter 20). For example, such signalling may enable a call to be transferred automatically to some distant exchange.

The above facilities are also possible with SPC analogue exchanges and many with non-SPC exchanges, though with less ease and greater cost. Chapter 15 gives examples of the SPC implementation of facilities by considering (i), (ii) and (iv) above.

3.3.1.3. Administration facilities. An SPC exchange offers an administration a wide range of operational facilities which would otherwise be expensive or labour-intensive to provide. Most of the day-to-day operations on the exchange involve the use of these facilities, accessed via computer terminals associated with the exchange, either locally or at remotely located operational-control centres (see Chapter 19).

Examples of administration facilities are:

(i) *Control of subscribers' facilities:* This enables the administration to change the list of facilities available to a subscriber.

(ii) *Change of routeing:* As referred to above, an operator may quickly change the choice of routes to be used by an exchange for calls to other exchanges, to meet short-term congestion problems or long-term changes to the routeing plan.

(iii) *Change of subscriber's number and trunk codes:* This can be undertaken by a simple instruction via the operator's terminal.

(iv) *Output of exchange-management statistics:* These include equipment occupancies (traffic loading) at specified times, call-destination data, details of congestion on routes, details of subscribers' calls, etc. The information may be available as print-outs or screen displays locally at the exchange and at remote network-management and operations-and-maintenance centres. In addition, they may be output on portable devices (e.g. cartridges or tapes) or on data links for processing at an administration centre.

(v) *Maintenance aids:* These comprise the automatic generation of tests and the recording of the results, alarm processing (i.e. interpreting and prioritising), diagnostic programs to assist fault location, error messages, etc.

Chapter 19 describes the principles of how network-management centres, together with the above-listed administration facilities, support the operations and maintenance of SPC digital exchanges.

3.3.2 The added advantages of digital technology
The use of digital (rather than analogue) switching within SPC exchanges adds the following features.

3.3.2.1. Speed of call set-up. The electronic hardware of the control element of SPC exchanges operates at high speed and at low-voltage levels (typically 5V DC). Thus, for SPC exchanges with electromechanical switches, which are by their nature slow and require relatively heavy operating voltages and currents, there is a significant mismatch in speed and power between the control and switching systems, and this must be overcome by appropriate buffering equipment. However, digital switches are entirely composed of semiconductor gates and stores, in integrated-circuit form, which operate at speeds and voltages compatible with the control systems, thus forming a fully electronic SPC exchange.

Connections can be established across digital switching systems very quickly (e.g. 250 µs). This, together with their economical high capacities and near-nonblocking characteristics, enables system-design economies to be gained by using the digital switches for sequential path set-ups between the various exchange subsystems (serial trunking) during the call-connection phase. In addition, automatic repeat attempts across the switches (e.g. to avoid congestion or outages in the network) can be made without incurring perceptible increases in post-dialling delay. The result is that digital switching with SPC enables complex call connections within an exchange to be established using relatively cheap and simple switch designs.

3.3.2.2. Accommodation savings. Digital switching systems are significantly smaller than analogue exchanges of equivalent capacity. This is due to both the miniaturisation achieved by the integrated circuitry and the large-scale time-division multiplexing used throughout a digital exchange. The latter is possible because of the high speeds of operation of the semiconductor technology used.

However, the savings in accommodation can be significantly reduced by the presence of analogue-to-digital interworking equipment required to terminate analogue lines. The need for such equipment, and hence the accommodation space, can be minimised by appropriate planning to maximise the proportion of digital circuits terminating at the exchange (see Chapter 21). The potentially high accommodation savings may also be reduced, to some extent, by the need to provide air-conditioning and a controlled environment for the exchanges. In some cases, additional power-supply equipment is required when SPC digital systems are installed in an existing exchange. Despite these additions, the overall space required for a digital SPC exchange is typically less than 25% of that for step-by-step or crossbar exchanges, and 50% of that for analogue SPC systems.

3.3.2.3 Ease of maintenance. The equipment used in digital SPC exchanges has a lower fault rate than that used in analogue SPC exchanges because of the absence of moving parts and the inherent reliability of semiconductor technology. In addition, unlike step-by-step and crossbar exchanges, digital systems do not require any routine adjustments. Diagnostic programs within the exchange-control system usually enable hardware faults to be located quickly to a particular module or plug-in unit. Where appropriate, the use of hardware redundancy enables the control system to restore service rapidly by automatic reconfiguration of the use of the equipment, replacing a faulty unit with a spare one. The exchange-control system then provides the necessary information for the maintenance staff to replace the faulty unit later, in a planned way. Faulty units are usually sent to specialist repair centres. Therefore, the hardware maintenance load is low in comparison with that for analogue exchanges.

Faults may occur in the exchange's software as well as in its hardware. Chapter 19 describes how software faults arise and the measures taken, both manual and automatic, to restore service in the event of failure. The software-maintenance process is eased by the range of diagnostic programs and error messages provided by the exchange-control system.

3.3.2.4 Quality of connection. There are three important transmission advantages with networks using digital switching and transmission (i.e. integrated digital networks). First, the overall transmission loss of a call connection through the network is independent of the number of digital switching units and transmission links. Furthermore, the overall loss is set by the analogue-to-digital conversion processes at each end of the connection. This enables the loudness to be optimised, taking into account preferred listening levels, consistent with adequate stability and echo control.[16] Secondly, because noise does not accumulate over digital transmission systems, users perceive significantly lower noise levels than with connections over analogue networks. Thirdly, digital local exchanges have line-interface cards permanently connected to 2-wire local lines. This enables improved impedance matching within the 2-to-4-wire conversion equipment, resulting in fewer instability problems compared to 2-wire-switched analogue networks. (Such improved impedance control is particularly important because, compared to analogue networks, digital networks tend to have lower end-to-end losses, but with increased delay, thus potentially worsening the echo performance.)

3.3.2.5 Potential for non-voice services. Digital transmission is an ideal medium for conveying the outputs from data terminals, computers, etc., which originate in digital format. Transmission of data, particularly at speeds above 4.8kbit/s, is cheaper and more efficient over digital systems than over analogue systems, because the signals can be carried directly, without the need for voice-frequency modems, with their attendant high cost and restriction on throughput. Digitally encoded analogue signals (e.g. speech and video) can be freely mixed with digitally-sourced traffic and carried over a common bearer without the power-spectrum constraints that exist when such a variety of signals is carried over analogue-transmission systems. Thus, digital exchanges, when associated with digital transmission, have the potential for the economic provision of a range of services in addition to telephony. The integrated services digital network (ISDN) is covered in Chapter 22.

3.3.2.6 Cost. In general, digital SPC exchange systems are economical to run compared with their analogue equivalents, and their capital costs can be significantly lower. However, the cost aspects of running telephone exchanges are varied and complex.[17,18] As well as the capital cost of the switching equipment, the initial investment involves costs resulting from accommodation, power, operation-and-maintenance support systems, spares holding, documentation, staff training, interconnection with the existing network. etc. When deciding on the choice of switching systems, administrations also need to consider the consequential running costs. These result from software support (provided by the manufacturer and the administration), operational management, maintenance, billing, etc. The ease of upgrading the system with new facilities may also be reflected in the cost evaluation. All these factors contribute to the 'whole-life cost' of an exchange system. Chapter 19 covers the operations and maintenance aspects and Chapter 21 describes how planning can minimise the accommodation and interworking costs of introducing such exchanges.

3.3.2.7 Installation time. The time to install digital SPC exchanges is less than that for analogue exchanges of equivalent capacity. This is due to the smaller physical volume and the modularisation of the digital equipment. Speedy installation is also attributable to the factory pre-testing and simple plug-in-unit construction of the equipment now employed in modern SPC exchanges.

3.3.3. Common-channel signalling

An important component of digital SPC exchanges is the common-channel signalling (CCS) which provides rapid communication between the control systems of such exchanges in a network. At the simplest level, people making telephone calls across a digital network using CCS enjoy very short call set-up times. In addition, CCS enables a range of subscriber facilities (see Section 3.3.1.2) to be extended across the network. However, it is in the realms of non-voice calls and network operation that CCS adds the greatest value to digital SPC exchanges. CCS has the capability to convey, between exchanges and from subscribers to the exchanges, a wide range of messages relating to telephone and data calls, as well as administrative and control information not directly associated with specific calls. Chapter 2 introduces the concept of CCS and Chapter 20 describes the CCITT standard system (SS No. 7) for digital networks. Chapter 20 also outlines the way that CCS is extended to the subscribers' premises in association with an ISDN.

3.4 Overview of an SPC digital exchange

Fig. 3.1 gives a block-schematic representation of a generalised digital SPC local exchange. It must be emphasised that the Figure shows the functional elements of an exchange rather than, necessarily, the physical units

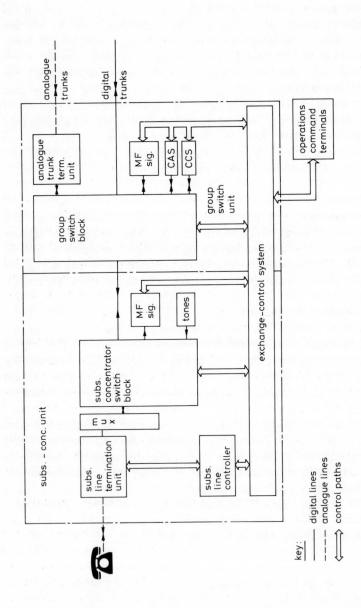

Fig. 3.1 *A generalised SPC digital local exchange*

key:

——— digital lines

– – – analogue lines

⇔ control paths

(subsystems) that might be used in any particular system. There are many types of digital SPC exchange system now being manufactured, each with a distinctive architecture. This results in different allocations of the functional elements to subsystems. However, Fig. 3.1 has been designed for good correlation between most of its functional elements and the subsystem functions of most of the available exchange systems. It should also be noted that the terms and principles used in this Figure, and adopted throughout this book, are general and descriptive, and are not intended to refer to any specific design of switching system.

The local exchange consists of two types of unit: one or more subscriber-concentrator units and a group-switch unit. Some of the subscriber-concentrator units can be remote from the group-switch unit (see Chapter 9), but, for simplicity, all units in Fig. 3.1 are collocated for the purposes of this description. The units contain digital-switching, line-termination, signalling and control equipment. The Figure shows a local exchange with only one subscriber-concentrator unit (SCU) and a group-switch unit (GSU); additional SCUs are connected to the GSU in a similar manner. Usually, control equipment in the SCU performs some of the call-control functions, in association with the main control equipment in the GSU. The degree of autonomy of the control equipment in the SCU depends on the design of the exchange system. Thus, the exchange-control system, which provides the SPC functions, is shown in Fig. 3.1 to embrace both of the exchange units. Digital SPC trunk exchanges do not terminate subscriber lines and thus consist of a GSU only (see Fig. 7.2 of Chapter 7).

Both of the exchange units contain digital switch blocks. (The term 'switch block' is used to describe a switching system which consists of several switching stages; see Chapter 2.) The subscriber-concentrator switch block switches originating calls from a large number of subscriber lines, with their low traffic loading, on to fewer, more highly loaded internal trunks leading into the group switch block. This provides interconnectivity between the trunks from subscriber-concentrator units and external trunk and junction routes. Calls terminating on an SCU are switched by the subscriber-concentrator switch block from the GSU trunks to the appropriate subscriber's line.

Digital switch blocks, by virtue of their digital semiconductor composition and time-division-multiplexed mode of operation, can deal only with formatted digital signals. Thus, any analogue lines terminating on the exchange must be converted to multiplexed digital mode (i.e. 24- or 30-channel PCM format, as described in Chapter 5) at the periphery of the switch block. This conversion is provided, for trunk lines, by the analogue trunk-termination unit at the periphery of the route switch block; the conversion for subscriber lines is provided by subscriber line-termination units (SLTU) and multiplexers at the periphery of the subscriber-concentrator switch block.

The SLTU also provides all the functions associated with subscriber lines.

These comprise the feeding of power for the telephone transmitter, the detection of the DC loop created by the telephone handset being lifted, the detection of dial pulses, protection of the switching equipment against high voltages on the line, conversions between the subscriber two-wire analogue line and the four-wire digital switching system, the feeding of ringing current to line, and certain test functions (Chapter 7). Economy in exchange design is achieved by minimising the equipment within the SLTU, because this is provided on a per-subscriber-line basis. Thus, some of the equipment providing the functions of subscriber-line termination is located in common units, the subscriber-line controllers, each serving many SLTUs. The line controllers provide the interface between the SLTUs and the exchange-control systems by acting as communication terminals. Thus, dial-pulse breaks detected by the SLTUs are converted into digits by the controllers.

A different form of SLTU is needed to terminate digital subscriber lines which convey a number of channels from special ISDN termination units or digital PABX units. Although digital SLTUs do not need to provide analogue-to-digital or 2-to-4 wire conversion, they must terminate the digital line transmission and undertake test and signal-extraction functions, as described in Chapter 7. Digital trunk and junction circuits in the standard 24- or 30-channel PCM format terminate directly on to the group switch block (see Chapter 6). However, digital routes carried over higher-order-transmission systems (see Chapter 5) must first be demultiplexed down to the standard PCM format at transmission stations associated with the digital SPC exchange before being terminated on the group switch block (not shown in Fig. 3.1).

With the exception of DC conditions (loops and loop disconnections), which are detected by SLTUs and their controllers, all signalling is handled by common pools of senders and receivers (see Chapter 8). Access between subscriber lines and multifrequency (MF) receivers is provided via the subscriber-concentrator switch block (see Chapter 14). This also provides access between a small pool of tone (and recorded announcement) units and the subscriber lines. Access between the trunk lines and the various pools of signalling senders/receivers, for MF signalling, PCM channel-associated signalling (CAS) and common-channel signalling (CCS), is provided by the group switch block.

Communication between the digital SPC exchange system and the administration's operational staff is provided by operations-command terminals using man-machine-interface software run on the exchange-control system. These terminals (e.g. VDUs and printers) may be collocated with the exchange or remotely located at distant operation-and-maintenance centres (see Chapter 19).

With Part I having introduced the concepts of telephony and digital SPC exchanges, the next two parts of the book concentrate on specific aspects: switching (Part II) and control (Part III). Finally, the network and operational aspects are considered in Part IV.

3.5 References

1 REDMILL, F.J. and VALDAR, A.R. (1984): *The Evolution of Telephone Exchange Systems,* African Technical Review, pp.64–66

2 MARTERSTECK K.E. and SPENCER, JNR. A.E. (1985): *The 5ESS Switching System: Introduction,* AT&T Technical Journal, **64** (6)

3 KEISTER W. *et al* (1964): *Special Issue on the 'ESS Switch,* Bell System Technical Journal, **43** (5)

4 FLOWERS T.H. (1976): *Introduction to Exchange System* (John Wiley & Son), Chap. 5

5 FLOWERS T.H.: Ibid, Chap. 6

6 HUGHES, C.J. (1986): *Switching — state-of-the-art,* British Telecom Technology Journal, **4** (1) pp.5–19

7 BENNETT G.H. (1983): *Pulse Code Modulation and Digital Transmission* (Marconi Instruments Ltd), Chap.1

8 CHAPMAN, K.J. and HUGHES, C.J. (1968): *A field trial of an experimental pulse-code modulation tandem exchange,* Post Office Electrical Engineers' Jounral, **61,** Part 3

9 GRINSEC (1983): *Studies in Telecommunications,* Vol 2, 'Electronic Switching', Part VII (Elsevier Science Publishers BV)

10 *Special Issue on the 4ESS switch,* Bell System Technical Journal, **56** (7), September 1977

11 OLIVER, G.P. (1980): *Architecture of System X: Part 3 — Local Exchanges,* Post Office Electrical Engineers' Journal, **73,** pp.27–34.

12 NILSSON B.A. and SORME, K. (1980): *AXE10 — A Review,* Ericcson Review, **57** (4), pp.138–148

13 KASSON, J.M. (1979): *Survey of Digital PBX Design',* IEEE Trans. Communications, **COM-27** (7) pp.1118–1124

14 SWAN, R (1983): *DMS-100 Family Evolution,* Telesis, **10,** (3) pp.2–5

15 HAYWARD, W.S. *et al.* (1985): *Special Issue on the 5ESS switching system,* AT&T Technical Journal, **64** (6) Part 2

16 HARRISON, K.R. (1980): *Telephony Transmission Standards in the Evolving Digital Network,* Post Office Electrical Engineers' Journal, **73,** pp.74–81

17 SHANTAI PAI, B.H. (1979): *Choice of Switching Systems,* Telecommunication Journal, **46** (IV), pp.219–231

18 REDMILL F.J. and VALDAR, A.R. (1985): *Selecting a System,* African Technical Review, pp.31–35

Part II
Digital Switching

Chapter 4

Time-division multiplexing

4.1 General

This Part of the book describes the principles of digital switching and considers some practical system implementations. The operation of digital switching systems relies on both digital technology and time-division-multiplexing techniques. Thus, these two topics are briefly reviewed in this and the following chapter before the subject of digital switching is tackled. Chapter 6 describes the principles of digital switching by focusing on the heart of digital exchanges: the digital switch blocks. Chapters 7 and 8 then expand the coverage by addressing the issues of line termination and the handling of signalling in an exchange. The various exchange systems currently available organise the switching, line termination, signalling and control functions differently. Thus, Chapter 9 considers the possible options for digital SPC exchange architectures.

The Part concludes with Chapter 10 describing an issue that is unique to digital exchanges, i.e. control of their internal timing and the synchronisation of a number of exchanges within a network.

4.2 The principles of time-division multiplexing

4.2.1 Introduction to multiplexing

The process of conveying a number of signals simultaneously over a single bearer, known as multiplexing, is widely applied across the field of electrical engineering. Signals may be multiplexed for a short distance of a metre or so, such as on the data and control buses within computers, or transmitted over distances of many thousands of kilometres, over an international route between two telephone exchanges, for example. Whatever the application, multiplexing is used to reduce the cost of transmitting or distributing a number of signals. A multiplexed system comprises the following elements: 'n' input signals (each from an 'input channel' or 'tributary') which are combined into

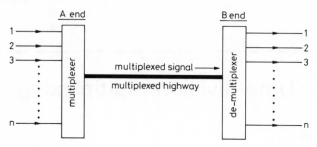

Fig. 4.1   *A multiplex system*

a single multiplexed signal; this complex signal may then be transmitted or processed, as required. If transmitted, the *n* individual signals are separated at the distant end and fed to the appropriate output channels. This configuration is illustrated in Fig. 4.1 for a system that multiplexes '*n*' channels, a multiplexing ratio of *n:*1.

For simplicity, Fig. 4.1 shows a unidirectional arrangement only. However, multiplexed systems used in telecommunications are usually bidirectional (also known as 'full duplex'). This is achieved by combining two opposing parallel unidirectional systems. The signals from the two systems are separated in space, time, or frequency, depending on the technology employed. For example, in digital exchanges and most cable and optical-fibre transmission systems, the two unidirectional signals are carried on different physical paths (i.e. use of spatial separation). The terminal equipment at each end of the bidirectional multiplexed system consists of a multiplexer for the 'transmit' (or 'go') direction and a demultiplexer for the 'receive' (or 'return') direction of signals. The combined multiplexer-demultiplexer terminal equipment is referred to as a 'muldex'; although, for convenience, it is most often called just a 'multiplexer' or 'mux'.

There are many multiplexing techniques, each of which exploits the ability of a system to generate and detect one particular parameter with sufficient accuracy. Multiplexing is achieved by the allocation of a dedicated range of that parameter to each channel. The most common examples of multiplexing techniques are frequency-division multiplexing (FDM) and time-division multiplexing (TDM). With FDM, each channel is allocated a dedicated band of frequencies, typcially 4 kHz wide for telephony. Recovery of each channel from the multplexed signal is achieved by selective bandpass-frequency filtering. TDM allocates dedicated periods of time to each channel, as described later. Another method of multiplexing, although not always recognised as such, is that of space-division multiplexing (SDM). Whilst not strictly a multiplexing technique, the term SDM does describe the way that physical (space) separation is used to route individual telephone calls through an analogue switch block (Chapter 2 covers this description). Fig. 4.2

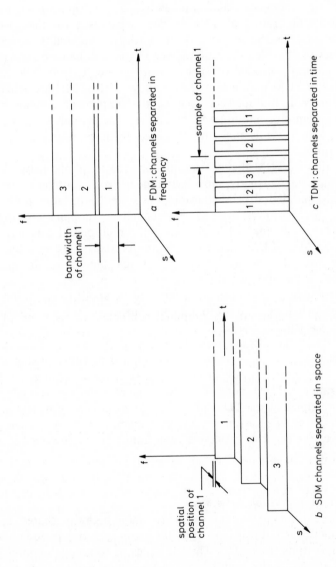

Fig. 4.2 *Illustration of SDM, FDM and TDM*

illustrates the three multiplexing techniques using the co-ordinates of space, time and frequency.

Multiplexing was originally introduced to telecommunications as a means of providing cost-effective long-distance transmission. The first application was to telegraphy, using TDM. However, the advent of telephony, with its 3100 Hz bandwidth of analogue signal, favoured the development of FDM systems.[1] Multiplexing ratios of 12:1 or 24:1 were used on the earliest telephony FDM systems. Since then, the progressive availability of better electronic devices has enabled the multiplexing ratios, and hence capacity, to increase. The current capacity of such systems is nearly 1000 telephony channels. Values of FDM multiplexing ratios above 12:1 or 24:1 are achieved by cascading levels of multiplexers: that is, the outputs of a number of multiplexers are themselves multiplexed, and so on. The pendulum has swung the other way and TDM is now the favoured form of multiplexing for telecommunications.

The introduction of TDM digital transmission into telephone networks is described in Chapter 3. Currently, most telephone networks employ a mixture of analogue FDM and digital TDM transmissions systems, the latter gradually replacing the former. However, prior to the recent advent of digital exchanges, all switching was at the channel level and so the multiplexed transmission links, whether TDM or FDM, had to be demultiplexed before entering the exchange. The use of fast semiconductor switch blocks within digital exchanges enables the economics of time-division multiplexing, which has for so long been exploited in transmission systems, to be applied successfully to public telephony switching. The profound consequences of this are described progressively in subsequent chapters.

Although reference has already been made to 'digital TDM' systems, time-division multiplexing is a process that can be applied to either analogue or digital signals. Furthermore, digital transmission and switching do not necessarily operate in a TDM mode. Therefore, the remainder of this chapter considers the principles of the TDM process in its analogue form, and gives an example of the early electronic TDM exchanges, which were analogue systems. It leaves the separate process of digitisation to be described in the following chapter.

4.2.2 A time-division-multiplexed system

A time-division-multiplexed system has a common highway shared by a number of channels, each of which occupies the highway during periodic slices of time, known as 'time slots' (TS). This multiplexing technique requires the transformation of each of the input signals into a sequence of samples, which are then interleaved and carried within appropriate time slots on the highway. The conversion of a continuous signal waveform, such as the analogue output from a telephone, to a sequence of discrete samples is produced by a sampling

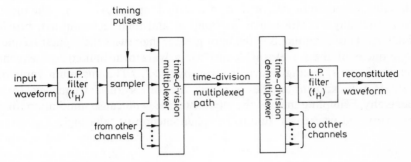

Fig. 4.3 *A time-division-multiplexed system*

system. The processes of sampling and time-division multiplexing are described separately below.

4.2.2.1 Sampling system. Fig. 4.3 shows the elements of a sampling system within an overall TDM system. The sampling processes comprise: lowpass pre-sampling filtering, sampling and low-pass recovery filtering. The sampling processes are applied to each of the constituent channels (or tributaries) carried over the TDM system. The required samples of each input waveform are derived from a periodic train of timing pulses which switch the sampler on and off, thus passing slices of the input waveform. These slices, known as samples, are in the form of a set of pulses with heights equal to the value (positive or negative) of the input waveform at the instants of the timing pulses. The profile of the sample heights approximates to the original waveform, as shown in Fig. 4.4.

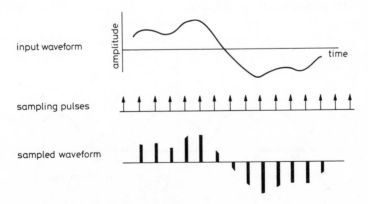

Fig. 4.4 *The sampling process (time-domain)*

The sampling theorem, attributed to Nyquist, enables an appropriate sampling system to be designed for a specified input waveform.[2] The theorem states that any waveform may be sampled and then reconstituted, provided that the rate of sampling is equal to or greater than twice the highest frequency component of the waveform. Thus, for a waveform limited to frequencies below f_H, the sampling rate f_S is given by $f_S > 2f_H$ The minimum value of f_S (i.e. $2f_H$) is known as the Nyquist rate or Nyquist frequency for f_H. Generally, sampling below the Nyquist rate will cause mutilation of the reconstructed waveform, due to the phenomenon of 'aliasing'.

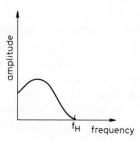

a the original waveform after low-pass filtering

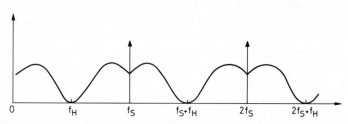

b the sampled waveform with $f_S = 2f_H$

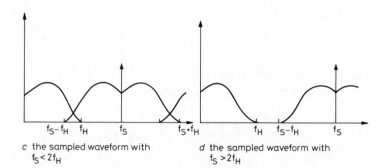

c the sampled waveform with $f_S < 2f_H$ *d* the sampled waveform with $f_S > 2f_H$

Fig. 4.5 *The sampling process (frequency domain)*

Aliasing is the term given to the overlapping of adjacent sidebands produced by the sampling process, the latter being essentially one of amplitude modulation. This concept is illustrated in Fig. 4.5, where the sampling of a waveform at a rate of f_S produces upper- and lower-sideband versions of the waveform centred at frequences of nf_S $(n = 1, 2, 3,$ etc.$)$. It can be seen that with f_S *less than* $2f_H$, the sidebands overlap and aliasing is produced. This representation demonstrates that the minimum separation of two adjacent harmonics nf_S and $(n + 1)f_S$ *must be* $2f_H$ if aliasing is to be avoided. Clearly, a sampling frequency greater than $2f_H$ would ensure no aliasing, but sampling at much in excess of the Nyquist rate is wasteful of bandwidth. (In practice, with certain types of waveform, sub-Nyquist-rate sampling may be possible, but this is a specialised subject beyond the scope of this book.)

It is interesting to note that Fig. 4.4 describes the sampling process in the time domain, while Fig. 4.5 describes the process in the frequency domain. Use of both domains is also helpful in considering the process of recovering a waveform from the sampling system. Thus, Fig. 4.5 shows that the frequencies associated with the input waveform can be retrieved by passing the sampled waveform through a lowpass filter with a cut-off frequency of f_H.

Fig. 4.6 illustrates the recovery process by reference to the time domain. Fourier analysis[3,4] shows that the temporal response of a lowpass filter to a single sample pulse, at time 't', follows a $(\sin x)/x$ characteristic centred on t (Fig. 4.6a). A reconstructed waveform, which approximates to the input

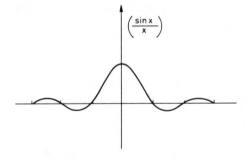

a filter response to one pulse

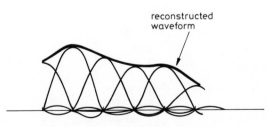

b result of filter response to the series of
pulses in a sampled waveform

Fig. 4.6 *Reconstruction of input waveform (time domain)*

waveform, is derived from the aggregate of the $(\sin x)/x$ responses resulting from the sequence of sample pulses passing through the low-pass recovery filter (Fig. 4.6b).

The elements of the practical sampling system shown in Fig. 4.3 can now be considered in more detail. The input signal is first passed through a low-pass filter (known as an antialiasing filter) to ensure that no frequencies greater than f_H are present. The sampler, driven by the timing pulses, produces a train of samples of the input waveform, which are then conveyed over the pulse path. Reconstruction of the original waveform, usually in attenuated form, is produced by an output lowpass filter. The sampling process introduces attenuation because only a proportion of the original waveform is passed by the sampler. The degree of attenuation is given by 20 log (T/d),

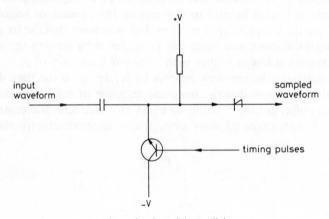

a sampler using transistor switch

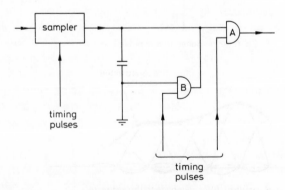

b sample-and-hold circuit

Fig. 4.7 *Sampling systems*

where T is the sampling period (inverse of sampling rate) and d is the width of the samples.[2]

The necessary rapidity of the sampling process means that some form of electronic gating must be used to produce the samples. Examples of such sampling devices are diode bridges and transistor switches.[5] Fig. 4.7a illustrates a transistor circuit that can sample the current from an input waveform. Ideally, a sample should represent an instantaneous value of the waveform, and thus it should have negligible duration. Practically, of course, the sample will have a duration, but this should be short enough for the slice of the waveform to be, essentially, rectangular. A 'sample-and-hold' circuit is used to ensure that the input waveform remains constant as the sample is taken (Fig. 4.7b). A capacitor (C) charges up to the maximum value of the input waveform during the sampling (for the duration of the timing pulse). This sample value is thus stored and may then be passed through the gate (A) for multiplexing or switching, etc. at the end of the sampling period. The capacitor is discharged via gate B just before a new sample is made, and the process repeated. Other timing pulses are required to operate gates A and B at the appropriate instants.

The action of producing a train of pulses whose height is equal or proportional to corresponding sample values results in the amplitude modulation of the pulses. Such a scheme is thus called pulse-amplitude modulation (PAM). There are three other means of analogue modulation of pulses by a set of samples,[2,4] namely pulse width, length or duration modulation (PWM, PLM or PDM), pulse-position modulation (PPM) and pulse-frequency modulation (PFM). These other modulation techniques are not considered in this book.

4.2.2.2 Time-division multiplexing. Provided that a waveform is sampled at or above the Nyquist rate, the duration of the pulses that represent the samples is arbitrary. Thus, the possibility exists of interleaving samples from a number of waveforms (input channels) on to the one common pathway. This can be done provided that the sets of samples from each channel are taken at different times, but at the same sampling rate. Fig. 4.8 illustrates how the interleaving of PAM samples from three channels enables the use of a common pathway, which is thus *time-division-multiplexed*. Note that each channel is first bandlimited by an antialiasing lowpass filter before being sampled by the appropriate set of timing pulses, P_1, P_2 or P_3. These trains of timing pulses all have the same height and repetition rate, but are staggered in time relative to each other so that the sampling gates operate in the sequence 1, 2, 3, 1, 2, etc. At the receiving end, an identical set of pulses (P_1, P_2 and P_3) applied to the corresponding sampling gates segregates the PAM samples output from the TDM highway into the constituent channels. The three output waveforms are reconstructed by passing the three trains of PAM samples through low-pass filters.

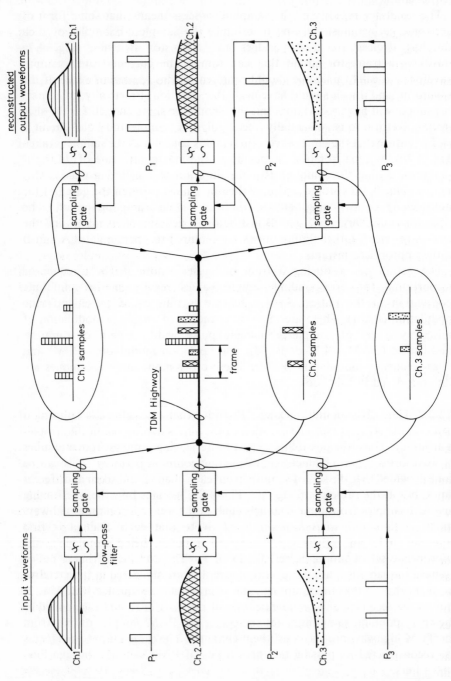

Fig. 4.8 Example of a 3-channel TDM system

Fig. 4.8 also illustrates the important concept of a TDM frame. A frame contains a number of time slots, each of which contains a sample from the corresponding channel. In the example given, the frame contains three time slots conveying the samples from three channels. The number of frames per second is equal to the sampling rate. The duration of the frames is given by the reciprocal of the sampling rate and is thus independent of the number of channels being multiplexed; however, the greater the number of channels, the smaller is each time slot.

In order for the above TDM process to work, it is essential that the train of timing pulses be applied at corresponding instants at the send and receives ends. If this is not done, the received PAM samples will be either mutilated or misrouted to other output channels. In this example, the two sets of timing pulses are assumed to originate from a common source and, hence, are in step. However, with a TDM transmission system, where the two ends are physically far apart, a means of aligning the two sets of timing pulses is required. In the case of a PAM/TDM system, this is achieved by the use of an additional time slot within the frame to carry a distinctive pulse. The detection at the receiver of this pulse, which has a constant value larger than the maximum height of the PAM samples for the speech channels, enables the start of the frame to be identified. The system is then said to be in 'frame alignment'. All TDM systems require a frame-alignment process. The concept is described further in the following chapter.

4.3 PAM/TDM switching

Early electronic exchanges used the concept of time-division multiplexing of PAM samples as a means of saving on switching hardware and exploiting the advantages of electronic systems. The use of TDM not only saved switch-block hardware but also enabled signalling and control equipment to be shared among many telephone users, as will be described in subsequent chapters. However, these PAM/TDM exchanges were still analogue, since the samples were direct slices of the input waveforms. The world's first such exchange to be in public service was opened with 600 subscriber lines at Highgate Wood, London, in December 1962.[6] Although the Highgate Wood prototype system demonstrated the technical practicability of PAM/TDM switching, the design, as it stood, had too many disadvantages to be a viable proposition and the exchange was soon closed. However, it is instructive to consider the principle of operation of such systems as a prelude to the fuller coverage of digital exchange systems in the following chapters.

The principle of operation of a PAM/TDM exchange is the controlled access to a common PAM/TDM highway (or bus) by each of the subscribers' lines involved in a call. The method of working can best be explained by

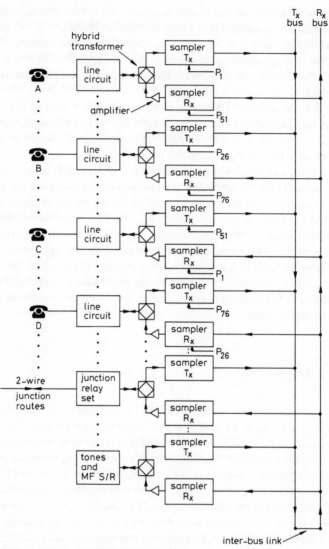

Fig. 4.9 *Basic trunking of a TDM/PAM exchange*

referring to Fig. 4.9. In the exchange, each call is allocated a pair of discrete time slots (one for each direction of transmission).

As with all electronic exchanges, each subscriber's line terminates on an appropriate interface unit (see Chapter 7). In the case of Highgate Wood, this unit was called a 'line circuit'. It performed the functions of extracting loop-disconnect dialling, providing ringing current, off-hook detection, test access, power feeding and overload protection. Thus, only the speech waveforms from the subscriber lines were passed to the samplers. The electronic gates used in the sampler were unidirectional devices, so the exchange operated as a 4-wire system, with separate transmit and receive channels. The signal from each two-wire subscriber line was split by a hybrid transformer into transmit and receive channels. The transmit channel passed through a transmit sampler, shown as 'sampler Tx' in Fig. 4.9. This comprised a low-pass filter and sampler gate and terminated on the TDM transmit (Tx) bus. The receiver sampler, shown as 'sampler Rx' in the Figure, comprised a low-pass filter and amplifier and was directly connected to the TDM receive (Rx) bus. Junction circuits, call-progress tones (e.g. dial tone) and MF signalling equipment were similarly connected to the two buses.

The sampling rate was 10 kHz, giving a frame length of 100 μs. There were 100 time slots on the exchange buses, each occupying 1 μs. To interconnect two subscribers, A and C in Fig. 4.9, for example, a pair of free time slots (one for each direction of transmission) was chosen by the exchange control. For the A-to-C direction, a pulse P_1 would be applied once every 100 μs to A's sampler Tx and C's sampler Rx. All other sampler gates were inoperative during the time slot relating to P_1. For control simplicity, the pulses for the transmit and receive directions of transmission were chosen to be half a frame, or 50 μs, apart. The return path was, therefore, established by the application of pulse P_{51} (i.e. P_1 plus $_{50}$) to C's sampler Tx and A's sampler Rx, as shown in Fig. 4.9.

A call set-up sequence began with the detection by the calling-line circuit of the off-hook condition. This was followed by the establishment of a connection between the dial-tone sender and the calling line (Fig. 4.9), using the technique described above. The loop-disconnect dialled digits were received by the line-circuit equipment and relayed to the exchange-control system over the bus (not shown in Fig. 4.9). The connection between the calling-line circuit and the dial-tone sender was cleared as soon as the first digit was received. Once the received digits had been analysed, the exchange-control system determined the identity of the called subscriber's line circuit, or the outgoing junction route in the case of an external call. The required connection between calling and called subscribers' lines was then established via the TDM buses (assuming an own-exchange call). In the case of a call involving an MF telephone, a unidirectional connection between the calling line and an MF receiver was established prior to the dial tone being sent. This additional connection across the exchange was cleared as soon as the cessation of MF signalling was detected.

The remainder of the 48 pairs of pulses could be used for the establishment of other connections through the TDM switch block. For simplicity, the Tx and Rx TDM buses are shown directly connected, in Fig. 4.9. In Highgate Wood, additional capacity was achieved by the use of a number of buses interconnected by gates using some of the available time slots on the buses.

Development of analogue TDM-exchange systems was hampered by the inherent susceptibility of analogue signals to noise, and the crosstalk and gain stability problems associated with PAM/TDM operation, particularly in public switched telephone networks (PSTN). However, these problems were minimised by the well controlled office environment of PABX systems with their use of short extension loops and provision of special telephone instruments. A number of analogue PAM/TDM PABX systems were manufactured during the late 1960s and 1970s, and these were highly successful.[7] Their method of operation was similar to that described above.

However, the inherent problems of analogue TDM switching for public telephony, and the emerging presence of digital TDM transmission systems, led the manufacturers and administrations to develop digital TDM exchange systems for the PSTN. The latest PABX designs have taken advantage of SPC digital developments, and both public and private exchange systems are now exploiting similar technology.

4.4 References

1 BYLANSKI, P. and INGRAM, D.G.W. (1980): *Digital Transmission Systems* (Peter Peregrinus Ltd), Revised edition, pp.11−15
2 BYLANSKI, P. and INGRAM, D.G.W.: Ibid. pp.38−43
3 SKILLING, H.H. (1966): *Electrical Engineering Circuits* (John Wiley and Sons, Inc), pp.501−508
4 SMITH D.R. (1985): *Digital Transmission Systems* (Van Nostrand Reinhold Co). Chap. 3
5 BENNETT, G.H. (1983): 'Pulse code modulation and digital transmission', written for Marconi Instruments Ltd, Second edition, pp. 23−25
6 HARRIS, L.R.S. *et al.* (1960): *the Highgate Wood experimental electronic telephone exchange. General introduction,* Proc. IEE, 107B, Suppl. 20, pp. 70−80
7 BELLAMY, J. (1982): *Digital Telephony* (John Wiley & Sons), pp. 13, 242, 243

Chapter 5

Principles of digital systems

5.1 Introduction

Chapter 3 described the worldwide trend towards the use of digital technology for the control, switching and signalling functions of telephone exchanges and the transmission links between them. An important element of this movement was the development of methods for conveying voice signals in digital format. The technique for converting analogue speech to a digital representation was invented by Reeves in 1937, but it was only the introduction of semiconductor technology that enabled practical digital transmission systems to be employed in telephone networks from the mid 1960s onwards.[1] Currently the most common method of digitising speech is pulse-code modulation (PCM), which is the focus of this chapter. However, there are several other techniques, which have been used in small PABX systems and digital leased-line applications, and these are also briefly described at the end of the chapter.

Unfortunately, the practical application of digital technology to telephony is confused: not only are there several different techniques for digitising speech waveforms, but there are two distinct internationally agreed standards for PCM and all of these methods are incompatible with each other. An additional complication is created by the different forms of digital signal representation used in digital transmission systems. Furthermore, digital technology is an area of rapid advance, with new forms of digitisation progressively being developed and internationally standardised. In this chapter, the reader is led through the jungle of the principles of digital telephony transmission so the fundamentals of digital switching, described in the remainder of Part II, can be more easily understood.

5.2 Pulse-code modulation

5.2.1 The concept

Pulse-code modulation is the common method of converting analogue signals into digital format (and vice versa) so that they may be conveyed over a digital

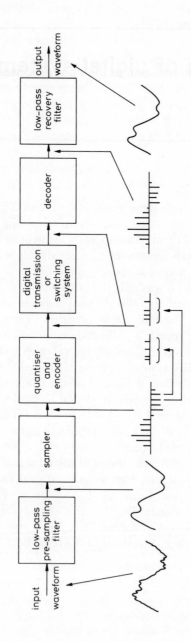

Fig. 5.1 The processes of PCM

line system or digitally processed (e.g. digitally switched or stored). The conversion comprises three processes: sampling, quantising and encoding. There is then a fourth process of time-division multiplexing a number of channels. However, PCM is not always a multiplexed system and is used in single channel format. The fundamentals of sampling and time-division multiplexing have been covered in the previous chapter.

Fig. 5.1 shows a simple representation of the PCM processes, as applied to a single channel. The Figure also shows how an input waveform is represented in each of the processes. The first stage of PCM is the sampling of the input voice signal which produces a sequence of analogue samples in the form of a pulse-amplitude-modulation (PAM) stream. These PAM samples have a continuous range of amplitudes within the working range. The next step is the division of this amplitude range into a limited number of intervals. All samples with amplitudes that fall within a particular interval are assigned the same quantam-level value. This is known as 'quantising'. Finally, within the encoder, the magnitudes of the quantised samples are representated by binary codes. The PCM process thus produces a stream of binary digits, grouped into 'PCM words', that represent the input speech waveform.

The binary stream may then be switched in a digital exchange. Alternatively, it may be carried over a digital transmission link after conversion to a suitable format (i.e. using a line code — see later). At the distant end, the binary code is converted back to a series of PAM samples by a decoder. Finally, an approximation of the original input waveform is derived from the PAM samples by a recovery (low-pass) filter.

It can be seen from Fig. 5.1 that the PAM signals at the outputs of the sampler and decoder mimic the filtered analogue audio input waveform, while the PCM signal is a binary representation of each sample. The following Sections consider the processes in more detail and give a description of the two internationally standardised practical PCM systems.

5.2.2 *Sampling*
The audio telephony signal has a power spectrum which is significant up to about 10 kHz. However, most of the power resides in the lower part of the range. Thus, for economy of bandwidth in multiplexed transmission systems, both FDM and TDM, telephony channels have traditionally been band limited to the range 300–3400 Hz. In practice, the use of relatively cheap pre-sampling filters means that there will be some attenuated power passed at frequencies higher than the nominal cut off of 3.4 kHz. To allow for this non-perfect filtering and hence keep aliasing negligible, a sampling rate of 8 kHz is internationally standardised for telephony PCM systems. This sampling rate, which corresponds to the Nyquist frequency for signals up to 4 kHz, produces a train of amplitude-modulated pulses, spaced at 125 μs intervals.

5.2.3 Quantisation

Quantisation is the process of producing a digital representation of a sampled analogue signal. After first limiting the amplitude to the working range of the equipment, each sample is compared with a set of quantum levels in turn and assigned to the one to which it approximates. By definition, all samples within an interval between two quantum levels are deemed to have the same value. The assigned value is then used within the digital switching or transmission system. Recovery of the original analogue signal requires the process to be reversed. Thus, an approximation to the original input signal is derived from a set of PAM samples which are generated to correspond to the quantum values in the stream of PCM samples. The effect of the quantisation process on a range of input PAM values is shown in Fig. 5.2. Fig. 5.2*a* shows how both

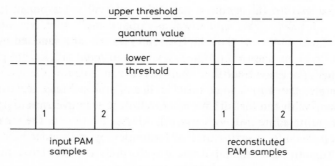

a the origin of quantisation error

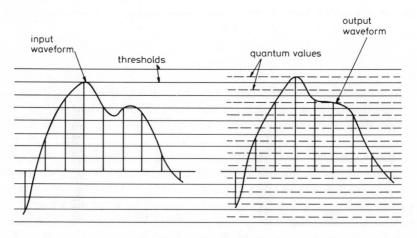

b example of quantisation distortion

Fig. 5.2 *Quantisation error and distortion*

input sample 1, which is close to the upper threshold, and input sample 2, which is close to the lower threshold, are represented by output PAM samples of the same height, as determined by their common quantum value. Thus, the quantisation process has imposed a rounding error on both samples. Fig. 5.2*b* shows that the quantisation process introduces some distortion to the input waveform, known as quantisation distortion. The series of errors in the reconstituted PAM samples is perceived by the listener as noise and is known as quantisation noise.

If the sample steps are equally spaced, the quantisation distortion is worse for small signals than for large ones. Since the distortion is perceived by the listener as noise, the ratio of signal (PAM sample height) to (quantisation) noise worsens as the input signal level is reduced. This problem is minimised by spacing the quantum levels logarithmically, so that large signals are allowed a large error and small signals can only have a small error, thus giving essentially constant signal/quantisation-noise ratios over the working amplitude range. An example of the effect of logarithmic quantisation on the signal/noise ratio is shown in Fig. 5.3. In the case of the linear quantum spacing, where the error E_1 with a small sample value equals the errors E_2 of a large sample, the signal-noise ratio is clearly much worse for the lower sample value. With the logarithmic quantising the errors E_3 and E_4 give the same signal/noise ratios for both sample values.

The process enables a wide range of amplitudes to be encoded by a given number of quantising levels. Since the effect is the compression of the higher amplitudes into fewer quantising levels, the logarithmic quantising technique is also known as 'companding'.

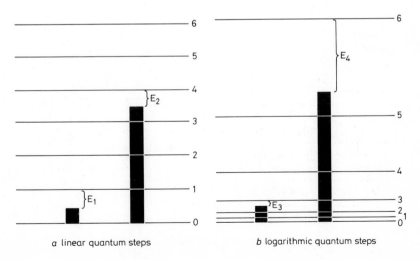

a linear quantum steps   *b* logarithmic quantum steps

Fig. 5.3 *Linear and logarithmic quantisation*

In addition to producing a constant signal/quantisation-noise ratio, companding offers system economies compared with linear quantisation. For a typical specification of a signal/quantisation-noise ratio of 40 dB over a dynamic range of − 35 dBm to + 3 dBm, linear quantisation would require some 8000 quantum steps compared to the 256 quantum steps required with logarithmic (companded) quantisation. Since each quantum step must be uniquely addressed by a binary code (which forms the PCM word), the linear system would require 13 bits compared to the 8 bits for the logarithmic system. Such saving is important, because the PCM word size has a direct effect on the cost and capacity of digital switching and transmission systems.

Two different companding laws have been internationally standardised: the 'A law' and the 'Mu (μ) law'.[2,3] Both laws have 256 levels, but they differ in their approximations to a logarithmic characteristic. Fig. 5.4 illustrates how the A-law compander uses 13 straight-line segments to represent a logarithmic curve; the Mu law uses 15 segments. The Figure plots the input sample voltage in a normalised form (so + 1 is the maximum positive voltage allowed) showing the positive quadrant in detail and the beginning of the negative quadrant, which forms a mirror image. A single segment covers the input voltage range between + 0.0156 and − 0.0156, which are encoded + 32 (00010000 in binary) and − 32 (10010000), respectively. The remainder of the positive and negative ranges are each represented by a series of six segments. The Figure illustrates how voltages in the top half of the range are allocated

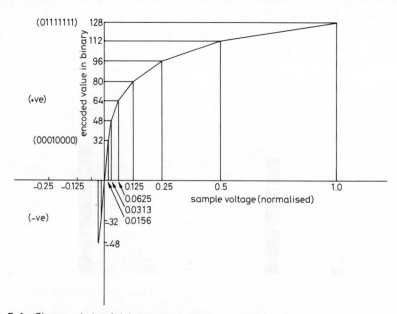

Fig. 5.4 *Characteristic of A-law companding*

only 16 quantum levels compared to the 32 quantum levels covering the lowest 1.6% of the positive voltage range.

5.2.4 Encoding

Each of the quantum levels is designated a binary number, e.g. an 8-bit number for a 256-level system ($2^8 = 256$). By convention, the first digit represents the sign of the sample: 0 for those sample values in the positive section of the range, and 1 for those values in the negative section of the range (see Fig. 5.4). The remaining seven digits then represent the magnitude; the first indicates the upper or lower half of the range, the second the upper or lower quarter, the third the upper or lower eighth, and so on. Each PAM sample representing one channel is thus encoded into a PCM word of 8 bits. Decoding, performed at the termination of the PCM system, is the process of reconstituting the PAM samples from the PCM words.

There are several practical methods of PCM encoding; Fig. 5.5 shows a schematic diagram of the commonly used successive-approximation system.[3] It consists of a comparator that compares a current, derived from an input sample, with a reference current. The latter is generated by an 8-step ladder network, the output of which represents the 8-bit (parallel) PCM word corresponding to the input sample. The name of the technique results from a sequence of successive comparisons between the sample current I_s and a decreasing range of reference current I_{ref}. The procedure is thus not instantaneous and the sample value must be held constant for the duration of the encoding process.

Referring to Fig. 5.5, the sequence of operations is as follows:

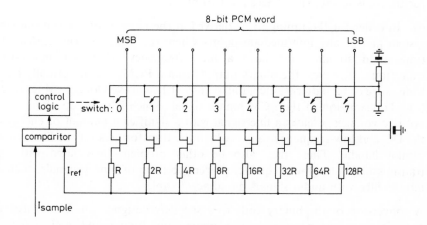

Fig. 5.5 *Successive approximation method of encoding*

(i) I_s is compared with I_{ref} when switch 0 is closed and all other switches are open.

(ii)   If the comparator output is positive, i.e. I_s is the greater, the switch 0 is kept closed, otherwise it is opened. The output from switch 0 is then either at a high or low voltage representing a digit 1 or 0, respectively; this forms the most significant bit (MSB) of the PCM word.

(iii) I_s is then compared with I_{ref} when switch 1 is closed (switch 0 open or closed as appropriate) and switches 2 to 7 are open.

(iv)   If the comparator output is positive, switch 1 is kept closed, otherwise it is opened.

(v)   This procedure continues until all switches have been tested in the closed position.

Since the currents from each switch 0 to 7 decrease by a factor of 2, owing to the value of the resistances in circuit, the summation of currents forming I_{ref} successively converges by, improving approximations, to I_s.

The decoder comprises the arrangement of eight current switches which are opened or closed according to the value of the corresponding bits in the input PCM word. The aggregate current from the switches forms the output sample current.

Although the successive-approximation encoder has been widely employed in PCM systems, recent advances in microprocessor technology have led to the emergence of new improved techniques. In particular, the special signal processors enable PCM encoding and decoding to be performed faster and for more complex coding algorithms such as ADPCM (see Section 5.4).

5.2.5 Line coding

The binary code generated for each sample by the PCM encoder is not suitable for transmission to line. There are two problems:

(i)   In order for the timing to be derived at the receiver of the transmission system and at any intermediate points where the signal is regenerated, the transmitted stream must have a high frequency of 1-to-0 and 0-to-1 transitions. However, the binary output from a PCM system normally has a wide range of bit patterns, including strings of 1s and even longer strings of 0s.

(ii)   Transmission of the PCM binary signal, which is made up of zero and positive voltages, creates a line signal with a significant amount of energy in the DC and low-frequency range, which the transformers on the line system cannot handle. There may also be other problems associated with the transmission of the low-frequency power spectrum of binary signals, such as interference with audio signals on adjacent cables.

A conversion of the binary code to a specially designed line (transmission) code is therefore necessary to overcome the timing and power-spectral problems. There are many line codes that have been devised which offer various degrees of bandwidth utilisation, performance and equipment complexity.[4] It is important to note that digital switch blocks work with binary signals, and so all digital-line systems terminating at the exchange require code conversion (Chapter 7). The most common line codes are briefly reviwed in the remainder of this Section.

5.2.5.1 ADI/AMI. This line code is a combination of two distinct techniques: alternate-digit inversion (ADI) to inject the necessary timing content and alternate-mark inversion (AMI) to eliminate the low-frequency component of the binary signal:

(i) ADI: In this process alternate digits are inverted (i.e. 0s become 1s; 1s become 0s). This gives an overall increase in the number of 0-to-1 and 1-to-0 transitions compared with the original PCM binary signal.

(ii) AMI: The AMI technique converts the ADI binary signal, with values of + 1 and 0, to a pseudo-tertiary (3-level) bipolar signal with values of + 1, 0 and − 1. This is achieved by changing alternate bits with a value of 1 in the input binary signal to − 1 in the output (line code) signal. (The term 'mark'

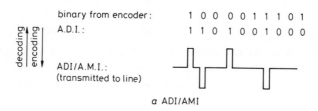

a ADI/AMI

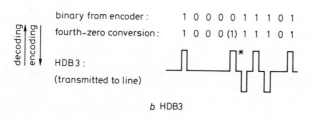

b HDB3

Fig. 5.6 *Example of two line codes*
Notes: (1) False mark to indicate 4th zero
*Indicated by a bipolar violation in transmitted signal

is used to indicate a binary value of 1.) The efficiency of this code conversion is given by the ratio of the information carried to the capacity of the line code. Thus, AMI has a code efficiency of just 67% since only a 2-level signal is derived from the three possible states of the tertiary line-code signal $[(1 \times 2)/(1 \times 3) = 67\%]$.

An example of ADI/AMI is given in Fig. 5.6. This shows the encoding process reading down the page and the decoding process reading up.

5.2.5.2. HDB3 (high density bipolar 3). The HDB3 line code uses AMI to produce a pseudo-tertiary bipolar signal with little low-frequency content. The name HDB3 relates to the way that a high-timing content is injected into the signal. The coding technique eliminates long strings of 0s by injecting false marks in place of each fourth consecutive 0 to ensure that the maximum number of consecutive zeros transmitted to line is limited to three. Each false mark is indicated by a 'bipolar violation', so that the original binary signal can be reconstituted at the far end. The bipolar violation is created by interrupting the normal sequence of $+1$, -1, $+1$, etc. for representing binary 1 (i.e. the AMI output), the false mark being indicated by repetition of the previous sign. Thus, for example, in a binary sequence of 100001, the line code comprises $+1000+1-1$; the second $+1$ symbol is interpreted by the receiver as a false mark representing a zero. This is illustrated in Fig. 5.6.

5.2.5.3 4B3T (4 binary to 3 tertiary). The coding produces a bipolar signal by converting blocks of four binary digits into blocks of three tertiary digits. This process gives a number of advantages in addition to removing the low-frequency content, namely:

(i)   The coding is designed to ensure that each 3-tertiary-digit block contains at least one mark, thus producing the necessary timing content.
(ii)   The scheme has efficiency of 89% since four bits of information (4×2 states) are carried by a 3-digit tertiary signal (3×3 states) $[4 \times 2)/(3 \times 3) = 89\%]$.
(iii)   The transmitted line rate is only ¾ of the PCM binary rate. Such a reduction enables transmission system parameters (e.g. crosstalk immunity, regenerater spacings, etc.) to be improved.

There are a number of variants of 4B3T coding, an example being MS43.[4]

5.2.5.4 6B4T (6 binary to 4 tertiary). By transcoding 6-bit binary blocks directly into 4-tertiary digit blocks, 6B4T achieves an efficiency of $(6 \times 2)/(4 \times 3) = 100\%$, together with a ⅓ reduction in line rate compared with the binary signal. The timing content is ensured by the presence of a mark in each tertiary block. However, the complexity and cost involved in producing 6B3T coders currently restricts this form of coding to the higher-level digital transmission systems (described briefly at the end of this chapter).

5.3 Operational PCM systems

The components of an N-channel multiplexed PCM line system are shown in Fig. 5.7. (This diagram is an elaboration of Fig. 5.1.) For clarity, only the components in the transmit direction are shown, the components in the receive direction being a mirror image. All PCM systems are 4-wire, that is, they have a pair of wires for each direction of transmission. Strictly speaking, the arrangement illustrated should be called a PCM/TDM system since N circuits are multiplexed on to a single 4-wire circuit. However, the TDM label is usually dropped unless it is necessary to differentiate it from single-channel PCM systems, such as are used in space-communication satellite links.[5]

Each audio channel terminates on a channel unit in the PCM terminal, as shown in Fig. 5.7. These units provide 2-to-4-wire conversion, extraction of DC signalling and pre-sampling filtering in the transmit direction of transmission, and low-pass recovery filtering and injection of DC signalling in the receive direction. Different types of channel units are used to terminate the other forms of analogue circuits, e.g. 4-wire audio without DC signalling. A description of PCM terminal equipment is beyond the scope of this book. However, similar functions are provided by the analogue line termination equipment in the digital exchanges, and these are described in Chapter 7.

Fig. 5.7 shows that the PCM system extends from the audio (analogue) channel inputs to the audio (analogue) channel outputs, and includes analogue-to-digital (A/D) conversion, multiplexing digital transmission, demultiplexing and digital-to-analogue (D/A) conversion. The digital line system refers only to the equipment between and including the digital line terminations. In some cases, digital line systems carry channels which originate in digital format; because no analogue-to-digital conversion is then required, such systems should be termed 'digital transmission' rather than 'PCM'.

There are now two types of operational PCM systems in the world: the North American 24-channel system (known as T1) and the CEPT 30-channel system. The 30-channel system is the CCITT international standard,[6] which has been adopted by many European countries and elsewhere for regional use, and all countries for international use. The 24-channel system is the alternative CCITT standard[7] for regional communications, and is used throughout the American continent and Japan. Digital exchanges are designed to operate in either 24- or 30-channel mode and, unfortunately, interworking equipment is necessary where there is a mismatch between the type of PCM system used by the digital exchanges and that employed by the digital transmission links.

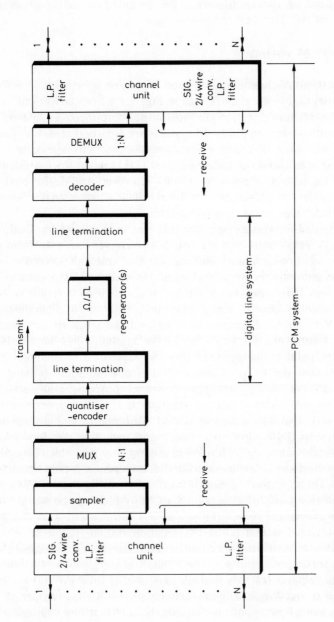

Fig. 5.7 *PCM-system components*

Although both systems use 8 kHz sampling, resulting in a 125 μs frame, they differ not only in the number of channels multiplexed but also in the signalling and frame-alignment format and the quantising laws employed. Emphasis in this book is placed on the 30-channel PCM system. For completeness, the 24-channel system will also be briefly described.

5.3.1 The 30-channel PCM system

5.3.1.1 Frame structure. Fig. 5.8 shows the CCITT standard frame format for the 30-channel PCM system.[8] The 125 μs frame contains 32 time slots: 30 of these are for speech, one is for signalling, and one is for frame alignment, line-system alarms and network-control signals. The time slots (TS) are numbered TS0 to TS31: TS0 is allocated to alignment and network-control signals; TS1 to TS15 are for speech channels 1 to 15, labelled Ch1 to Ch15; TS16 is used to carry either channel-associated or common-channel signalling (CAS or CCS); TS17 to TS31 are for the remaining 15 speech channels, labelled Ch16 to Ch30.

Each time slot within the frame occupies 125 μs/32 = 3.9 μs. Eight-bit encoding using A-law companding is used, giving 256 levels to represent the speech samples. Thus, each bit occupies 3.9 μs/8 = 0.488 μs, which is made up of a 0.244 μs pulse and 0.244 μs separation (using a 50% duty cycle). The gross-bit rate of the PCM system is 8 kHz × 8 bit × 32 timeslots = 2048 kbit/s or 2.048 Mbit/s, which is often abbreviated to '2 Mbit/s' for convenience.

5.3.1.2 Frame alignment. Reference has already been made to the term 'frame alignment' which, in the 30-channel PCM system, is associated with TS0. The function of frame alignment is best described by considering the needs of the termination at the receiving end of a PCM system. At the termination, a stream of binary bits is received at the rate of 2048 kbit/s. However, this stream of bits is meaningless unless they can be allocated into the right 8-bit time slots, enabling the contents of each channel to be identified. This allocation is achieved by the sending terminal injecting a distinguishable pattern in TS0 which the far end searches for within the stream of received bits. Once the pattern is detected, bit 0 of TS0 can be located and hence all the following 255 bits of the frame identified; the receiving terminal is then in 'frame alignment' with the sending terminal. The function of frame alignment is sometimes termed 'frame synchronisation'

Table 5.1 shows the standard format of 8 bits of TS0 in a 30 – channel PCM system used to convey the frame-alignment pattern. The process of achieving frame alignment requires the use, in TS0, of a unique bit pattern that has a very low chance of occurring in the remainder of the frame. This objective could be met by using a very long pattern, say of 32 bits, with 8 bits occurring in each TS0 of successive frames, thus requiring four frames to transmit the pattern once. However, the longer the pattern, the greater the time that is required in searching for it and hence the time for a PCM system to achieve alignment. Since the PCM system is out of service during states of lost-frame alignment, the search time must be minimised. The compromise adopted for the 30-channel PCM system is for a single 7-bit pattern of '0011011', known as the 'frame alignment signal' (FAS), to be carried in TS0 of each odd frame. Frame alignment is achieved when the sequence 'FAS-absence of FAS-FAS' is detected during three successive frames.

Loss of frame alignment is determined by the receipt of three consecutive frames in which the FAS is not detected. This threshold is a compromise between avoiding unnecessary re-alignment action as a result of transmission impairment of the received signal and unnecessarily delaying corrective action when there is genuine loss of alignment. Re-alignment is achieved by searching successively for the FAS and then checking for the 'FAS-absence of FAS-FAS' sequence, as described above. When the receive terminal detects loss of frame alignment, an alarm is signalled back to the transmitting terminal by setting bit 3 of the non-FAS word from 0 to 1 in the TS0 of the return-transmission link.

Table 5.1   *Bit allocations for time slot 0*

Odd Frames	Even Frames
Frame alignment signal (FAS)	Non-FAS word
Y0011011	Y1*XXXXX

Key

X: digits not allocated by the CCITT to any particular function and normally set to 1.[8]
Y: reserved for international use (normally set to 0).
*: digit normally 0 but changed to 1 when loss of frame alignment occurs and/or system-fail alarms occur.

5.3.1.3 Signalling. In the 30-channel PCM system, TS16 is a separate dedicated channel for the conveyance of either channel-associated or common-channel signalling for a group of dependent speech channels. It is important to note that these signalling methods are mutually exclusive and cannot be mixed on one PCM system. Both of these forms of signalling are described in Chapter 2, while this Section considers how they fit within the 30-channel frame format, in order to appreciate their influence on digital switching.

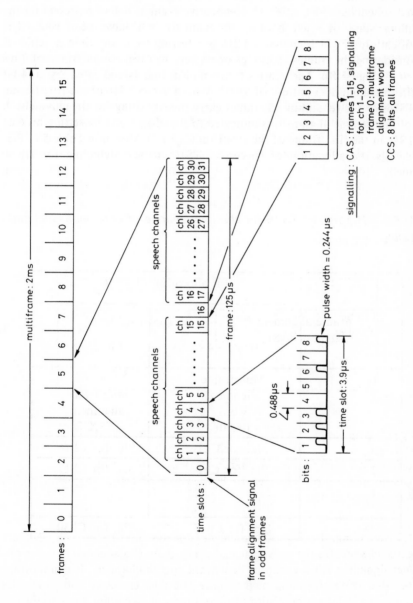

Fig. 5.8 *The 30-channel PCM frame format*

(i) Channel-associated signalling: For channel-associated signalling (CAS), TS16 is used to convey a 4-bit representation of the in-band 10 p.p.s. (line and selection) signals of all 30 speech channels of the PCM system. During each frame, the 8 bits of TS16 are assigned to two particular channels according to a fixed schedule. Thus, after 15 consecutive frames, 4 bits representing the signalling state of each of the 30 channels will have been sent. The identification of which channels TS16 is referring to at any time is achieved by considering the frames as groups of sixteens to form a multiframe of 2 ms duration (Fig. 5.8). The start of the multiframe is indicated by a 4-bit multiframe-alignment pattern of '0000' carried in the TS16 of the first frame; the TS16s of the remaining 15 frames carry the signalling for the channels. A loss of multiframe alignment is indicated to the distant end by setting bit 6 of the TS16 in the first frame of the multiframe to 1, as shown in Table 5.2. Fig. 5.8 details the allocation of successive TS16s to the signalling for all 30 channels.

Table 5.2 *The use of TS16 for the conveyance of channel-associated signalling*

Frame number	Bit allocation of TS16	
	Bits 1 – 4	Bits 5 – 8
0	Multiframe alignment word (0000)	Loss of multiframe alignment (X*XX)
1	Sig. for Ch1	Sig. for Ch16
2	Sig. for Ch2	Sig. for Ch17
: :	: :	: :
15	Sig. for Ch15	Sig. for Ch30

Key:

*: digit normally 0 but changed to 1 when loss of multiframe alignment occurs.
X: digit not allocated to any particular function and set to 1.

For a particular channel, the same 4-bit pattern in TS16 (known as 'ABCD') is repeated until the signalling state changes. For example, the break condition during a loop-disconnect dialling sequence is represented by the pattern '1010' every 2 ms. Since the break condition lasts about 33.3 ms, the 1010 bit pattern will be repeated some 17 times. Although this repetitive use of the ABCD bits does not fully exploit the potential of the 2 kbit/s signalling capacity available to each channel (4 bits every 2 ms), it does give a simple, cheap and robust method of conveying DC signalling over PCM.

(ii) Common-channel signalling: The concept of common-channel signalling (CCS) is introduced in Chapter 2 and described in detail in Chapter 20. With CCS between two exchanges linked by 2 Mbit/s digital line systems, TS16 is used to convey the CCS messages in a standard format, 8 bits at a time in successive frames. Thus, the CCS is conveyed at the rate of 64 kbit/s. No multiframe arrangement is used because there is no fixed relationship between the TS16 contents and individual channels; rather, each signalling message has a label indicating to which channel the signals relate.

5.3.2 The 24-channel (T1) system

This system has a 125 μs frame with 24 8-bit time slots allocated to 24 speech channels. Speech is normally encoded into 8 bits using Mu-law companding. The frame and multiframe alignment patterns are carried by a single bit at the beginning of each odd and even frame, respectively. Thus, the frame contains $1 + 24 \times 8 = 193$ bits. The gross-system-bit rate is 193×8 kHz $= 1544$ kbit/s, or 1.544 Mbit/s. which is abbreviated to '1.5 Mbit/s' for convenience. Fig. 5.9 illustrates the frame and multiframe format of the 24-channel system.[7]

(i) Channel-associated signalling: Channel-associated signalling for each channel is conveyed in every sixth frame, using the least-significant bit of each corresponding time slot, a technique known as 'bit-stealing'.[8] This means that for frames 1 to 5 and 7 to 11, 8 bits carry the encoded speech for each channel, while for frames 6 and 12 only 7 bits of encoded speech are carried. The perceived degradation to the quality of transmission is negligible. The bit-stealing technique provides a 1.33 kHz (i.e. 8kHz/6) signalling capacity for each channel within its time slot. The signalling bits for each channel in the sixth and twelfth frames are known as the 'A bit' and 'B bit', respectively. DC signalling is represented by AB (2-bit) patterns. As with the 30-channel system, these patterns indicate the signalling state and are repeated for the duration of the state.

(ii) Alignment: The 12-bit frame-alignment pattern is carried, one bit at a time, at the beginning of each odd frame. Similarly, the 12-frame 1.5 ms multiframe is identified by a 12-bit multiframe-alignment pattern carried in the first bits of even frames (Fig. 5.9).

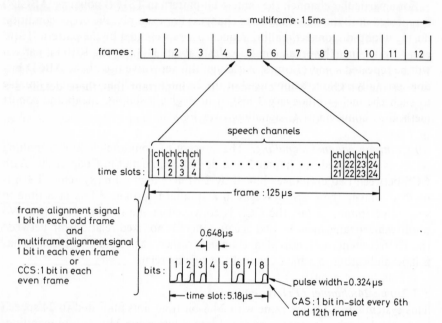

Fig. 5.9   *The 24-channel PCM frame format*

(iii) Common-channel signalling: Since a multiframe is not required for common-channel signalling, the first bit of successive even frames is used to convey CCS on a T1 system. This gives only a 4 kbit/s signalling capacity.

However, the T1 system can be modified to allow 64 kbit/s CCS to be carried transparently. This requires the elimination of the bit-7-zero-code-suppression process normally provided in T1 systems.[9] The suppression process involves the setting of bit 7 to 1 for any channel which has 8 zeros in a frame. Although this occasional changing of bit 7 has no perceivable effect on speech transmission, it does prevent the use of time slots for carrying 8 bits of data. The T1 system is thus sometimes referred to as having 'nonclear' channels. With the necessary suppression, the T1 system can carry in its 'clear channels' not only 64 kbit/s CCS but also any other 64 kbit/s data stream. The resulting loss of the zero suppression, with its consequent reduction in timing content, should not affect the performance of the newer line-transmission systems. With clear-channel operation, the T1 system carries one CCS channel of 64 kbit/s and 23 traffic channels of 64 kbit/s.

5.3.3 Comparison of the two PCM systems

There are significant differences between the 30-channel and 24-channel PCM systems. Apart from the number of speech channels within the frames and the companding laws used, the systems employ radically different methods of carrying signalling. The 30-channel system uses a separate dedicated time slot in a bunched format for channel-associated and common-channel signalling, while the 24-channel system uses a dispersed format with a bit-stealing within-time-slot method for CAS. Common-channel signalling on the 24-channel PCM (T1) system may be conveyed over a single-bit separate channel or within one of the 64 kbit/s 8-bit channels. It is important that these details are appreciated when considering the processes of digital switching in the following chapter. A summary comparison of the two PCM systems is given in Table 5.3.

Table 5.3   *Comparison of PCM systems*

Parameter	PCM system	
	CCITT CEPT	CCITT Regional (T1)
Number of 8-bit time slots	32	24
Number of speech channels	30	24
Number of encoding bits for speech	8	7/8*
Encoding law	A	Mu
Signalling (channel-associated)	Bunched in TS16 4 bits per channel every 16 frames (2 kbit/s)	1 bit per channel every 6 frames (1.3 kbit/s)
Signalling (common-channel)	8-bit bytes in TS16 (64 kbit/s)	1 bit in every even frame ** (4 kbit/s)
Frame-alignment pattern	7 bits bunched in TSO of odd frames	1 bit spread over odd frames
Bits per frame	256	193
Bit rate	2.048 Mbit/s	1.544 Mbit/s
Line code	HDB3 or 4B3T	ADI/AMI

*: 7 bits when channel-associated signal bit is used every 6ht frame.
**: Unless modified (see 5.3.2. (ii)).

5.4 Other digital encoding techniques

Although currently the PCM 64 kbit/s encoding techniques, using either the A or Mu law, are the only internationally standardised methods for digital telephony exchanges, several other encoding methods are employed in specialist applications. These other methods perform speech encoding at rates lower than the 64 kbit/s of PCM and hence obtain greater utilisation of digital transmission capacity. Inevitably, the lower encoding rates are achieved at the expense of transmission quality, particularly because of quantisation noise and frequency distortion. They have therefore been constrained to specific applications, usually on private or military networks without any interconnection with the public switched telephone network (PSTN).

Examples of the low-bit-rate voice-encoding systems include:

(i) CVSD (continuously variable slope delta modulation): This technique[10] is a derivative of delta modulation, in which a single bit is used to code whether each PAM sample is bigger or smaller than the previous one. Since it is not restricted to 8-bit words, the coding can be operated at a variety of speeds down to around 20 kbit/s. It is widely used in military networks (where the use of trained operators enables quality to be sacrificed in order to gain easily deployable robust communication). CVSD is also used for the storage of digitalised speech where economy in memory capacity is vital.

(ii) ADPCM (adaptive differential PCM): This is a derivative of standard PCM where the encoded difference between successive samples, rather than the acutal encoded samples, is transmitted to line. The CCITT has recommended a standard for 32 kbit/s ADPCM encoding of speech.[11]

Research is continuing into the design of acceptable speech-encoding techniques at rates as low as 16 kbit/s. However, the indications are that the ubiquity of the 64 kbit/s channel in both digital switching and transmission throughout the world favours the continuing dominance of standard PCM for the immediate future. Undoubtedly, the picture will change as lower-bit-rate speech encoding becomes cheaper and the quality improves. Finally, it should be noted that the 30-channel PCM standard was based on the requirements of a mixed analogue and digital network; allowing for up to 18 A/D conversions in a CCITT hypothetical international connection. As the network becomes progressively more digital, the number of A/D conversions encountered in call connections decreases and the need for the 64 kbit/s PCM standard diminishes. Thus, once the vestiges of analogue plant are eliminated, the alternative low-bit-rate encoding techniques can be deployed, with consequent plant economies.

5.5 Higher-level digital transmission systems

The PCM systems may be further time-division multiplexed on to higher-capacity digital transmission systems. The CCITT has defined a TDM digital hierarchy based on the 30-channel 2 Mbit/s block,[11] analogous to the existing FDM analogue hierarchy which, in turn, was based on the 12-channel 48 kHz group. The hierarchy contains higher-level digital line rates at 8, 34 and 140 Mbit/s. A fifth level at 565 Mbit/s is used but is not a CCITT standard.[12] Table 5.4 summarises the features of each hierarchical level, including the 64 kbit/s channel capacities available.

A similar hierarchy of higher-level digital transmission formats has been defined for 1.5 Mbit/s based networks.[13] This contains levels at 6 Mbit/s (carried on T2 transmission systems) and 45 Mbit/s (carried on T3 transmission systems).

Table 5.4 *CCITT (CEPT) TDM Digital transmission hierarchy*

Level	Gross rate Mbit/s	64 kbit/s channel capacity	Transmission systems available
1	2.048; (2)*	30	symmetric pair cable transverse screened copper cable micowave radio
2	8.448; (8)	120	carrier copper cable optical fibre microwave radio
3	34.368; (34)	480	coaxial cable optical fibre microwave radio
4	139.264; (140)	1960	microwave radio coaxial cable optical fibre
5	e.g. 563.992;** (565)	7840	optical fibre

Notes

* Figures in brackets give the nominal Mbit/s value.
** Not standardised.

As far as the influence on digital exchange design is concerned, most digital exchanges terminate digital line systems only at the 2 Mbit/s or 1.5 Mbit/s block level. However, some digital switching systems also terminate 8 Mbit/s lines directly (e.g. Fujitsu Fedex 150[14]); this offers a measure of TDM transmission-cost saving by eliminating the need for 2-to-8 Mbit/s multiplexers. The indications are that, in the future, digital exchanges will increasingly provide ports to digital transmission systems at the higher-order rates in order to provide cost savings in network plant.

5.6 References

1 CATTERMOLE, K.W. (1979): *The History of Pulse Code Modulation,* Proc. IEE, **126** (9), pp. 889–892

2 CCITT (1984): Red Book (VIIIth Plenary Assembly, Malaga — Torremolinos, October 1984), Recommendations of the Series G: Volume III — Facicle III.3, Rec. G711

3 BENNETT, G.H. (1983): *Pulse Code Modulation and Digital Transmission,* Marconi Instruments Ltd (White Cresent Press Ltd), Second edition, Chap. 2

4 BYLANSKI, P. and INGRAM, D.G.W. (1980): *Digital Transmission Systems* (Peter Peregrinus Ltd), Revised edition, Chap. 11 and Appendix (Bibliography of some transmission codes)

5 EVANS, B (1987): Satellite Communications Systems (Peter Peregrinus Ltd) pp. 19, 63

6 CCITT Rec. G732, Op. cit.

7 CCITT Rec. G733, Op. cit.

8 WELSH, S. (1981): Signalling in Telecommunications Networks (Peter Peregrinus Ltd), Chap. 6

9 RUFFALO, D.R. (1987): *Understanding T1 Basics: Primer Offers Picture of Networking Future,* Data Communications, pp. 161–169

10 SMITH, D.R. (1985): *Digital Transmission Systems* (Van Nostrand Reinhold Co), Chap. 3

11 CCITT Rec. G721, Op. cit.

12 CCITT Rec. G702, Op. cit.

13 CCITT Rec. G743, Op. cit.

14 KUTSUKAKE, S., HATANAKA T. and MATSUURAY (1987): *A Digital Switching System For Central Office,* Paper 32B4, Proceedings of the International Switching Symposium, Phoenix, 15–20 March 1987

Chapter 6

Digital Switching

6.1 Introduction

Digital switching is the process of interconnecting time slots between a number of TDM digital transmission links. This enables digital routes at 2 Mbit/s or 1.5 Mbit/s, from other exchanges or digital PABXs, to be directly terminated on the digital switch, without the need for their conversion to the constituent audio channels for switching, as is necessary in an analogue exchange. The absence of per-channel equipment gives digital switching considerable cost and accommodation advantages. Of course, any analogue circuits terminated on the digital exchange whether subscriber lines, trunk or junction circuits, must be converted to PCM format prior to entering the digital switches. Similarly, circuits leaving the exchange on analogue bearers must also be converted from digital to analogue at the periphery of the switch block. These analogue-to-digital (A/D) and digital-to-analogue (D/A) conversions, together with any necessary signalling conversions, are undertaken by 'interworking equipment'.

The role of interworking equipment is illustrated in the simple generalised structure of Fig. 6.1. The Figure shows that digital (PCM) streams enter the switch block directly at the multiplexed level, while the analogue circuits terminate at the individual circuit level on the interworking equipment. Interworking equipment, therefore, because of its per-circuit provision, adds a penalty of cost and accommodation to a digital exchange. For exchanges in a predominately analogue transmission environment, this can be substantial. The high interworking cost, resulting from the expensive A/D and D/A conversion technology then available, constrained the first working applications of digital switching to trunk and junction exchanges during the late 1960s and early 1970s.[1] Such exchanges acted as tandem nodes switching between the PCM line systems then being increasingly deployed in the junction and trunk networks. The development of cost effective digital local exchanges had to await the availability of cheap speech-encoding systems from the late 1970s onwards in order to minimise the interworking cost per subscriber line, as described in Chapter 3.

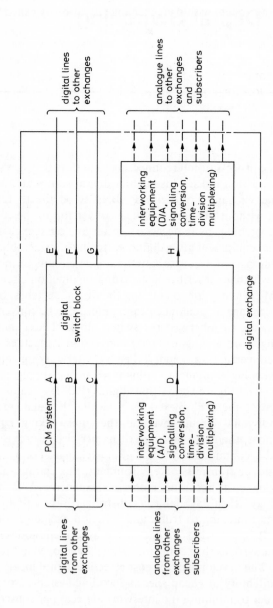

Fig. 6.1 *The role of a digital switch block*

This chapter concentrates on the mechanism of digital switching and the construction of practical digital switch blocks. The peripheral functions of terminating subscriber and trunk lines are covered in Chapter 7, and signal handling within the exchange is described in Chapter 8.

Before considering digital switching, a brief consideration of terminology is required. The switching system within an exchange is variously called 'switch', 'switching network', 'switching-centre network' or 'switch block'. To avoid any confusion with the terms used to describe telephone networks, this book uses the term 'switch' to describe a single switching element, and 'switch block' to describe composite groups of switches, such as the subscriber concentrator switch block or group switch block.

A digital switch block provides connections between a number of PCM systems, each comprising 30 (or 24) channels in a time-division-multiplex frame. The PCM systems terminate at the switch block on TDM digital 'highways' or buses'. Chapter 5 describes how the samples from each channel form 8-bit PCM words, which are conveyed within the time slots of a PCM frame. Thus, digital switching involves the transfer of the PCM words relating to a channel in a time slot on an inlet bus to a time slot on an outlet bus. Although the terms 'channel' and 'time slot' are distinct, they are often used synonymously in the literature. For clarity, this description considers digital switching simply as a means of time-slot interconnection, leaving the description of the various forms of channels (e.g. traffic, signalling, test, supervision, etc.) that can be conveyed via the time slots to later.

It is useful at this stage to consider a simple example of a connection through a digital switch block. Referring to Fig. 6.1, consider the example of a call, carried in time slot 6 (TS6) of PCM system A, which requires connection to an exchange over the trunk route of PCM system F. If TS6 is free on PCM system F, then the connection could be established by linking the two PCM systems during the period when TS6 occurs simultaneously on both systems. This process, which is a simple connection in space, is known as *digital SPACE switching*.

However, sole reliance on space switching within a digital switch block gives rise to serious blocking problems due to the high probability of two or more calls competing for the same outgoing time slot. For example, blocking will occur if TS6 of PCM system F is already engaged (say, by connection to TS6 of PCM system C) and so is unavailable for connection to TS6 of system A. This blocking can be avoided by choosing a time slot other than TS6 on system F for the call from system A. Normally, this is possible because any free time slot on a route to an exchange is suitable for carrying the call. The connection between TS6 of system A and some other time slot on system F involves not only digital space switching between the two PCM systems but also digital *TIME switching* between the different inlet and outlet time slots. Practical digital switch blocks normally use a combination of time and space switching.

In the following Sections, digital space and time switches are separately described before their combination into switch blocks is considered.[2-5] These descriptions assume that all PCM systems terminating on the switch block are aligned (Chapter 7) and synchronised (Chapter 10) so that all corresponding time slots occur simultaneously throughout the switches. Thus, in Fig. 6.1, TS1 of PCM system A coincides with the TS1s of PCM systems B,C,D,E,F,G,H, and the TS1 of the switch block. It is further assumed that the connections through the switches have been established; path set-up is not considered until Section 6.5.

6.2 Digital space switching

6.2.1 The technique

A digital space switch consists of a time-division-multiplexed matrix with inlets and outlets conveying PCM systems. Thus, to connect any time slot in an incoming PCM system to the corresponding time slot (i.e. same TS number) of an outgoing system, the appropriate crosspoint of the space-switch matrix must be operated for the duration of that time slot, and whenever that the time slot reappears (i.e. once per frame), for the duration of the call. During other times, the same crosspoint may be used to switch the contents between time slots which relate to other calls. In addition, the remaining crosspoints in the matrix may be used for other connections. Hence, TDM digital space switching is essentially the rhythmic establishment of a set of connections through the matrix at the rate of once per frame, each connection lasting for the duration of one time slot. Because a normal telephone call lasts for the duration of many PCM frames, typically of the order of one million, some kind of simple cyclic control of the connection pattern is needed. This is achieved by local connection memories associated with the space switch.

Fig. 6.2 illustrates the principle of a TDM digital space switch. The switch comprises a square matrix, of size $n \times n$, whose rows carry the incoming PCM systems and whose columns carry the outgoing PCM systems. The crosspoints in each column are controlled by a connection memory (CM) which stores as many words w as there are time slots per frame. A unique binary address is assigned to each crosspoint in a column. The appropriate address is then used to select the required crosspoint to establish a connection between inlet and outlet buses. These selection addresses are stored in the CM in time-slot order, according to the current connection schedule. That is, for column 1, the address of the crosspoint to be closed during TS1 is stored in memory cell number 1 of the CM for column 1, the crosspoint address to be closed during TS 2 is stored in cell 2, and so on.

The word size of the CM must be sufficient to hold a binary address for each of the n crosspoints, plus one address that will keep all crosspoints open (i.e. unconnected). Thus, $(n + 1)$ addresses are required, each identified by a binary number of $\log_2(n + 1)$ bits. Once each of the CMs have been loaded with the

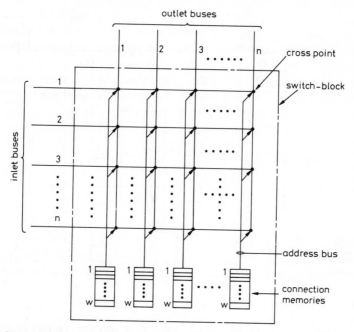

Fig. 6.2 *Digital TDM space switch*

crosspoint addresses for their columns, the switching-control process consists of reading the contents of each CM cell during the appropriate time slot and using the addresses to select a crosspoint which will remain operated for the duration of that time slot. This process continues until each cell in the CM has been read and the appropriate crosspoint operated. The procedure is then repeated, beginning with the first cell of each CM. Each cycle occupies one frame duration, during which a PCM word from each inlet time slot may be 'switched' to the corresponding outlet time slot.

The process of digital space switching may be best appreciated by reference to the example in Fig. 6.3. The space switch has a 4×4 matrix of crosspoints. Input buses A to D are connected to the rows of the matrix and output buses E to H are connected to the columns. In this simple example, each PCM system (one per bus) is assumed to have just three time slots. Therefore, there are three cells in each CM. For this 4×4 matrix a CM word size of 3 bits (.i.e. $\log_2 [4 + 1]$) is required for crosspoint addressing. Fig. 6.3 shows the binary address '011' being applied by CM-H during TS3 to the 3-line address bus in order to close the crosspoint at the junction of row B and column H. The address-decoding logic located at each crosspoint consists of a 4-input digital logic AND-gate, with inverters, as required, on the inputs so that each bit on the address bus presents a '1' to three of the AND-gate inputs. Different decoding logic is required at each crosspoint, so that only the appropriate

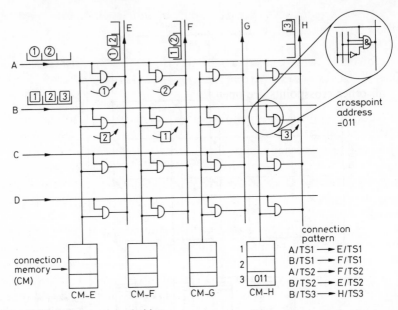

Fig. 6.3 *TDM digital space switching*

address primes the AND gate of the selected crosspoint. In this primed state, a '1' on the input B row passes through the crosspoint formed by the AND gate to the output bus H; a '0' is similarly transferred. An address of '000' on the address bus causes all crosspoints in the column to remain open (unconnected).

In the example of Fig. 6.3, time slot 1 of inlet bus (A/TS1) is to be connected to time slot 1 of outlet bus E (E/TS1). Other connections to be made are B/TS1, to F/TS1, A/TS2 to F/TS2, B/TS2 to E/TS2 and B/TS3 to H/TS3. Once the connection memories CM-E and CM-F have been loaded with the appropriate crosspoint addresses relating to time slots 1 and 2, the sequence of operations is as follows:

(i) During TS1:
(a)   crosspoint A,E is closed connecting A/TS1 to E/TS1 (shown as a 1 within a circle in Fig. 6.3);
(b)   crosspoint B,F is closed connecting B/TS1 to F/TS1 (shown as a 1 within a square in Fig. 6.3);
(c)   all other crosspoints are open, although those in columns G and H could be used for connections to rows C and D.

(ii) During TS2:
(a)   crosspoint A,F is closed connecting A/TS2 to F/TS2 (shown as a 2 within a circle in Fig. 6.3);
(b)   crosspoint B,E is closed connecting B/TS2 to E/TS2 (shown as a 2 within a square in Fig. 6.3);

(c)   all other crosspoints are open, although those in columns G and H could be used for connections to rows C and D.

(iii) During TS3:

(a)   crosspoint B,H is closed connecting B/TS3 to H/TS3(address 011 is applied to the crosspoint decode logic, as shown in Fig. 6.3);

(b)   all other crosspoints are open.

This sequence of operations continues, once every frame, until the content of a CM is changed to enable a call to be initiated or cleared-down. New connections across the digital space switch are established, or existing connections cleared, by the insertion or deletion, respectively, of the crosspoint addresses in the relevant CM cells. During each time slot, when a crosspoint is closed, the 8 bits of one PCM word are passed sequentially (i.e. in serial mode) from inlet to outlet bus.

It is apparent that, because the matrix highways and crosspoints are time-shared, far fewer crosspoints are required compared to a space-divided (non-TDM) matrix. For example, considering one direction of transmission only, the interconnection of 32 input and 32 output channels requires a 32×32 space-divided matrix but only a single TDM crosspoint using 32-time-slot digital input and output buses. A 1600-crosspoint matrix (40×40) is required to interconnect 40 PCM systems (1280 time slots in and out) terminating on each side of the space switch. But to provide similar capacity on a single space-divided matrix requires $1280 \times 1280 = 1\,638\,400$ crosspoints, again considering only one direction of transmission. However, these figures are only indicative since two crosspoints are required for each telephony connection (to allow both directions of speech); also, a single square space-divided analogue space switch of such size would not be cost effective.

6.2.2 Practical digital space switches

The preceding description of digital space switches was necessarily simplified to explain the principle of operation. The following Sections introduce the various additional features of practical digital space switches.

6.2.2.1 Control-memory orientation and crosspoint implementation. The example of Fig. 6.3 has the crosspoint control organised column-wise; that is, the address given by the CM selects the input row for its column. The control could alternatively be organised on a row basis, with the contents of the CM selecting the output column for its row. The alternative configurations are illustrated in Fig. 6.4, using digital logic multiplexers and demultiplexers. A digital-logic multiplexer is a device, usually in integrated-circuit form, which provides connection to its single output from one of its *n* inputs, according to a binary address applied to its control leads. A digital-logic demultiplexer provides a connection between the single input and one of the *n* outputs, as indicated by a binary address.

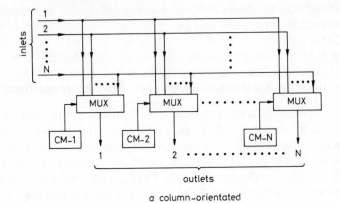

a column-orientated

b row-orientated

Fig. 6.4 *Practical space-switch construction*

With column orientation, the $N \times N$ space-switch matrix is produced by a row of N digital-logic multiplexers whose outputs form the N outlets of the matrix (Fig. 6.4a). The N inlets to the matrix are applied to the corresponding inputs to each of the multiplexers. The connection memory selects which of the input rows is connected to the outlet columns. With row orientation, the matrix is composed of N digital-logic demultiplexers, whose outputs are commoned to form N switch outlets (Fig. 6.4b). The connection memories select which of the output columns are connected to the inlet rows.

The choice of control orientation depends on the switch-block configuration and the degree of control integration with adjacent time-switch stages, as discussed later in this chapter.

6.2.2.2 Super-multiplexing. The capacity of a space switch may be enlarged by increasing the number of time slots on each bus. This is achieved by time-division-multiplexing several PCM systems on to each of the inlet buses, with corresponding demultiplexing of the signal from the output buses into its constituent PCM systems. The technique is known as super-multiplexing. Since the time slots on a super-multiplexed bus must be proportionally shortened, which means that the bits within them must be transmitted faster, the extent of the increase to the switch capacity is limited by the operating speed of the digital logic forming the crosspoints.

The required speed of the crosspoint digital logic may be reduced, by a factor of eight, by operating the space switch in parallel rather than serial mode. In this case, the 8-bit serial transmission of each PCM word carried on a single-line bus is converted to 8-bit parallel transmission on an 8-line bus. The space switch them comprises a matrix of 8-line parallel buses, interconnected by 8-line crosspoints. Parallel-to-serial conversion is required on the output side of the switch block before the PCM can be transmitted to line in the normal serial format.

Space-switch highways for N 30-channel (i.e. 32-time-slot) PCM systems will operate at $N \times 2048$ kbit/s in serial mode, or $n \times 256$ kbit/s (i.e. eight times slower) in parallel mode. In general, to switch P incoming 32-time-slot systems to P outgoing systems requires $[P/N]^2$ crosspoints, operating at $N \times 2048$ kbit/s in serial mode. The corresponding figures for parallel mode are $8[P/8N]^2$ crosspoints, operating at a speed of $N \times 256$ kbit/s. These figures should be compared with the equivalent space-divided analogue space switch of $[32P]^2$ crosspoints to appreciate the high capacities possible with digital TDM space switches.

A typical practical digital space switch accommodates 16 30-channel PCM systems (512 time slots) super-multiplexed on to each of the parallel highways of the matrix. With each time slot having a duration of 244 ns, the crosspoints have to operate at 4096 kbit/s (i.e. 16×256 kbit/s).

6.2.2.3 Modularity. As with all square matrices, the size of a digital space switch increases with the square of the number of input highways. Even with the maximum possible degree of super-multiplexing, an exchange capable of growing to about 20 000 erlangs capacity requires a digital space switch of about 96×96 ports. Currently, switches of such capacity are impossible to construct on one printed-circuit board, even using VLSI chips. Another aspect in the construction of digital space switches is the economic penalty associated with fully equipping new exchanges with their ultimate required capacity. This imposes a cost burden of spare plant which might last many years. In practice, therefore, digital space switches are constructed using conveniently sized modules in order to achieve switch blocks of adequate capacities and which are easily extendable.

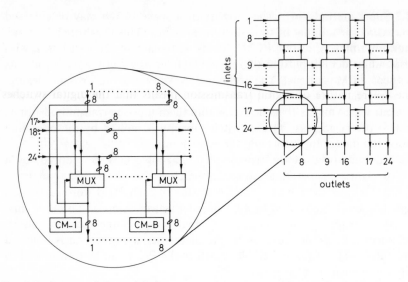

Fig. 6.5 Space-switch modularity

Fig. 6.5 illustrates the modular concept for a digital space switch. The switch is constructed from a square array of modules, each being an 8×8 matrix. In this example, the 8×8 column-oriented matrix modules contain eight digital-logic multiplexers, each with its connection memory (labelled CM-1 to CM-8). The smallest space-switch provision is one module, giving an 8×8 switch. This can be extended to 16×16 by the provision of three more modules, then further extended to 24×24 by the provision of five further modules, and so on.

6.3 Digital time switching

6.3.1 The technique
Time switching has already been defined as the transfer of the contents of one time slot to another, non-coincident, time slot. The process involves the introduction of an appropriate delay within a time switch. Thus, for example, to switch between TS3 on the inlet bus to TS8 on the outlet bus a delay of 5 time slots' duration is required. Fig. 6.6 illustrates the process of time switching by considering the relationship of time slots within the PCM frames.

In Fig. 6.6a the simple case of switching the contents of TS3 to TS8 is shown for two consecutive frames, the top line representing the input sequence of frames and the bottom line representing the output sequence. Because TS8 is later in time than TS3, the input and output-channel contents (i.e. one PCM word) are kept within the same time frame. However, where the time switching is between a time slot and one that is earlier in the frame, the necessary delay means that the contents of the input time slot are delayed until the next frame. Thus, in this case, frame integrity is not possible. Fig. 6.6*b* illustrates the case of time switching TS8 to TS3. Here, the incoming channel must be delayed until TS3 appears in the following frame: that is, a delay of $(32-8)+3 = 27$ time slots' duration, assuming a frame of 32 time slots.

The 4-wire nature of digital transmission means that the digital switches must also be 4-wire, with separate 'transmit' and 'receive' paths. Fig. 6.6*c* shows the time-slot and delay arrangements for a typical 4-wire connection through a digital time switch. A call is shown switched between TS3s in the transmit and receive directions of PCM system A, carried on buses A_T and

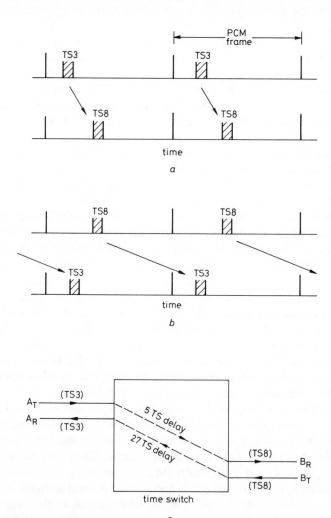

Fig. 6.6 *Time-slot changing*

A_R, respectively, and TS8s of the transmit and receive directions of PCM system B, carried on buses B_R and B_T, respectively. As previously explained, the connection of A_T/TS3 (i.e. bus A_T Ts3) to B_R/TS8 requires a delay in the time switch of 5 time slots' duration, while the connection of B_T/TS8 to A_R/TS3 requires a delay of 27 time slots' duration. Because of the need to delay B-to-A transmission beyond the frame boundary, corresponding time slots in the A buses will contain PCM words from two adjacent frames, unlike the PCM words on the B buses which relate to the same frame.

For clarity, the remainder of this Section on time switching will consider only one direction of transmission. Both-way transmission will be discussed again during the description of complete digital switch blocks, later in the chapter.

A number of mechanisms have been used to create the necessary delay for time switching, including electromagnetic tape and ultrasonic delay lines.[1] One of the earliest digital exchanges, the British Post Office 'Empress' junction tandem, in service between 1968 and 1978, used magnetostrictive delay-line time switches.[6,7] However, the advances in digital-logic technology have resulted in fast, cheap semiconductor storage, which is now used by all modern digital exchanges for the delay components of their time switches.

The time switch contains a speech memory (SM), where the PCM words for each of the input time slots are stored and thus delayed for the required time (Fig. 6.7a). Since a delay of one full frame length would not give a net shift of time slots, the delays required range from the durations of one time slot up to one frame less one time slot. The writing of the PCM words from the incoming time slots into the speech memory is sequential and controlled by a simple counter. Thus, the content of time slot 1 is written into SM cell 1, the content of TS2 is written into SM cell 2, etc. The reading out of the SM is directed by a connection memory (CM). As with the digital space switch, this CM has as many cells as there are time slots per frame; during each time slot it directs the reading of a specific cell in the SM. The effective delay is the difference in time slots between writing into and reading out of the SM. It is interesting to note that the time switch is not normally time-divided. This is because the same SM cell is used exclusively by a single call for its duration. Therefore, a digital exchange contains the paradox that its space switches are time-divided, whereas its time switches are space-divided![3]

Fig. 6.7a illustrates the principles of a storage-type time switch with a simple example of a 4-time-slot system. As is characteristic with time switches, this example has a single input and output bus. The writing into SM is in time-slot order, under the control of a 4-stage cyclic counter. Thus, the content of input TS1, i.e. PCM word A, is written into SM cell 1. Similarly, the content of TS2, i.e. PCM word B, goes into SM cell 2, and so on. The read addresses are stored in the CM in the order of the required output sequence. In this example, the content of SM cell 3 is to be read during each time slot 1; thus,address '3' has been stored in the first location of the CM. The second location in the CM

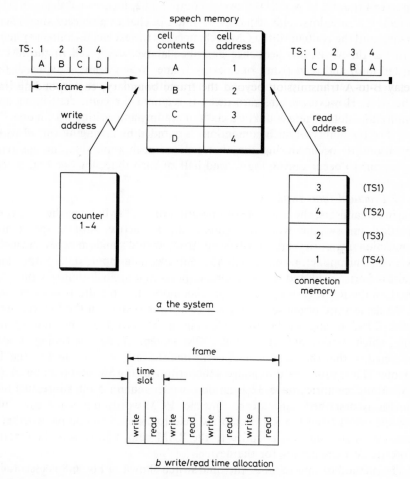

a the system

b write/read time allocation

Fig. 6.7   *The principle of time switching using storage*

contains the address '4', indicating which SM cell is to be read during time slot 2. Similarly, CM cells 3 and 4 contain the SM addresses '2' and '1', respectively.

In operation, during each time slot 1 within the switch block, the content of TS1 on the incoming bus (i.e. PCM word A) is written into cell 1 of the SM, using the address from the counter. Also during TS1, the content of SM cell 3 (i.e. PCM word C) is read on to the output bus, using the address from the CM. Similarly, during time slots 2, 3 and 4, PCM words D, B and A, respectively, are read out of the SM. Thus, the content of input TS1 (i.e. PCM word A) experiences $4 - 1 = 3$ time slots' delay; input TS2 (i.e. PCM word B) experiences a delay of $3 - 2 = 1$ time slot. Input TS3 (i.e. PCM word C) moves to output TS1, incurring a delay of TS3 to TS4 to TS1 = 2 time slots;

input TS4 (i.e. PCM word D) moves to output TS2, incurring a delay of TS4 to TS2 = 2 time slots. This sequence of time-slot changing repeats, once every frame, until the content of the CM is changed by the exchange-control system.

In the time-switching sequence described above, every cell in the SM is written into, and read from, once every frame. When delay is required, the writing and reading operations on an individual cell occur during different time slots. However, where zero delay is required (i.e coincident input and output time slots), there could be a clash of reading and writing within one SM cell. The possibility of such contentions is avoided by arranging for all write operations to occur during the first half of each time slot and all read operations to occur during the second half of each time slot (see Fig. 6.7*b*).

6.3.2 Practical time switches

Having established the principle of operation of time switching, the details of practical time switches will now be discussed. In practice, both the speech and connection memories are constructed from digital random-access-memory (RAM) semiconductor chips. A RAM chip has data-input, data-output and address ports, read and write command ports and a clock-timing input. All ports on the RAM chip operate in parallel mode. Fig. 6.8 illustrates how two RAM devices are organised to form a basic time switch. In the Figure, the 2 Mbit/s clock timing for the two RAM chips is extracted from the input speech bus, which carries one 30-channel PCM system. Time-slot timing is also extracted so that the beginning of each time slot may be indicated to the TS counter. This is then used to address the writing of the SM and reading of the CM. Both memories have 32 locations, which require 5-bit addressing. As previously described, during each time slot PCM words are first written into and then read from the RAM chips. A selector is used to control whether a 'read' or 'write' address is aplied to the RAM chip, and so sets the read or write (R, W) commands for the device.

The method of operation of the time switch shown in Fig. 6.8 is essentially the same as the simple example of Fig. 6.7. Thus, the input 8-bit PCM words are written sequentially into the SM and read out on to the output speech bus, under the control of the CM. The SM read address is output from the CM. The required connection pattern through the time switch is determined by the control system, and is established by writing the appropriate contents into the cells of the CM, as described later in this chapter. When no calls are being set up or cleared down, the CM is in the read-only state.

In order to construct economical switches of high capacity and to exploit the speed potential of semiconductor systems, time switches, like space switches, usually employ some degree of super-multiplexing and work in the parallel mode. The super-multiplexing of several PCM systems can conveniently be incorporated into the digital time switch rather than provided by separate multiplexing equipment. In early designs, eight PCM systems were combined in a single super-multiplexed time switch. This was achieved using eight sets

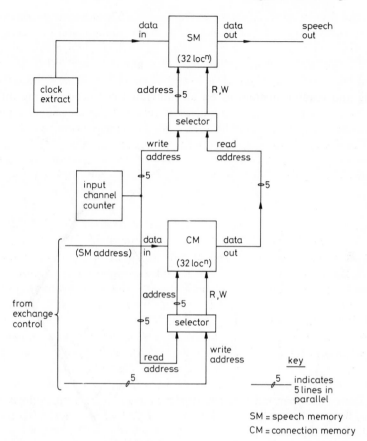

Fig. 6.8 *Basic time switch using RAM*

of SMs with their outputs connected to a common 256 (8×32) time-slot bus. A single 256-cell CM could then be used to control the super-multiplexed time switch.[5]

The advent of higher-speed bipolar semiconductor memories of suitable size in the early 1980s made possible a further improvement in the design of time switches. Instead of providing an SM per PCM system, it became practicable to perform the function for all the PCM tributaries with one time switch using one set of integrated-circuit chips. This led to a greatly simplified control mechanism for the SMs with improved reliability and fault diagnostics. It also enabled a significant size reduction to be achieved.

Fig. 6.9 shows a modern time switch acting as a first stage of a digital switch block. The time switch terminates 16 30-channel PCM systems on the inlet side and one super-multiplexed bus on the outlet side. Each inlet PCM signal is first converted from serial to parallel mode and frame aligned with the exchange timing by a digital line-termination unit (DLTU), provided at the periphery of

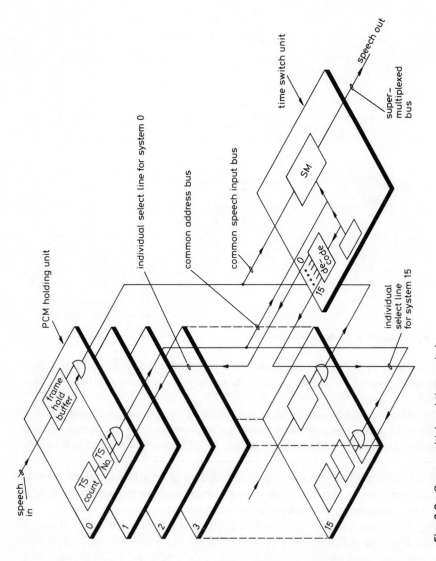

Fig. 6.9 *Super-multiplexed time switch*

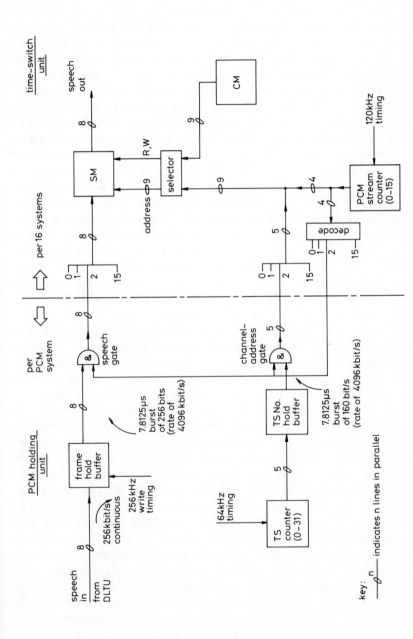

Fig. 6.10 *Modern input super-multiplexed time switch*

the switch block, as described in Chapter 7. The overall time-switch system comprises one PCM system holding unit (a frame store) for each of the 16 PCM systems and a single super-multiplexed time-switch unit.

The principle of the super-multiplexing operation is that the content of each PCM stream is continuously written into its frame-hold buffer (within each holding unit) and these are read periodically, in sequence, by the time-switch unit. Thus, the writing into the frame-hold buffers is continuous, while the reading into the super-multiplexed time-switch unit is in bursts of one frame of PCM words. Selection-control wires from the time-switch unit prime each holding unit in turn, so that single bursts from each of the 16 PCM systems may be interleaved within each super-multiplexed frame. The outputs from the 16 frame-hold buffers are thus joined to form a common, super-multiplexed input to the SM, as shown in Fig. 6.9. Similarly, the output buffers within each holding unit that store the time-slot numbers, as described below, are also common on to the inputs to the digital logic which produces the SM addresses.

The operation of the super-multiplexed time switch may be described by reference to Fig. 6.10, which shows the digital logic within the two groups of time-switch equipment in more detail. The Figure represents a view from the top of Fig. 6.9 showing just one of the PCM holding units and the time-switch unit. In Fig. 6.10 the equipment provided for each of the inlet PCM systems is shown on the left of the chain line, while the equipment on the right of the chain line is shared by the 16 PCM systems. The 30-channel PCM systems write into their frame-hold buffers at the continuous rate of 256 kbit/s (8-bit-parallel mode), while the reading from this buffer is in bursts lasting 1/16 of a frame duration, i.e. 7.8 μs. During the bursts, the line rate on the 8-bit-parallel buses out of each frame-hold buffer is 16×256 kbit/s = 4096 kbit/s. As with all frames in the PCM digital exchange, the super-multiplexed frame duration is 125 μs; thus, each of its 512 time slots occupies only 244 ns.

The procedure for selecting which of the 16 PCM holding units should write into the single SM is initiated by a 'PCM-stream counter' stepping from 0 to 15 once every frame. This counter produces a 4-bit binary number, representing the chosen PCM holding unit, which, when decoded, is used to prime one of 16 control lines. (The line to the holding unit of PCM system 2 is primed in the example of Fig. 6.10.) The control line then primes two sets of digital logic AND gates within the holding unit. When primed, each of the speech gates in turn passes the single 256-bit frame contents from their associated frame-hold buffers to the super-multiplexed bus and hence into the SM of the time-switch unit. Simultaneously, the channel-address gates pass a sequence of 5-bit binary numbers representing each of the 32 time slots. These numbers are concatenated, five bits at a time during each 244ns time slot, with the 4-bit PCM-system numbers to make a 9-bit write address for the SM. In this way, each of the 512 locations of the SM is addressed sequentially by a 9-bit number, whose first four bits represent the PCM system number (0–15) and the remaining five bits represent the time slot within that PCM system (0–31).

Time switching is performed by reading from the SM under the control of a CM, as described earlier in this chapter. For clarity, the inputs to the CM have not been shown in Fig. 6.10.

6.4 Structures of digital switch blocks

At the beginning of this chapter, the blocking problem of a single time-divided space switch was described. This serious blocking probability precludes the use of digital switch blocks composed of only space switches. However, high-speed time switches, as described in the preceding Section, can offer nonblocking interconnectivity to all the channels of the PCM systems terminating on their highways. For example, with 16 PCM systems multiplexed on to a single time switch, all 512 time slots can be freely interconnected. If a switch block is composed of just one such time switch, the capacity is actually a half of 512 time slots, so that the unidirectional switch could interconnect PCM channels in both directions of transmission. A single-stage time switch is thus a practical switch block for switching 256 circuits (with transmit and receive channels). However, in a local exchange application, allowing for signalling and control paths and junction routes, this would support only about 200 subscribers. Increasing the capacity of the time switch is limited by the technology of the RAM devices and the associated control logic. Thus, additional subscribers or trunks could be supported only by adding extra time switches; these would then require a space switch for their interconnection.

The conclusion is, therefore, that low-blocking digital exchanges of practical capacities must employ a combination of space and time switches. The remainder of this Section considers the variety of space- and time-switch combinations, and their merits and applications. It is common practice to label the switch-block configurations 'S' and 'T' for space and time switch stages, respectively.

6.4.1 The T-S switch block

A T-S switch block comprises a time switch on each of the input buses of a single space switch. An example with three time switches and a 3×3 space switch is shown in Fig. 6.11. In this configuration, the writing into the time switches is cyclic, under the control of a counter, and the reading is acyclic, under the control of a connection memory. For clarity, the address logic associated with the CMs, SMs and crosspoints of the switches have been omitted. The Figure shows the connection memories, CM-A1 to CM-A3, for the time switches and speech memories, SM-A1 to SM-A3, for the switch-

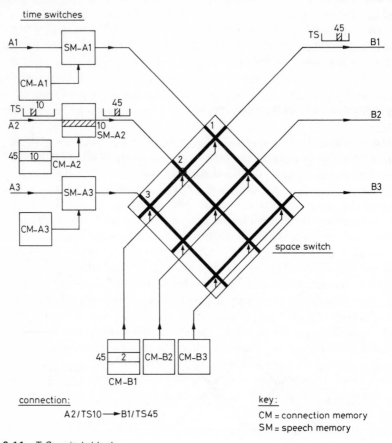

Fig. 6.11 *T-S switch block*

block-inlet buses A1 to A3. The outlets from the space switch are labelled B1 to B3 and the column-orientated space-switch connection memories are labelled CM-B1 to CM-B3, correspondingly.

With call connections provided by a T-S switch block, the time switches shift PCM words between the incoming and desired outgoing time slots, while the space switch interconnects the inlet and outlet buses. This activity can best be explained by considering the example shown in Fig. 6.9. Assuming a super-multiplexing ratio of 8:1, each bus carries 256 (8×32) time slots in a 125 μs frame. Thus, the SMs and both sets of CMs each contain 256 locations. The Figure shows the necessary CM contents for a connection between bus A2 time slot 10 and outgoing bus B1 time slot 45 (written: 'A2/TS10 to B1/TS45'). The way that the exchange-control system determines the contents of the CMs is discussed in the following Section.

In order for the time switch associated with A2 to shift the content of TS10 to TS45, CM-A2 has the address '10' stored in cell 45. Thus, during TS45 the content of CM-A2 cell 45 is read and used as the address of the A2 speech memory cell to be read. The PCM word in SM-A2 cell 10 is then transferred on to the space-switch inlet bus A2. Also during TS45, the content of cell 45 of CM-B1 is taken to prime the space-switch column B1 crosspoint 2. As a result, the content of cell 10 of time switch A2 is transferred through the space switch on to the outlet bus B1 during TS45. This connection between A2/TS10 and B1/TS45 will repeat every frame until the contents of the connection memories CM-A2 and CM-B1 are changed.

Although the T-S switch block enables larger capacities to be achieved than a single T-stage, it does suffer from the inherent blocking of corresponding time slots on the outlet buses from the space switch. For example, in Fig. 6.9 with the connection A2/TS10 to B1/TS45 established, A2/TS15 is blocked

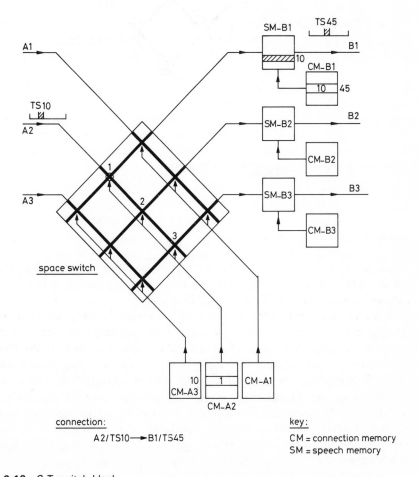

connection:

A2/TS10 ➞ B1/TS45

key:

CM = connection memory
SM = speech memory

Fig. 6.12 *S-T switch block*

from accessing B2/TS45 or B3/TS45 (although time slots on A1 and A3 are not blocked).

6.4.2 The S-T switch block

The characteristics of an S-T switch block are similar to those of a T-S switch block, except that the space switch interconnects the inlet and outlet buses first and then the time switch undertakes the necessary time-slot delays. Fig. 6.12 illustrates the CM contents required for the A2/TS10 to B1/TS45 connection example used in the T-S explanation. In the S-T case, the connection through the space switch is made during time slot 10; thus, CM-A2 contains the crosspoint address '1' in cell number 10. With cyclic write and acyclic read, the content of A2/TS10 passes through the space switch and is stored in cell 10 of time switch SM-B1. The speech samples are then read during TS45 under the control of CM-B1, which has address '10' stored in cell 45.

The S-T combination offers capacity and blocking advantages compared to the single- S-stage. However, the S-T switch block does have an inherent blocking characteristic, namely only one of the inlets to the space switch may access an outlet bus during any of the time slots. Thus, in the example of Fig. 6.12, with the A2TS10 to B1/TS45 connection established, neither A1/TS45 nor A3/TS45 can access B1, irrespective of how many of the 255 remaining time slots on the outlet B1 are free. This problem will clearly worsen as the size of the time switch is increased and more time slots become blocked. The blocking characteristics of S-T and T-S switch blocks are overcome by the addition of a third stage, S or T, respectively.

6.4.3 The S-T-S switch block

In an S-T-S switch block, the input space switch connects the input bus with a time switch during the input time slot, and the output space switch connects the time switch with the output bus during the output time slot. This arrangement is illustrated in Fig. 6.13 for the connection example of A1/TS10 to C1/TS45. With the inlet bus/time slot and outlet bus-time slot specified, the control system is able to select any time switch that has both TS10 input and TS45 output free. In Fig. 6.13, time switch C3 has been chosen and the CM contents for the three stages are as shown. Bus A1 is connected to time switch B3 during TS10 via crosspoint 3 of row A1 in the inlet space switch. Therefore, cell 10 of CM-A1 contains the crosspoint address '3'. Time switch B3 is required to shift the PCM word from inlet TS10 to outlet TS45; thus, with cyclic write and acyclic read, cell 45 contains the address '10'. The outlet space switch C1 connects the outlet from B3 time switch to the outlet bus C1 during TS45 via crosspoint 3.Thus, CM-C1 contains the crosspoint address '3' in cell 45. It can be seen that control symmetry for the S-T-S configuration is achieved if the input space switch is row-oriented and the output space switch is column-oriented.

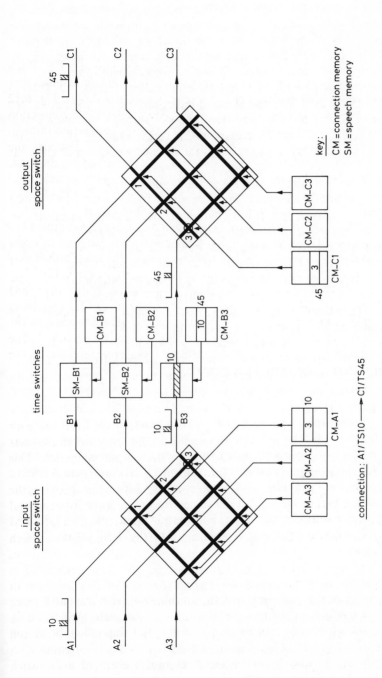

Fig. 6.13 *S-T-S switch block*

6.4.4 The T-S-T switch block

With a T-S-T switch block, the input time switch connects the input time slot to any free time slot on the bus to the space switch input, while the output time switch connects the chosen time slot from the space switch to the required outgoing time slot. Thus, call connections through the switch block may be routed through the space switch during any convenient time slot. The arrangement is shown in Fig. 6.14 for the connection example A2/TS10 to C1/TS45. With the inlet bus/time slot and outlet bus/time slot specified, the switch control is able to select any free time slot through the space-switch crosspoint A2, C1 (C1 column, crosspoint 2). In the example of Fig. 6.14, space-switch time slot 124 is assumed and the necessary contents of the CMs are shown.

The CM of the input time switch A2 contains address '10' in cell 124; thus, with cyclic write and acyclic read, input time slot 10 is connected to output time slot 124. The space switch has column-oriented connection memories. The crosspoint address '2' is held in CM-B1 cell 124 so that the output of the A2 time switch is connected to the input of the C1 time switch during each switch-block time slot 124. For reasons of control symmetry explained in the following Section, there are advantages in organising the output time switches in acyclic-write, cyclic-read mode. The content of CM-C1 in cell 124 is the address '45'. The output from the space-switch column B1 during time slot 124 is thus transferred to output bus C1 during time slot 45.

This sequence, which is repeated once every frame until the contents of the relevant connection memories are changed, creates a path from A2/TS10 to C1/TS45. In order to establish a both-way connection, a corresponding path is also required for the transmission of speech from C1/TS45 to A2/TS10. These two paths through the switch block may be established independently for each call or established as a pair. The former method offers more flexibility in the utilisation of the switch block, but the second method simplifies the control process because of the connection symmetry achieved. Fig. 6.15 illustrates the practical arrangement of a T-S-T network providing a 4-wire connection (transmit and receive paths) between A2/TS10 and C1/TS45. As the Figure shows, it is convenient to organise the T-S-T switch block with the corresponding input and output time switches grouped together. (A similar arrangement is applicable to the T-S, S-T and S-T-S configurations.)

6.4.5 Comparison of the switch-block configurations

At the beginning of this Section it was shown that, although a single S-stage digital switch block is impractical because of the extremely high blocking probability, a single T-stage can be used as a non-blocking switch block for capacities up to about 250 lines. The higher capacities normally required by exchanges can be met only by switch blocks comprising a combination of S and T stages. The two-stage, S-T or T-S, switch blocks enable small-to-

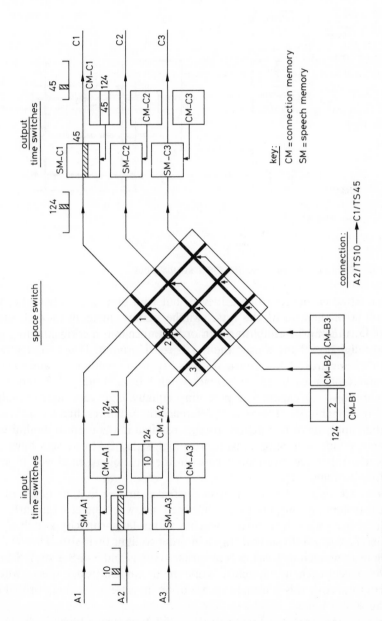

Fig. 6.14 *T-S-T switch block*

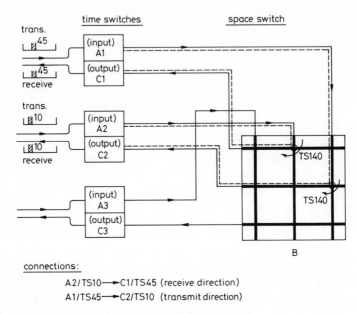

connections:

A2/TS10 ──► C1/TS45 (receive direction)
A1/TS45 ──► C2/TS10 (transmit direction)

Fig. 6.15 *Four-wire routeing through T-S-T switch block*

medium sized capacities to be achieved, but with blocking probabilities that increase with the size of the time switches. Thus, medium-to-large switch blocks must comprise three stages in order to achieve low blocking.

Early digital exchanges used the S-T-S configuration. However, from the late 1970s onwards, T-S-T configurations began to predominate and they are now invariably used in all multistage switch blocks. Both S-T-S and T-S-T configurations are capable of providing equally high capacities with low-blocking probabilities. The earlier preference for S-T-S resulted from the high cost then applying to fast digital storage, resulted in the need to minimise the amount or time switching. Advances in high-speed RAM devices have now eliminated this price differential between the technology employed in space and time switches.

The space switch, being a matrix, increases in size as the square of the number of input or output buses, while the time switch increases linearly with the number of time slots. Space switches within large-capacity switch blocks are therefore split into several stages in order to limit their size. This leads to configurations such as S-S-T-S-S in earlier systems and T-S-S-T or T-S-S-S-T in more recent exchange systems. While the use of multistage space switches reduces total cost, this is at the expense of an increase in blocking probability for the switch block.

An important factor in favour of the T-S-T type switch block is the ability to incorporate some of the peripheral functions into the input and output time switches. Such functions include alignment of PCM systems with the exchange frame start, and super-multiplexing. The inclusion of these functions into a time switch is illustrated in Figs. 6.9 and 6.10.

For simplicity, the description of space and time switches has, until now, tacitly assumed an equal number of inlet and outlet time slots in a time switch, and that the space switch is square; thus there is no traffic concentration. However, subscriber-concentrator switch blocks are constructed from T, T-S or T-S-T configurations with traffic concentration being provided by the T-stage having fewer output than input time slots for one direction of transmission and vice versa in the other direction. An adequate grade of service (GOS) is achieved provided that each time switch is evenly loaded with input traffic within the planned range. This requires the spreading of the high-traffic generating subscriber lines across a number of time switches. (Connection between a subscriber line and the input of a concentrator switch block is provided by a wiring frame, the MDF, as described in Chapter 7.) For example, with a T-S concentrator switch block with a concentration ratio of 8:1, a good GOS is obtained if the average loading per input line is a maximum of 0.1 erlangs. This will result in an average loading on the output channels of 0.8 erlangs, a typical figure for digital exchanges.

In theory, non-concentrated T-S-T switch blocks can be guaranteed non-blocking if the number of internal time slots across the space switch is $2N-1$, where N is the number of external time slots in the input and output sides of the peripheral T-stages.[5] However, even if the number of internal and external time slots are equal, the grade of service through a T-S-T switch block may be as low (i.e. good) as 3.1×10^{-17} for 0.7 erlang per circuit, increasing to 4.7×10^{-8} for 0.8 erlang per circuit. Furthermore, because not all of the time slots on the PCM systems are allocated to traffic, there is usually a greater number of internal time slots available for routeing through the space switch than there are external time slots. Hence, even with the high-traffic loading of 0.8 erlangs per circuit, the grades of service through non-concentrated T-S-T switch blocks can be around the low values of 8^{-10} to 10^{-10}. However, the grade of service will degrade rapidly with increasing traffic loading[8] (see Section 6.5.3).

6.5 The control of digital switch blocks

6.5.1 Introduction

So far in this Chapter the description of the way that a path is maintained across a digital switch block has assumed that a steady-state condition exists; that is, the path has already been selected and established. It has been shown that the path will be created through the space and time switches for the duration of one time slot once every frame, for as long as the connection memories hold the appropriate information. The function of establishing or

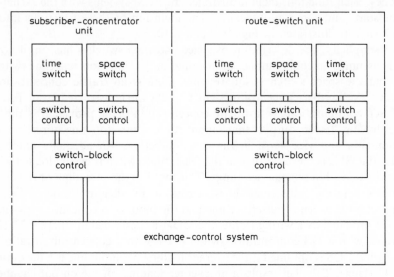

Fig. 6.16   *The three levels of digital switch-block control*

clearing a path through the switch block therefore entails the insertion, or deletion, of the relevant addresses within the connection memories. This activity is the result of the interaction between the exchange-control system and special control units associated with the switch blocks.

The control of a digital switch block is functionally split into three components, namely the exchange-control system, switch-block control and switch control (see Fig. 6.16). The exchange-control system provides the overall control of the exchange, including the call-processing function, as described in Chapters 14 and 15. Although this control system is shown as a single entity in Fig. 6.16, it may be physically realised in a distributed rather than a fully centralised architecture, as discussed in Chapter 17.

An exchange may comprise just one switch block, for example in a purely trunk exchange, or many switch blocks, as in a local exchange with a group switch block and several subscriber concentrators. Each switch block has its own switch-block control. Within the switch block, each switch has a switch control consisting of a connection memory and associated control logic. The following Sections discuss the way that these control systems establish connections through exchange switch blocks.

6.5.2 Switch-block control
The switch-block control must provide the management of all the paths through the switch block. Typically, this management covers:

(i)   Set-up of a path.
(ii)   Clear-down of a path.

(iii)   Reservation of a path.
(iv)   Tracing of a path.
(v)   Checking of a path.
(vi)   Interrogation of a path's status (.i.e. free, busy or reserved).

The paths through the switch block are normally both-way, but uni-directional paths may also be established for the conveyance of supervisory, control or alarm information. Thus items (i) and (vi) above must relate to both uni-directional and bi-directional connections. As indicated above, the switch-block control is concerned only with the task of managing paths through the switch block and is not in control of the overall call connection. This is because the complex call-processing activity is confined to the exchange-control system, while the primitive switch block path-management activity is delegated to the switch-block control.

The prelude to any request for a path to be established through the switch block is therefore the call-processing activity within the exchange-control system. An example of this activity is the recognition of a particular subscriber lifting his telephone handset and thus requiring a path through the subscriber-concentrator switch block to a multifrequency (MF) signalling receiver (see Chapter 8). Another example of the call-processing activity is the request to the group-switch-block control for a path between a subscriber-concentrator output and a junction route, based upon the translation of the subscriber's dialled digits and the routeing rules for that exchange. (The full range of call-processing work is described in Chapters 14 and 15.) Irrespective of the form of call processing activity involved, the exchange-control system will command the relevant switch-block control to set up (or clear down, etc.) a path between specified inlet and outlet time slots.

6.5.2.1 Communication between exchange and switch-block controls. The command from the exchange-control system to the switch-block control is usually in a high-level message format so as to achieve an efficient control structure which maximises the use of the high-speed and high-level signal intelligence of the exchange-control processors. An example of the format of such a message is shown in Fig. 6.17 (labelled type I). Like any message-based signal, the format is based on specified fields. Although the actual format of type I messages will differ according to the individual manufacturer's design, the following fields will always be required:

(i)   Operation code: This indicates whether a path is to be set up, cleared down, reserved, etc. If the 12 operations described above ((i) to (vi) for both uni- and bi-directional paths) are to be coded, a field of 4 bits is required.
(ii)   Input-time-slot field group: This group of fields gives the inlet-channel address in the form of switch number, PCM system number and time slot number within the PCM system. Although a single address, this does identify

message No.	operation code	input time slot			output time slot			error detecting code
		time switch No.	PCM system No.	TS No.	time switch No.	PCM system No.	TS No.	

a type I message (exchange–control system to switch–block control)

message No.	information field	reference message No.	error detecting code

b type II message (switch–block control to exchange-control system)

input time switch				space switch				output time switch			
time switch No.	CM contents	B	P	column No.	CM contents	P		time switch No.	CM contents	B	P	

c type III message (switch–block control to switch controls)

Fig. 6.17 Switch-block control message formats

both transmit and receive time slots at the inlet switch. The sizes of the first two fields depend on the number of input switches and the number of the PCM systems super-multiplexed on to a switch, respectively. The time-slot number field is 5 bits for both 32- and 24-time-slot PCM systems.

(iii) Output-time-slot field group: This gives the outlet-time slot address in a similar format to the input-time-slot field group.

(iv) Error-detecting code: This enables the switch-block control to detect any corruption of the message which may be incurred during transmission from the exchange-control system. Such a code might be simply parity check sums or the result of a more elaborate polynomial-division process.

(v) Message number: Each message is labelled with a simple number to facilitate referencing in subsequent messages. The use of a message number enables the switch-block control to advise the exchange-control system that a particular message had been received in error and hence request its retransmission.

On receipt of a type I message, the switch-block control executes the instruction. In the case of a path set-up request, the switch-block control performs the path-search procedure, as described later in this chapter, and selects a path through the switch block. The exchange-control system must then be advised that a path has been found. Alternatively, the switch-block control must advise the exchange-control system that no path is available. Similarly, messages will need to be sent to the exchange-control system indicating that paths have been cleared, reserved, and so on. The messages back to the exchange-control system will need to contain the fields shown in Fig. 6.17 (type II), namely:

(vi) Reference-message number: This field contains the identity of the message from the exchange-control system to which this return message relates.

(vii) Information field: This contains the message information being sent to the exchange-control system. The message is sent using a repertoire of codes, covering, for example, 'path established', 'no path available', 'path reserved' and 'forward message received in error'.

(viii) Message number and error-detecting code: These fields are similar to those in the type I messages.

In the case of a call path set-up, the switch-block control will need to determine the necessary addresses that are to be written into each of the connection memories involved, so that the time and space switches will provide the required path. The switch-block control must then inject the required contents into the specified cells of the connection memories.

The time-switch control consists of the connection memory and time-slot counter logic (as described in association with Fig. 6.8), in addition to a switch-control unit. The space-switch control comprises the connection memories (as described in association with Fig 6.4) together with a switch-control unit.Before considering how the necessary information is injected into the appropriate connection memories, the format of the messages involved is first described.

6.5.2.2 Communication between switch-block and switch controls. The information that must be sent from the switch-block control to each switch control, whether a space or time switch, is:

the address of the CM,
the address of the CM cell,
the data content to go into the CM cell.

This suggests that, for a three-stage switch block, three fields are required in the message for each of the three switch controls involved in a path set-up. However, this need not be the case. In the example of the T-S-T network in Fig. 6.14, a connection was made between A2/TS10 and C1/TS45 using time slot 124 through the space switch; it will be noticed that the cells of the three CMs controlling that connection are all at location 124. (In fact, the output time switch has an acyclic write/cyclic read arrangement, the reverse of the input time switch, in order to achieve this commonality.) Advantage can be taken of this feature by providing only one field to address all three CM cells. However, an even more elegrant approach is to send the message to the relevant CMs during the time slot corresponding to the CM cell's address: in our example, during TS 124. In such a case, the message(s) from the switch-block control does not require an address for the CM cells, thus saving three address fields.

A possible format for (type III) messages between a switch-block control and the T, S and T switch controls is shown in Fig. 6.17c. The message is composed of three groups of fields, addressing the controls of the input time switch, the space switch and the output switch, respectively.

(i) Input-time-switch field group: The first field in this group identifies the particular input time switch. The second field contains the content that is to be inserted into the connection memory. There are two additional 1-bit fields associated with the input-time-switch control, namely the 'busy' bit and a 'parity' bit, shown as B and P, respectively, in Fig. 6.17c. The parity bit is set by the switch-block control, using an odd or even parity rule, so that simple transmission errors introduced on the address and CM-content buses may be detected by the recipient.

The busy bit is used by the switch-block control to register the busy/free status of the outlet time slots in the input time switches. This information is used in the search for a free time slot for routeing through the space switch to an output time switch, as is described later. Rather than establish a separate memory in the switch-block control to map the status of the output time slots,

input time switch   space switch   output time switch

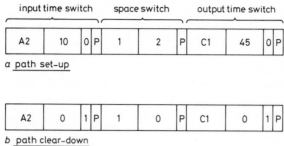

a path set-up

b path clear-down

Fig. 6/18   *Example of type III message*

the busy bit can conveniently be added to the contents of the input-time-switch connection memories. Thus, when loaded by a type-III message, each connection-memory cell within the input time switches will contain either the address of the SM cell and busy bit set to 0, or zero address and busy bit set to 1, according to whether a call using its output time slot is inprogress or not, respectively.

(ii) Space-switch field group: The first field in this group identifies which space-switch column and associated connection memory are being addressed. (If the space switch is controlled on a row basis, as described in Section 6.2.2.1, the first field will contain a row address.) Where the switch block contrains several space switches, the first part of the column number will need to indicate which switch is being addressed. A parity bit is provided, as described above.

(iii) Output-time-switch field group: The four fields in this group are similar to those for the input-time-switch control, except that the busy bit relates to the busy/free status of the input, rather than output, time slots.

An example of the use of type III messages between switch-block control and the three switch controllers is shown in Fig. 6.18. For continuity, the example of a connection across a T-S-T network shown in Fig. 6.14 (A2/TS10 to C1/TS45 using TS124 via space-switch crosspoint 2 of column B1) will be considered. As previously explained, the appropriate information needs to be deposited in cell 124 of the relevant time- and space-switch connection memories. Thus, the message shown in Fig. 6.18a is sent during the next occurrence of time slot 124 to the three switch controllers, in order to set up the required path through the switch block. The relevant parts of the message are extracted by the three switch controllers, as indicated in Fig. 6.18a and described later.

The message necessary to clear down this path through the switch block is shown in Fig. 6.18b. This message is sent during the next occurrence of time slot 124, so that the 124th cells of the connection memories of input time switch A2, space-switch column 1 and output time switch C1 are loaded with

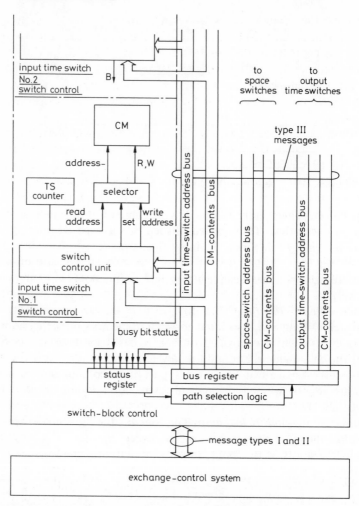

Fig. 6.19   *Use of bus and control wires in switch-block control*

zero addresses. The busy bits of the 124th cells of the three CMs are set to 1
to indicate that internal time slot 124 can be used for other connections.

Fig. 6.19 illustrates the uses of buses and control wires to link the various
switch-control units with the switch-block control. For clarity, only the switch
control for input time switch 1 is shown, thus omitting controls for the space
switch and the other time switches in the T-S-T switch block. The
communication is provided by three pairs of buses (address and CM contents):
one pair linking all the input time switches, one pair for the space switches and
one pair for the output time switches. In addition, a single wire links the busy-
bit fields from all the input and output time-switch connection memories to
a status register within the switch-block control.

The input-time-switch address buses are connected to all the input-time-switch control units, as shown in Fig. 6.19. Each control unit continuously searches the contents of the bus for an address of one of its time switches. When an appropriate address is found, that address is used to route the data from the CM contents bus into a connection memory cell. A similar arrangement applies to the bus connections to the space switch and output time switch.

The three pairs of buses terminate on a bus register within the switch-block control. This enables the various fields of type III messages held in the bus register to be output to the relevant buses. The necessary messages are generated by the switch-block control, based upon commands from the exchange-control system and, for new path set-ups, using the output of the path-selection logic. (Fig. 6.19).

6.5.3 Path selection

The path-selection procedure for a T-S-T switch block consists of finding a free internal time slot across the space switch. This means that a time slot must be chosen that is free on the outgoing side of the input time switch and on the incoming side of the output time switch. The busy/free status of each of these two sets of time slots is indicated by the busy bit in the respective CM cells. The path-selection method is, therefore, a simple process of seeking for coincident free time slots by examining the pairs of busy bits from the two sets of time switches.

The mechanism for path selection is contained within the switch-block control. This system, which is usually microprocessor-based, may perform the search using a number of digital logic processes. The most common method, the use of a mask, is briefly described below.

The busy/free status of each time-switch connection memory cell is stored during each time slot within the status register of the switch-block control (Fig. 6.19), as described above. This register contains two fields: one holds the status for the input time switches (0 for busy, 1 for free); the other holds the status for the output time switches. During each time slot, all the corresponding busy bits are presented in the status register; for example, during time slot 150, all the busy bits from cell 150 of all the CMs are present. The path selection, however, is concerned with coincident free time slots only in the input time switch and output time switch specified by the exchange-control system. Thus, the switch-block control constructs a binary word, equal in length to the status register, which contains zeros in all bit positions except those two corresponding to the connection memories of the required input and output time switches. This word, known as a mask, is held in a selection register. During each time slot, the contents of the status register are ANDed on a corresponding-bit basis with the contents of the selection register. When a free time slot (status 1) appears in the two CMs under test, the result of the AND process places a 1 in both halves of a result register. An example of this

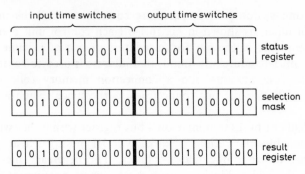

Fig. 6.20 *Path-search example*

process is shown in Fig. 6.20. Thus, presence of a free path through the space switch is identified by testing for a non-zero result in both fields of this register during each time slot.

The chosen internal time slot is then used to create the necessary type III message to set up the path through the switch block. Alternatively, the switch-block control may be requested to just reserve the path. In this case, a type III message will be produced which contains 0s in the two B fields and merely zero content in the three CM-content fields.

The manner in which the switch-block control uses the mask to seek a free path influences the grade of service achieved. There are several path-search algorithms employed by digital exchanges, each of which offers different traffic characteristics and control complexity. An outline of three of these algorithms, as applied to T-S-T switch blocks, is given below.[9]

6.5.3.1 Sequential search, random start. With the algorithm, the search for a free internal time slot across the S-stage begins randomly and hunts sequentially through the time slots. This has the effect of spreading the occupied time slots across the range: for example, between TS0 and TS511 in a frame of 512 time slots. The resulting grade of service (GOS) is around 2.5×10^{-4} for an offered traffic of 0.825 erlangs per circuit.

6.5.3.2 Sequential search, fixed start. In this method, the path search always begins with the same internal time slot (the 'home' time slot) and then hunts sequentially through the range. Such an algorithm has the effect of packing the occupied time slots, because more calls will be set-up in the lower-numbered time slots than in the higher-numbered time slots. This results in an improved grade of service compared to the random start method. (For example, a GOS of 1.46×10^{-4} for 0.825 erlangs per circuit loading.)

6.5.3.3 Repeat attempt. When a path search has failed to find a free time slot across the time switch, a repeat attempt at searching, initiated some time later, may enable a vacant time slot to be found. The delay between searches affects the improvement in grade of service; the longer the wait, the greater is the likelihood that some calls will have cleared. For example, there is a 50% chance of finding a path after a 1 s wait, assuming traffic offered at 0.8 erlangs per circuit with 3 min mean holding time. However, a 1 s wait for a repeat attempt can add noticeable delay to the call set-up times, which are normally only 0.5 to 1 s through a digital exchange.

6.5.4 Multiple paths

Control techniques are also required to set up multiple paths across a switch block in which n inlets are connected to 1 outlet (or vice versa). Such connections may be used for multiple access to tone or recorded-announcement equipment, as described in Chapter 8. The techniques are similar to those described above for point-to-point connections, except that the content of the common speech-memory cell in the inlet or outlet time switch is read n times. This is achieved by storing the SM-cell address in the appropriate n cells within the controlling connection memory.

6.6 Switch-block security

The switch block and its associated control equipment must be adequately secured against failure. Public telephone exchanges are required to have an average time between total system failures of 20 to 50 years, depending on an administration's specification. Because it is only one of the exchange components, the switch block itself must contribute only a small proportion of this overall system failure rate. In addition, most telephone administrations stipulate a low probability of any one fault affecting more than a small proportion of circuits. This requires that considerable attention be paid to the reliability of digital switch blocks because of the high number of circuits that are carried over the time-division-multiplexed hardware. For example, a hardware fault on a single bus between a time and space switch could affect up to 512 traffic-carrying circuits. Digital switch blocks, therefore, employ some degree of equipment replication to protect against such an eventuality.

The simplest and most effective protection is the complete duplication of the switch block, as shown in Fig. 6.21. This means that the switch block comprises two T-S-T (or S-T-S, etc.) planes with each call established simultaneously with parallel paths through the planes. Although not essential, the control is simplified if the two paths through the switch-block planes are identical. Other replication methods, such as 'm in n' sparing of time switches, and multiple appearances of time switches on the space switch, produce far more complexity for little improvement in security compared with the simple full duplication of the switch block. The latter is, therefore, the most commonly used method.

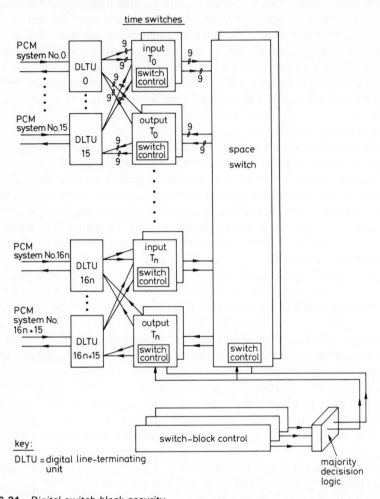

Fig. 6.21 *Digital switch-block security*

In addition to replicating the switching stages, it is necessary to have a mechanism to detect failure so that the faulty equipment can be isolated. The method of duplicating the switch block enables the application of a simple parity check to both sets of parallel paths to indicate the faulty plane. This parity checking is most effective when applied to each inlet and outlet channel. A parity bit is added to each 8-bit PCM word within the transmit direction of each 2 Mbit/s system terminating on the switch block. The stream of 9-bit words is then duplicated and fed to the corresponding inlet time switches on both planes. In the receive direction, each 9-bit word from both planes is examined for parity errors. Normally, the output from either one of the planes is passed to the PCM line system once the parity bit has been stripped off.

However, in the event of a persistent parity error being found the offending plane is considered fualty and the output from the switch block is automatically taken from the other plane. The functions of adding the parity bit to the transmit 2 Mbit/s stream, checking parity and choosing one of two received 2 Mbit/s paths is usually performed by digital line-termination units (DLTU). Further discussion of this equipment is given in Chapter 7.

The use of a complete duplication of the switch block also enables planned work, such as adding extra switch modules, to be carried out on one plane while the other carries live traffic. Obviously, in this situation there is no second plane to carry the traffic if a failure occurs; so planned outages of the switch-block plane should be kept to a minimum.

The duplication normally includes the switch controls, since they are primitive and intimately associated with their switches.However, the switch-block control will tend to be physically remote from the T and S switches and it undertakes some relatively complex processing functions. The switch-block control is, thus, not normally included in the duplicated planes, and so is separately secured.

The required security can be provided by constructing the switch-block control as a duplicated or triplicated system. In either case, some form of decision logic is needed to select an output from the replicated switch-block control. Chapter 18 considers the various forms of control-system architecture and their resilience to hardware and software errors. The block schematic diagram of Fig. 6.21 shows a duplicated switch block and triplicated switch-block control with majority-decision logic.

6.7 References

1 HUGHES, C.J. (1986): *Switching — State of the Art,* British Telecom Technology Journal, **4** (1) pp.5 – 19

2 FLOWERS T.H. (1976): *Introduction to Exchange Systems* (John Wiley & Sons) Chap. 10

3 HOSHI, M. (1979): *Introduction to Exchanges with Stored Program Control — Part II, Telecommunication Journal,* **46**, pp.204 – 218.

4 RONAYNE, J. (1986): *Digital Communications Switching* (Pitman), Chap.5

5 SLANA, M.F. (1983): 'Time-Division Networks', in *Fundamentals of Digital Switching,* ed. McDONALD J.C. (Plenum Press), Chap. 5.

6 CHAPMAN, K.J. and HUGHES, C.J. (1968): *A field trial of an experimental PCM tandem exchange',* Post Office Electrical Engineers' Journal, **61**, Part 3, p.2

7 ROSE, D.J. (1971): *Pulse code modulation tandem exchange — field trial model,* Electrical Communication, **46**, p.246

8 BELLAMY, J. (1982): *Digital Telephony* John Wiley & Sons), pp.252 – 264

9 TOMLINSON, P.N. and CHIA, C.W. (1977): *Teletraffic Aspects of Digital Switching,* Post Office Electrical Engineers Journal, **70**, Pt. II, pp.102 – 107.

Chapter 7

Line terminations at digital exchanges

7.1 Introduction

Earlier chapters have covered the fundamentals of converting speech on telephone lines to PCM format and the time and space switching of PCM channels. These processes are at the heart of an SPC digital exchange. The following two chapters consider the equipment that needs to be associated with the digital switches in order to fulfil the telephony functions of an exchange. Fig. 7.1 gives a block schematic representation of the range of subsystems required in a local exchange covering the functions of line terminations, signalling and control. Note that the Figure is an elaboration of the basic structure of Fig. 3.1. This chapter concentrates on the fundamentals of the subsystem functions and how they are realised, using Fig. 7.1 as a representation of a logical configuration. However, there are many variations in the way that the digital exchange systems now on the market organise the subsystems; such architectural options are considered in Chapter 9.

Fig. 7.2 shows how toll, trunk and international telephone SPC digital exchanges require only a subset of the local exchange subsystems because of the absence of subscriber lines. Such exchanges, therefore, would not include subscriber concentrator units. Although trunk and international exchanges are not encumbered by the control of subscriber lines, their role within the telephone network does involve additional control complexity covering routeing and network-management functions. In addition, international exchanges need international call-accounting capabilities.

This chapter considers the fundamentals of the various digital-exchange subsystems concerned with line termination, under the following groupings:

(i)   Analogue subscriber-line termination.
(ii)   Digital subscriber-line termination.
(iii) Analogue trunk termination.
(iv)   Digital line termination (external and internal).

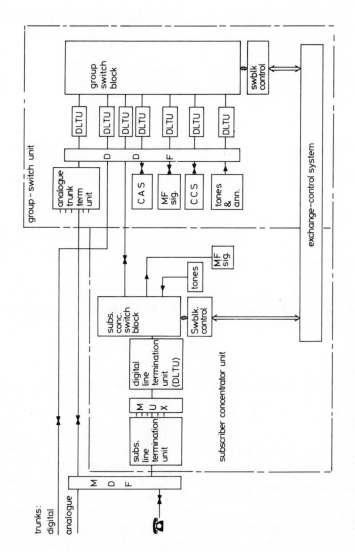

Fig. 7.1 *Digital local exchange subsystems*

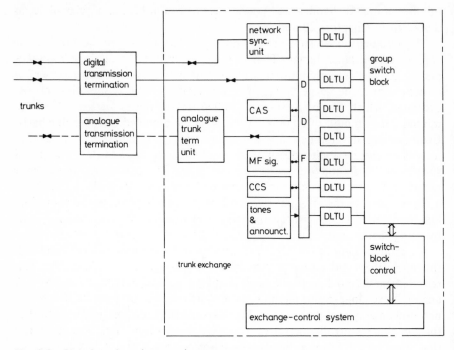

Fig. 7.2 *Digital trunk-exchange subsystems*

The following chapter considers the subsystems, shown in Figs. 7.1 and 7.2, concerned with the handling of subscriber and inter-exchange signalling.

7.2 An overview of the termination of subscriber lines

Undoubtedly, the most difficult hardware function to realise in a digital exchange system is the termination of subscriber lines. Such lines have a wide range of lengths, convey a variety of signalling systems, require ringing current and line-power feeding, and have a need for protection and test features. The difficulty becomes even greater when other forms of subscriber lines are considered: for example, PABX lines and public call-box lines. Currently, the vast majority of subscriber lines are analogue, using twisted pairs of copper wire. However, the single pair of wires may provide access for a number of subscribers where subscriber-carrier systems are used. Typically, these systems multiplex two subscriber lines, in the form of transmit and receive channels, thus achieving a saving of line plant (known as 'pair gain'). Furthermore, the implementation of an integrated services digital network (ISDN) introduces the need to terminate digital subscriber lines at exchanges. Such lines carry a number of digital channels and a new form of common-channel signalling, as described in Chapters 20 and 22.

The termination of subscriber lines represents a high proportion of the capital cost of the exchange. Thus, a major challenge for any local-exchange system designer is to minimise the cost of coping with the variety of subscriber terminations. The large number of subscriber lines terminating on an exchange, which can be as high as 50 000, forces the exchange designer to minimise the amount of equipment that must be provided on a per-line basis. Evolution and development of exchange systems is therefore directed towards increasing the flexibility of common equipment to cope with a variety of subscriber lines, and hence reduce the range of equipment types, while reducing the cost for each line. As will be described in Chapter 9, there are a variety of exchange-system architectures used to implement the subscriber line termination functions, together with a continuing drive to minimise costs by using semiconductor and VLSI techniques.[1-5]

A digital exchange must interface with not only the new modern terminals that subscribers may attach to their lines but also a range of old systems. This requirement is most onerous when a digital exchange replaces an existing analogue exchange. Some administrations chose to minimise the burden on the new exchange systems by reducing the range of telephone instruments and thus the signalling-control procedures that have to be supported. This can mean that some of the telephone instruments, attachments or PABXs have to be replaced by modern systems when a digital exchange is installed. A typical catalogue of types of subscriber lines/terminals that need terminations on a digital local exchange is as follows:

(i)   Analogue lines:
  (a) Direct exchange line:
  standard telephone instrument with bell
    — LD signalling (dial or push button)
    — MF signalling (push button).
  (b) PBX exchange line:
    — LD signalling (earth or loop calling)
    — MF signalling (earth or loop calling).
  (c) Coinbox:
    — LD or MF signalling and a variety of signalling methods to convey coin
      information (e.g. register loop, 50 Hz pulsing).
  (d) Auxiliary:
    — key-controlled transfer
    — subscriber's private meter.

(ii)   Digital lines:
  (a) Direct line exchange:
  ISDN access over a single subscriber line (CCITT 'basic access')
  (b) PBX exchange line:

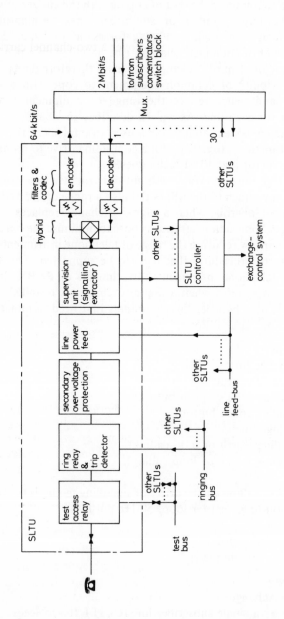

Fig. 7.3 *Analogue subscriber-line termination at a digital local exchange*

ISDN access at 1.5 Mbit/s or 2 Mbit/s over 4-wire digital transmission systems (CCITT 'primary access').

The list above is not affected by the conveyance of subscriber circuits over pair-gain transmission systems. This is because the systems are transparent and still present each circuit to a subscriber line termination unit (SLTU) as if two exclusive local lines had been used. Thus, for example, where two standard telephones with dial and bell are connected over a two-channel carrier system (known as '1 + 1') on a single cable pair in the local network, the two subscriber circuits are demultiplexed at the exchange and each terminated on an individual analogue SLTU in the normal way.

The termination of digital subscriber lines differs significantly from that for analogue subscriber lines. The termination of analogue subscriber lines, which until the mid-1990s will represent the vast majority, is described in this Section first, followed by a consideration of the SLTUs for digital lines.

7.3 Termination of analogue subscriber lines

The functions required to terminate an analogue subscriber line at a digital local exchange are summarised by the acronym 'BORSCHT', namely: battery feed, over-voltage protection, ringing, supervision, coding (i.e. encoding and decoding between analogue and digital mode), hybrid transformer (i.e. 2-to-4-wire conversion) and testing. Fig. 7.3 illustrates the allocation of these functions to equipment provided for each subscriber line, the SLTU, together with common equipment shared by many lines. The principles of providing each of the BORSCHT functions are now described.

7.3.1 Battery feed

The microphone used in the standard telephone instrument (traditionally carbon gradule, but increasingly using transducers such as charged electret foil) requires an energising current to be fed from a central source at the local exchange.[6] This current, of between 20 mA and 100 mA, is provided over the subscriber line by a central battery at a voltage of about -50 V DC with respect to earth. The negative voltage prevents corrosion of the copper pair which would result from ions causing plating-off in water solutions formed with any leakage to earth.[7]

In a conventional analogue exchange, power is fed to the two subscribers involved in a call connection on the same local exchange via a transmission bridge within the switch block. The transmission bridge consists of a power feed to the subscriber A and B lines via choke coils, whose inductance prevents crosstalk between lines by blocking the transmission of speech frequencies to the common central battery. The transmission bridge is completed by the use of capacitors in series with the connection path,[4,8] These give DC-voltage

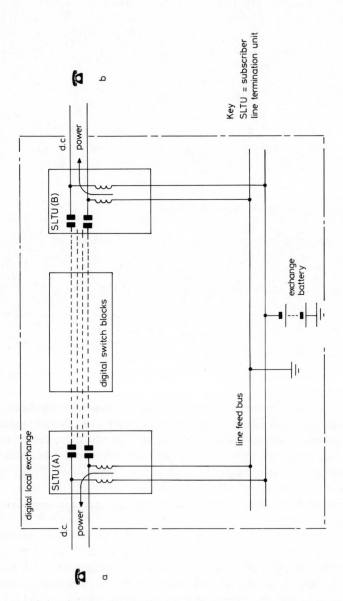

Fig. 7.4 Transmission bridge and power-feed mechanism for an own-exchange call connection

isolation between the lines, which is necessary to ensure the separate control of power feeding to each subscriber line and detect its on-hook and off-hook conditions.

However, unlike an analogue exchange, a digital local exchange may pass voltages and currents only at digital logic levels (e.g. 5 V), and not the massive 50 V required for the power feeding of subscriber lines. Thus, the basic transmission-bridge function is provided by the SLTUs at the periphery of the exchange, and not from within the switch block as in analogue exchanges. The arrangement for an own-exchange call is shown in Fig. 7.4 in which each of the SLTUs performs half of the transmission-bridge function.

A key requirement for the termination of analogue subscriber lines is the ability to cope with the variable loss introduced by differing lengths of line. The current flowing through the telephone instrument depends on the exchange voltage and the loop resistance. The latter is made up of the resistances of the subscriber line and the telephone itself. In any local-exchange catchment area there will be a range of subscriber-line resistances, the shorter circuits having lower loss than the longer circuits. This variation is kept within a working range by the use of cable of different gauges or, in the case of very long lines, by the use of amplifiers or loading coils.[7] The working range of subscriber-loop resistances is set by the power-feed requirements of the telephone instruments, the sensitivity of the loop-detection equipment at the local exchanges, and the transmission loss permitted by the national transmission plan. Typical values of maximum loop resistances range between 1250 and 1800 Ω. Examples of the planned maximum attenuation in the local line are 8 dB at 1 kHz in the USA, and 10 dB at 1.6 kHz in the UK (15 dB for lines parented on digital exchanges). Where appropriate, amplification is implemented in the SLTU.

Many telephone instruments (e.g. the BPO type 706 in the UK and the model 500 in the USA) incorporate a regulator which compensates, to some extent, for the potential reduction in power-feed currents resulting from the higher resistances of longer lines. These regulators take the form of varistors (non-linear resistors) or forward-biased rectifiers which increase their resistance as the current falls, thus increasing the sensitivity of the microphone circuit. The effect of the regulators is to vary the telephone-set resistance around the nominal 100 to 200 Ω.

The power supplied to line by the SLTU in a digital local exchange can be in the form of a constant-voltage or a constant-current source. Where the telephone-instrument regulators are exploited, the SLTU needs to supply a constant voltage to line, giving a current range of between about 20 mA and 100 mA. This is the method favoured in North America.[9] In the UK the specification for digital local exchanges requires a constant-current source capable of supplying a 25 mA current to loop resistances of up to 1800 Ω. This arrangement has the advantage of maintaining an even load per circuit on the exchange battery and it reduces the power loss in the subscriber line. However, because the regulator in the telephone instrument is thus disabled and cannot compensate for line loss, some other mechanism is required. One possibility, as used in the UK System X exchanges, is to use the variable voltage to control an automatic-gain-control system in the SLTU to keep the transmission loss constant, as described later in this chapter. It is interesting to note that disabling the regulator stabilises the transmission parameters on a subscriber line, which enables a better balance to be achieved in the 2-to-4-wire conversion system (e.g. hybrid transformer) in the SLTU (see Section 7.3.6).

7.3.2 Over-voltage protection

Any type of telephone exchange requires protection from the dangerously high voltages and currents that might occasionally be present on subscriber lines or the metallic paths of junction or trunk circuits. Such protection is for both the exchange equipment and the staff that work with it. Two main categories of electrical hazard need to be protected against: those due to lightning and those due to electrical power-distribution effects.

Lightning, with its massive short-duration electrical discharges measuring thousands of amperes and millions of volts, can cause considerable damage to external telephone cable. On overhead pole routes, even with external protection on the line, around 5000 V may be presented to the exchange during lightning attacks. Thus, it is standard practice for all physical circuits likely to be at risk to terminate on lightning protectors located at the main distribution frame (MDF) in the exchange building. These protectors consist of an air-gap device or a gas-discharge tube connected to each wire of the circuit, which short-circuits the line to earth at voltage in excess of about 750 V.

Hazardous voltages can also be present on the lines due to accidental contact between overhead power cables and overhead telephone lines, or by induction from faulty unbalanced power cables in the vicinity of underground telephone lines.[10] In addition, mains-powered terminals attached to telephone wires at the subscriber's premises (e.g. answering machines and switchboard systems) can apply hazardous voltages to the line under fault conditions. Protection from these possible electrical power problems is usually provided in the form of a fuse in each wire of the lines. There are several types of fuse used: some are designed to blow immediately a high current is present; others blow only after a period of time beyond which continuous flow of slightly excessive currents would have caused damage to exchange equipment by overheating. Fuses are often located at the MDF, together with the lightning protectors.

The particular vulnerability of digital electronic exchanges which, unlike electromechanical exchanges, cannot recover from even momentary voltage surges, necessitates the inclusion of secondary protection against excessive voltages in each SLTU, in addition to the primary protection at the MDF. The protection within the SLTU, which is directed against electrical-mains voltage levels, normally comprises fuses in each wire of the subscriber line and shunt circuits consisting of power resistors, varistors and Zener diodes.[3,7]

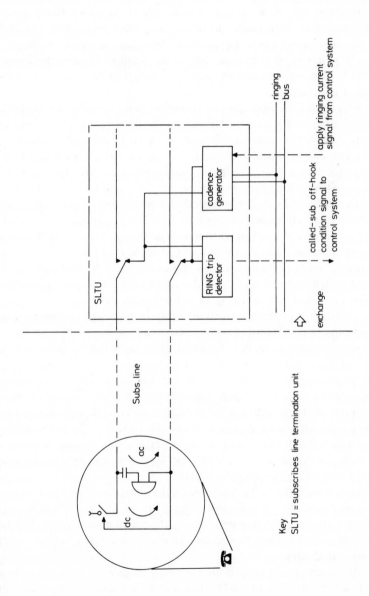

Fig. 7.5 *Application of ringing current*

7.3.3 Ringing

An intermittent alternating electrical source, typically of 75 V to 80 V and 200 mA at about 20 Hz (values of between 16⅔ Hz and 25 Hz are used in various countries), needs to be applied to a subscriber line by the digital exchange in order to ring the bell of a conventional telephone. This relatively massive electrical signal must be injected on to the subscriber line within the SLTU between the line entry and the analogue-to-digital conversion stage. In addition, the response of the subscriber to the ringing, namely the lifting of the handset ('off-hook' condition), must be detected by the SLTU to intitiate the cessation of ringing.

Fig. 7.5 illustrates the principles of ringing-current injection and detection of the called-subscriber's answer. The basic ringing current is produced by an appropriate alternating-current generator located centrally within the exchange. The characteristic bursts of ringing current, known as 'ringing cadence', (typically 2s on, 4s off) are usually produced by interrupting the continuous output from the generator. As in analogue exchanges, the interruption can be made by a mechanical cam mechanism associated with the generator, which produces several outputs for different groups of SLTUs with the on-off periods interleaved, so as to maintain an even load on the generator. Alternatively, the cadence can be introduced locally within the SLTU when ringing current needs to be applied, as shown in Fig. 7.5. In the latter case, the random distribution of SLTUs in the ringing state is sufficient to ensure an even loading on the central ringing-current generator.

When a subscriber's telephone is to be rung, the exchange control instructs, via an appropriate distribution (e.g. a control and address bus), the relevant SLTU to initiate ringing. As Fig. 7.5 shows, this results in the subscriber line being connected to the ringing-current bus via the cadence generator. The cadenced alternating ringing current then operates the bell in the subscriber's telephone instrument. The called subscriber's answer is detected by the presence of the DC loop created when the telephone receiver is lifted off-hook. The ring-trip detector must be sensitive enough to detect the loop condition during both the ringing bursts and the quiet period.[9] On sensing the loop, the ring-trip detector disconnects the subscriber line from the cadence generator and notifies the exchange-control system.

The devices within the SLTU that switch the subscriber line to the ringing-current supply need to pass a substantial 75-to-80 V 200 mA signal. This requirement is traditionally implemented by the use of electromagnetic relays, or increasingly by the use of special high-voltage discrete semiconductor devices, usually diodes.

A further design option is to locate the ring-trip detector with the common equipment serving a group of SLTUs (i.e. the SLTU controller), as described in Section 7.3.9.

7.3.4 Supervision

All local telephone exchanges are required to supervise each subscriber circuit continuously so that any changes of state can be quickly identified and an appropriate response made. With analogue subscriber lines using conventional telephones, all supervision is achieved by monitoring, at the exchange, the presence or absence of the electrical loop over the metallic path comprising the subscriber line and telephone set. The appropriate interpretation of this simple binary-loop condition, together with the use of timing and sequence parameters and a knowledge of the previous condition, enables the seven basic states of the subscriber line to be identified. Table 7.1 lists the various subscriber states that can be deduced at the exchange from the loop conditions and a knowledge of the preceding line states for that subscriber.

For digital local exchanges, supervision of the subscriber's loop must be performed at the periphery of the exchange because digital switches do not offer a metallic through path. Each subscriber line is therefore supervised within the analogue portion of the SLTU. The supervision function, at its simplest, is the detection of the presence or absence of a loop on the subscriber line and the flagging of this state in a form suitable for interpretation by the line-scanning function of the exchange-control system. The supervision device within the SLTU consists of a loop detector capable of working within the required range, typically with a minimum of 20 mA at about 50 V DC on the subscriber line. The presence or absence of a loop is indicated directly to the line scanner within the exchange-control system at the appropriate digital-logic level, typically 5 V DC. Fig. 7.3 illustrates the arrangement. Chapter 14 describes how the exchange control scans the output from the SLTU loop detectors and interprets the results according to the state of progress of the call.

The SLTU has to monitor the duration of the loop conditions in order to detect DC signals on the subscriber line, as described in Chapter 2. There is a clear design option between providing some measurement of duration within the SLTU loop detector or by incorporating all timing and persistence checking within the scanning equipment of the exchange-control system. The implementation of such functions within the SLTU requires a cheap and simple solution, which in early exchange systems was realised using conventional relay logic. The alternative method of exploiting the scanning process within the common equipment of the exchange control is now usually favoured. This is because of the flexibility offered by the scanning software to cope with a variety of timings and tolerances, enabling the exchange to be adapted easily to different network specifications. Also the speed of operation provided by microprocessors enables the function to be handled for a large number of lines by just a few centralised systems.

Fig. 7.3 shows the same scanning-point output from the loop detector being used to indicate both on- and off-hook conditions, as well as each dialled digit. If the subscriber's telephone uses MF signalling, then all the selection

Table 7.1 *The supervisory status resulting from the various subscriber lines states*

Subscriber's action	Loop condition	Supervisory status
Calling subscriber		
1 Lifts handset	'Off-hook' sustained loop initiated	Subscriber calling
2 Dials	Series of short makes and breaks of the loop	Calling subscriber dialling digits derived from train of makes and breaks
3 Replaces handset	'on-hook' sustained absence of loop	Calling subscriber clears
Called subscriber		
4 Handset on	Continuous absence of loop	Called subscriber free
5 Lifts handset	'off-hook' sustained loop made in the presence of ringing current	Called subscriber answers
6 Handset off	Continuous loop	Called subscriber busy
7 Replaces handset	'on-hook' sustained absence of loop	Called subscriber clear

signalling passes through the SLTU within the speech band and is detected by a common pool of MF receivers associated with the subscriber concentrator switch block, as described in the following chapter. The loop detector then indicates only the line signalling, i.e. subscriber actions 1, 3, 4, 5, 6 and 7 from Table 7.1.

7.3.5 Coding

The principles of PCM encoding and decoding have been described in Chapter 5. These functions form a key element of a digital exchange. The PCM primary multiplexer has historically used a common digital encoder and decoder (codec) for all of the 24 or 30 channels in the group format. This architecture was set by the high cost of realising PCM codecs using the available descrete logic devices. However, the overall economics of PCM transmission for routes between exchanges does not easily transfer to subscriber lines, with their low traffic loading and greater numbers. Two options were pursued by the designers of the early digital-exchange systems to meet this problem. The first was the use of analogue subscriber concentrators in order to concentrate traffic on to currently available PCM-group encoders, as described in Chapter 3. The E10 local exchange introduced in 1973 used this hybrid approach, as did the early versions of AXE10 and System X. The other option was to develop a single-channel codec and hence employ digital subscriber concentrators from the start. This option was adopted in the DMS100 introduced in 1980, the first all-digital local exchange system, based on development by Bell Northern Research.[11] The first single-chip integrated circuit PCM codecs became available in 1976, followed by single-chip integrated filters in 1977. However, the major breakthrough that gave the impetus to the all-digital local exchange was announced in 1978, namely a cheap integrated-circuit version of a single-channel filter and codec.[3,7] Progressively, from the early 1980s onwards, the increasing availability of these devices has been exploited by many manufacturers to produce all-digital local exchange systems.

Single-channel codecs are now available as VLSI devices in which, typically, the transmit and receive filters and the control logic are all contained on one chip.[7] As well as providing economical A/D conversion for each subscriber line at an exchange, these integrated-circuit codecs are now used on a per-channel basis in PCM transmission systems in place of the group codec used in the first-generation equipment. Recent developments in chip design have been directed at reducing power consumption and cost by the inclusion of digital signal processing. The new technique reduces the cost and complexity of the pre-sampling (anti-aliasing) filters by sampling the analogue channel at rates much higher than the Nyquist minimum of 8 kHz. The 64 kbit/s signal is then recovered by use of digital filtering techniques.[3,12]

7.3.6 Two-to-four-wire conversion

Chapters 5 and 6 have described how the use of semiconductor devices to constitute the digital switch block means that the transmit and receive paths through the digital exchange are on separate highways, i.e. they form a 4-wire circuit. Although within the telephone set the separate transmitter and receiver form a 4-wire circuit, the subscriber's circuit to the local exchange is normally provided over a single pair of wires (i.e. a 2-wire circuit). The necessary 2-to-4-wire conversion at the telephone set and at the exchange is provided by hybrid transformers or by a solid-state amplifier system.[9] Fig. 7.6 shows the circuit arrangement for an own-exchange call. The design of the hybrid transformer in the SLTU has to meet two essential criteria, namely:

(i)   The stability of the 4-wire circuit.
(ii) The suppression of echo.

The stability of the circuit is dependent on the degree of matching between the 2-wire line and the balance impedance within the hybrid transformer.

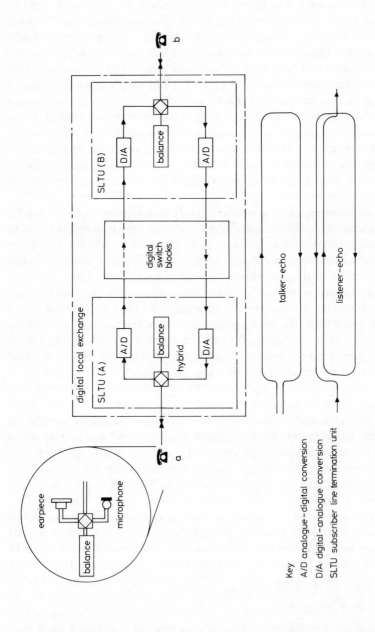

Fig. 7.6 *The four-wire path within a digital exchange*

Key

A/D   analogue–digital conversion

D/A   digital–analogue conversion

SLTU subscriber line termination unit

Typically, a simple balance impedance comprising resistors and capacitors is used in SLTUs to provide a reasonable match to their particular subscriber's line.[3]

Echo is a phenomenon encountered in all telephone networks as a result of a component of the sent speech signal being reflected from the distant end back to the talker. This return of signal occurs at the distant 4-to-2-wire conversion point. Echo becomes a problem when the distance of the call connection is sufficiently long for the propagation delays to create a perceived lag between speech sent and echo received. Two echo paths are possible: talker-echo path and listener-echo path, as shown for the speech direction A to B in Fig. 7.6. Current techniques for controlling echo use echo-suppressors, which automatically disconnect the return path according to which end is speaking. Alternatively, the transmission loss of the circuits may sufficiently attenuate the echo. Normally, some form of echo control is required for long-distance calls in large countries, such as the USA, and for intercontinental calls. In most small countries, which have predominantly analogue networks, the inherent transmission losses are sufficient to suppress the small amount of echo encountered on national connections. However, the overall reduction in transmission losses resulting from the introduction of integrated digital networks, especially for long-distance calls, means that even small countries will in future be required to control echo on national connections.

Both stability and echo requirements are met by the introduction of adequate loss into the transmission path formed by the transmit-and-return loop between the two 2-to-4-wire conversions in a connection. The minimum loss between 2-wire points required to ensure stability is 3 dB, giving a 6 dB stability margin around the loop.[13] However, a flat loss on connections between subscriber lines of just 3 dB could be inadequate for the control of echo; also, calls would be perceived as too loud and thus uncomfortable. Consequently, national transmission plans provide for loss across the digital network of at least 6 dB.[14] This loss will also usually be sufficient to suppress the talker-and-listener echo paths and eliminate the need for extra echo control for national connections in small countries.

7.3.7 Testing

A fundamental requirement of any local exchange system is the ability to test each subscriber line. In analogue exchanges, this requirement is met by test equipment which can access any line, often via the subscriber-concentrator switch block. However, such an arrangement is not possible with digital exchanges because the digital switch block does not provide the continuous metallic path required for subscriber-line testing. Thus, access must be provided at the periphery of the digital switch. The means of access is part of the subscriber-line-termination function, and is usually achieved by a test-access relay within the SLTU. Fig. 7.3 shows the position of this relay at the front of the SLTU, which switches to a common test bus or to inlets on a test switch.

The realisation of the subscriber-line test arrangement differs between switching systems, but the following basic principles apply:

(i)   Each subscriber line must be accessible and the metallic path must be extended to the test equipment for the duration of the tests.
(ii)   Tests may be made as a routine or on demand when required.
(iii)   Access between the SLTU and the test equipment may be via a bus or a small separate switch block.

The use of a test bus to link SLTUs to the test equipment is the simpler solution for (iii) above. However, it does have the limitation of allowing access to only one of the SLTUs parented on the bus at a time. Thus, the number of SLTUs dependent on one bus must be limited to a manageable group, the size of which is determined by such factors as the probable fault rate, maintenance procedures, duration of tests and the degree of automatic routine testing.

Where automatic routine testing of subscriber lines (e.g. every night) is carried out, the use of a simple test-switch-block access may be favoured. Such a switch block needs to provide a continuous low-resistance metallic path; this restricts its implementation to either standard analogue switches (e.g. reed relay) or special solid-state switches (e.g. PN diode). The small test switch block requires some element of control, which for operational security is usually physically separate from the exchange-control system.

Where the subscriber-concentrator unit is remote from the route-switch unit (see Chapter 9), the option exists of locating the test switch block either at the remote concentrator or at the parent exchange. The latter arrangement minimises the cost and complexity of the remote concentrator, but it does require the use of several junction pairs to extend the metallic path between the remotely located SLTU and the test switch block at the parent exchange.

7.3.8 Practical realisation of analogue SLTUs

All digital local exchanges need to provide the BORSCHT functions in their SLTUs according to the principles described above. However, the realisations differ between different manufacturer's systems, depending on the architecture and technology employed. In practice, the BORSCHT functions are split into three groups for physical implementation, as follows:

(i)   Test, ringing and (secondary) overvoltage protection.
(ii)   Supervision, battery feed and hybrid transformer.
(iii)   Codec and audio-transmission functions.

Group (i) functions are provided by a single module. Group (ii) functions are provided by a single module, commonly referred to as the 'subscriber-line-interface circuit' (SLIC). Group (iii) functions are either provided as a single module, e.g. the 'subscriber-line audio-processing circuit' (SLAC) in System X,[12] or are combined with the group (ii) functions to form a single entity, e.g. the 'channel circuits' in the ESS No. 5 switch.[15]

The SLIC is commonly implemented using transformers. The supervisory function relies on a threshold circuit which is set to distinguish between the minimum current flowing due to the off-hook condition and the maximum due to leakage through the cable insulation. The threshold circuit usually incorporates a low-pass filter to discriminate against the medium-to-high-frequency line noise. This form of SLIC has the great advantage of being robust and not easily damaged by high voltages.[7] However, it does suffer from the size and weight of the transformer necessary to pass the loop currents of up to 100 mA and transmit frequencies as low as 300 Hz. A number of design variants of the transformer-based SLIC have been aimed at reducing the size and cost of the transformer and minimising the heat dissipation when high line currents flow to short subscriber lines. Many designs now eliminate the use of transformers entirely by the use of operational amplifiers to implement the SLIC function.[3,4,12]

7.3.9 SLTU controller

The SLTU controller is a piece of common equipment that acts as the interface between the exchange-control system and a group of SLTUs. The number of SLTUs parented on a controller depends on the design of the exchange system; typical values range from 32 to 128. As emphasised at the beginning of this chapter, the key objective of the design of the subscriber-line termination is minimum per-circuit cost. Thus, the SLTU controllers tend to provide as many of the subscriber-line-termination functions that can be centralised, as well as the control interface functions. The SLTU controller's functions include the control of the five items described below:

(i) Supervision control: The supervisory unit in each SLUT raises a logic level '1' on its output to indicate the presence of a loop on the subscriber line. The SLTU controller is able to monitor the status of each of its subscriber lines by regular scanning of the outputs from their SLTU supervisory units. The controller usually processes low-level signals derived from the scanning, which indicate seizure, dialled digits, clear down, etc. This information is then transferred to the exchange-control system in a higher-level format, typically messaged-based.

(ii) Ringing control: Activation of the process of injecting ringing current on to a subscriber line is provided by the SLTU controller. The relay (or equivalent) in the relevant SLTU is operated by a low-level control-line signal from the SLTU controller, which has previously received a command from the exchange-control system. As previously described, the cessation of ringing current is under the control of the ring-trip detector. The SLTU controller may

also incorporate a common pool of cadence generators and ring-trip detectors for the dependent SLTUs.

(iii)   Test-access activation: The controller activates the test-access relay (or equivalent) in the appropriate SLTU upon receipt of a command from the exchange-control system. Cessation of the test access results from a further command from the exchange-control system.

(iv)   Power-switch activation: Power consumption within the SLTUs is minimised by switching power to the active components only during call connections. This means that power to the codec and audio transmission circuitry is switched off when the subscriber line is idle, i.e. the condition operating for most of the time. Power-switch activation by the SLTU controller may be initiated by its detection of the relevant subscriber's line off-hook condition, or on receipt of a command from the exchange-control system.

(v)   Activation of software-selectable options: Many exchange systems offer the ability to switch in and out optional features for an individual subscriber line under command from the exchange-control system, usually refered to as 'software-selectable' features. The initiation command from the exchange-control system to the SLTU controller may result from call-processing activity or from operational inputs via the man/machine-interface terminals. Examples of such options include:

(a)   short or long-line power-feeding system
(b)   earth or loop calling for PABX lines
(c)   shared or exclusive-line ringing
(d)   activation of a private subscriber's meter.

All the activations (i) to (v) above require the receipt by the SLTU controller of a command from the exchange-control system. Activations (i) and (ii) also require message signalling from the SLTU controller to the exchange-control system. The various messages need to convey two basic pieces of information, namely, the SLTU identity (e.g. equipment number) and the instruction signal or information (e.g. power-on, operate test relay, digit 3 received, etc.). The control messages between the SLTU controllers and the exchange-control system may be conveyed via a variety of transport mechanisms. These include a separate message bus, spare time slots across the concentrator switch block, and a separate message-switching network. Chapter 9 considers the various exchange architectures involved in routeing the control messages.

7.3.10 SLTU mutliplexing

The 64 kbit/s transmit channels for each SLTU are multiplexed into standard 30 or 24 channel. PCM frame formats ready for transmission over the exchange highway into the subscriber-concentrator switch block; the 64 kbit/s receive channel is similarly demultiplexed from the exchange highway (Fig. 7.1). This multiplexing equipment provides a permanent association between a particular SLTU and a time slot on the bus to the subscriber concentrator switch block.

7.4 Termination of digital subscriber lines

Although the vast majority of subscriber lines terminating on digital exchanges are currently analogue, there is a small, but rapidly growing, number of digital lines. These are provided over existing telephony distribution cable pairs in the form of ISDN 'basic access'. The concept of an ISDN is described in Chapter 22. However, for the purpose of describing a digital SLTU (D/SLTU), it is sufficient to define an ISDN basic access as a digital path from the subscriber's premises to the digital local exchange which is capable of supporting a range of non-voice (e.g. data) services, as well as telephony. The digital path provides two independent 64 kbit/s circuits for traffic use and common-channel signalling at 16 kbit/s. The traffic circuits are separated at the exchange within the D/SLTU and are then routed through the switch block to individual destinations on the exchange or via the external telephone network. Fig. 7.7 presents a block-schematic diagram of the role of the D/SLTU in association with a network-termination unit (NTU) and terminal adaptor (TA) at a subscriber's premises.

Consider first the use of one of the circuits over the digital local access for a simple telephone connection, as shown in Fig. 7.7. The major difference from an analogue SLTU is that, because a digital local line system is used, some of the BORSCHTfunctions are provided by the NTU and subscriber attachments at the subscriber's premises, rather than in the SLTU. Thus, the hybrid transformer and codec are located in a TA attached to the NTU. (Telephones designed specifically for an ISDN, known as 'digital telephones', may incorporate the TA function.) The remainder of the BORSCHT functions (i.e. battery or power feed, test, and over-voltage protection) are provided in the D/SLTU.

The subscriber common-channel-signalling system carries all supervision and address signalling for both the 64 kbit/s traffic channels, as described in Chapter 20. The subscriber-line-supervision function is performed in the TA rather than the D/SLTU. In the D/SLTU, the 16 kbit/s signalling channel is extracted and passed to the signalling-system logic (Fig. 7.7). Ringing current is not passed by the exchange to the telephone; instead, a message is sent via the common-channel signalling system to the TA. The TA then provides the ringing current.

Fig. 7.7. also shows, as an example, the second 64 kbit/s traffic circuit in use for a data call. This requires some additional software within the exchange-control system to handle the non-telephony call processing. The NTU offers a standard data interface to the terminal (e.g. CCITT X21, X21 bis or I420). No hybrid transformer or codec is required within the NTU for

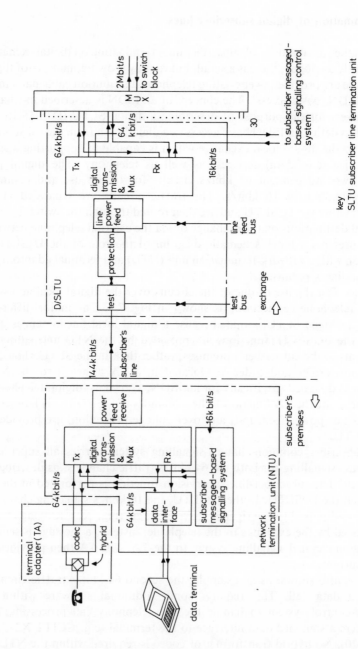

Fig. 7.7 *Digital subscriber-line termination*

data terminals because the latter directly presents a digital signal over discrete transmit and receive wires at the interface to the terminal.

7.5 Termination of analogue trunk lines

In a network comprising digital trunk and junction transmission and digital SPC switching (known as an integrated digital network 'IDN', as described in Chapter 21), there is no need for analogue trunk-termination equipment in either the local or trunk exchanges. However, most exchanges will be installed in networks which are progressively being converted from analogue to digital transmission. Thus, there will usually be some requirement for analogue trunk and junction routes to be terminated at the digital exchange. The termination for analogue trunks is provided by interworking equipment. Such equipment tends to be expensive because it usually deals with the termination of individual circuits (unlike digital routes which terminate on the exchange at the 24- or 30-channel group level). (However, there are systems, known as transmultiplexers, which convert analogue FDM supergroups directly into 2 Mbit/s digital format.[16]) Furthermore, the interworking equipment is redundant once the analogue transmission is replaced by digital systems. Therefore, administrations aim to minimise the need for such interworking equipment by suitable network planning (see Chapter 21).

The functions of the analogue-trunk-termination equipment are similar to those of the analogue-subscriber-line termination, namely test access, over-voltage protection, power feeding, digital encoding/decoding, signalling and multiplexing. However,there are some important differences, and these are described below.

Fig. 7.8 shows a block-schematic diagram of a typical analogue-trunk-termination unit (ATTU) supporting up to 30 trunk lines. The analogue trunk or junction circuits are routed via the exchange MDF in 2- or 4-wire form. Trunk routes using FDM high-capacity transmission systems are terminated at the nearest transmission terminal station, where the individual 4-wire audio circuits are derived by FDM multiplexers and extended to the exchange MDF.

7.5.1 Signalling
The provision of signalling senders and receivers on a per-line basis is both expensive and inefficient, particularly where discrete components or relay-logic systems are used. However, the use of high-speed semiconductor logic, together with microprocessor control, enables a single sender-receiver system to be time-shared by a group of circuits. Thus, the handling of analogue signalling within a digital exchange is centralised within a common pool of equipment. The primitive DC signalling from 30 trunk circuits is converted into standard PCM channel-associated TS16 format within the 2 Mbit/s stream produced by the ATTU. The signalling is then handled, along with

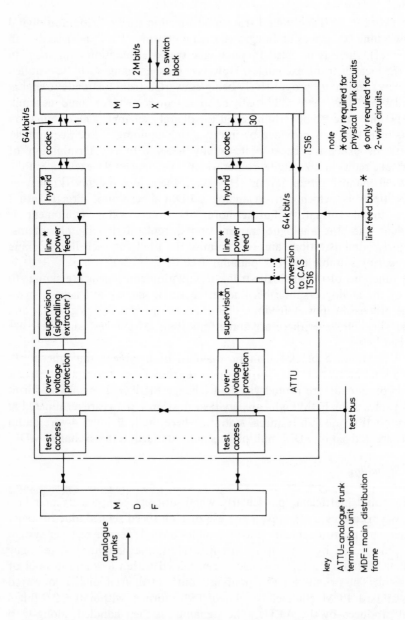

Fig. 7.8 *Analogue trunk termination*

channel-associated signalling from PCM trunks, by the pool of channel-associated signalling (CAS) exchange equipment, as described in the following chapter. Any single-voice frequency (1VF) or multifrequency (MF) signals within the analogue trunk circuits pass unaffected through the DC signalling extractor. Such signals remain within their individual channels and are echoed as though they were speech. As explained in Chapter 8, these PCM-encoded signals are later extracted and processed by common equipment within the exchange.

7.5.2 Power feed

This facility is required by trunk and junction circuits using DC signalling. These would normally be 2- or 4-wire physical circuits, but may also include links carrying outband signalling between an exchange and FDM terminal equipment housed in transmission stations. Analogue trunks using FDM transmission systems must employ voice-frequency signalling because DC conditions cannot be transmitted. The powering of such FDM systems is usually performed by the trunk transmission station, and so is not part of the exchange function.

7.5.3 Hybrid

This facility is required only for 2-wire trunk circuits terminating on the ATTU. The hybrid transformer is similar to that used in an SLTU, except that only a simple resistive balance impedance is used.

7.5.4 Multiplexing and control

The multiplexing arrangements are similar to those for the SLTU, except that the ATTU deals with up to 30 channels. TS 16 is fully utilised for channel-associated signalling for the 30 traffic channels. Thus, any control signals between the control system and the ATTU can be carried in spare capacity in TSO or via a separate control highway. A further option is for control signals to be carried over one of the PCM channels; this, of course, reduces the traffic capacity of the ATTU to 29 channels.

7.6 Termination of digital trunk lines

The structure and nature of a digital switch block and the specific requirements for connections to its ports were described in Chapter 6. The digital line-termination unit (DLTU) performs the function of interfacing the binary, time-phase-critical, secured digital switch blocks to the outside world. The range of necessary digital line terminations encompasses:

(i) External digital trunk and junction lines (from other exchanges).
(ii) Internal digital line connections from other subsystems within the exchange.

Figs. 7.1 and 7.2 illustrate how the digital switch blocks of both local and trunk exchanges are connected to the internal and external digital lines via DLTUs, one per 2 Mbit/s (or 1./5 Mbit/s) system.

Although the transmission of 2 Mbit/s or (1.5 Mbit/s) over an external line has different requirements compared to transmission between subsystems within the close confines of the exchange, there are advantages in using a common design of DLTU for all digital terminations. The major advantages result from the use of a standard interface at the switch ports, at the subsystem ports, and to the 2 Mbit/s highways interconnecting them. This not only provides flexibility in the way that switch ports are allocated to 2 Mbit/s highways, enabling a variable mix of internal and external lines, but it also provides the added advantage of a common design of equipment. The approach is particularly attractive for local exchanges that support remote concentrator units (see Chapter 9) because the same interface can be used between the group switch block and all its subscriber concentrators, irrespective of whether they are collocated or remote. The disadvantage of using a common interface is that all subsystems within the exchange need to incur the complexity of presenting a standard 2 Mbit/s interface suitable for external transmission, even if this is not warranted by the length of the link within the exchange.

The function of the exchange digital line-termination unit comprise:

(i)   Line-code-to-binary conversion.
(ii)   Frame alignment;
(iii) Serial-to-parallel conversion.
(iv)   Switch-block security.
(v)   Channel-associated signalling injection and extraction.
(vi)   Transmission-system termination.

The functions are fully described in other chapters; this Section only gives an overview of the functional components of the DLTU to enable its role to be appreciated. Fig. 7.9 gives a block-schematic diagram of the functional elements of a DLTU provided for one 2 Mbit/s digital line system.

7.6.1 Line-code-to-binary conversion

The digital line system terminating on the switch block has a line code suited to the type of transmission used (see Chapter 5). For cable systems this is some form of bipolar code (such as HDB3, or 4B3T or AMI), while for optical-fibre systems a unipolar over-speed code (such as 4B5B) is used. All these line codes require conversion to the simple binary format of the digital switch block. Rather than design a range of equipment to terminate all the different line codes that might be used, the DLTU usually offers a single standard interface. For exchanges based on 2 Mbit/s format, this is typically the interface according to CCITT G 703, which uses the common HDB3 code. Most digital transmission systems now also offer this standard interface, irrespective of the

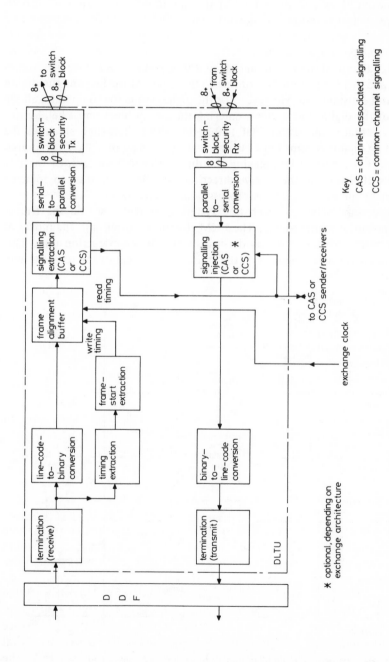

Fig. 7.9 *Digital trunk-line termination*

Key
CAS = channel-associated signalling
CCS = common-channel signalling

* optional, depending on exchange architecture

actual line code used. Thus, a single design of DLTU can terminate a wide range of line systems on to the switch block. Transcoding between the binary of the switch block and the HDB3 at the line interface is required on both directions of transmission, as shown in Fig. 7.9.

7.6.2 Exchange frame alignment

Each digital line system terminating at the exchange will have its frames starting at a different instant according to the length of the line and the timing source at the distant end. However, for the digital switching to operate, all the frames of the digital line systems entering the switch block must be aligned. This means that at the instant that TS0 begins within the switch block, all the terminating digital systems must also be at the beginning of TS0. The required realignment is achieved by delaying the digital signal from each of the line systems sufficiently, in the receive direction, for the frame start to coincide with that of the exchange. This process is illustrated in Fig. 7.10.

The necessary delay for each line system is provided within the frame-alignment buffer of the SLTU. The PCM stream is written into the buffer at the rate of the received line, as derived by the timing-extraction unit. This writing into the buffer begins with the frame-start instant of the received line,

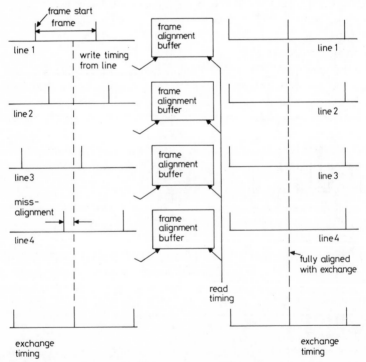

Fig. 7.10 *Re-alignment of digital-line frames to exchange-frame timing*

which is also derived by the timing-extraction unit from the received frame-alignment pattern (see Chapter 5). The content of the frame-alignment buffer is then read out at the rate of the exchange clock, beginning at the instant of the frame start of the exchange. The reading-out is performed simultaneously from all the DLTUs in the exchange. This process requires the frame-alignment buffer to have a capacity of at least one frame (i.e. 256 bits for a 2Mbit/s system) to cope with the maximum amount of misalignment between a digital line system and the exchange.

Frame realignment is not necessary in the transmit (to line) direction because the timing and alignment for all lines are directly generated by the exchange.

7.6.3 Serial-to-parallel conversion
Serial-to-parallel (S/P) conversion of the input digital line is achieved by writing each PCM word (or 'octet') sequentially into an 8-bit buffer store at the exchange rate (e.g. 2 Mbit/s) and reading the word's bits simultaneously on to an 8-bit parallel bus. The rate on the parallel bus will be an eighth of the input serial rate, i.e. 256 kbit/s for a 2 Mbit/s input. The timing for the reading and writing actions is derived from the exchange-waveform generator, as described in Chapter 10. It is important that writing into the serial-to-parallel buffer is aligned with the exchange-frame start, so that each of the 8 bits grouped for conversion represents an integral PCM word and not two parts of adjacent PCM words.

7.6.4 Switch-block security
Where the security of the switch block depends on its replication, the DLTU acts as the common peripheral points for signals entering and leaving the various planes of the switch block. As described in Chapter 6, the most common form of security replication is full duplication of the switch block. In this case, the DLTUs, which are provided on the basis of one per 2 Mbit/s (or 1.5 Mbit/s) line system, perform the following functions for the two identical parallel planes of the switch block:

(i)   In the switch-security transmit unit:
   (a) generation of check codes (e.g. adding a parity bit),
   (b) splitting the transmit signal into two identical bit streams.
(ii)   In the switch-security receive unit:
   (a) monitoring the check codes on the received paths from the switch block,
   (b) selecting and switching between received paths,
   (c) raising the required alarms for the control systems indicating changeover from one plane to the other, the working status of each plane, etc.

7.6.5 Signalling injection and extraction
Chapter 8 describes how the DLTU is used to inject and extract signalling to and from digital line systems terminating on a switch block. Whether or not

the DLTU performs this function depends on how the signalling is conveyed on the PCM line systems and the exchange-system architecture adopted. In the case of 24-channel PCM systems, channel-associated signalling is carried within the time slot of the channel to which it relates. Therefore, this signalling needs to be extracted by the DLTU at the periphery of the switch block and passed to the appropriate signalling sender/receivers, as shown in Fig. 7.9. The DLTUs for 1.5 Mbit/s-based exchanges will also extract/inject common-channel signalling where this is carried in the first bit of even frames of the 24-channel PCM systems (see Chapter 5). However, where signalling is carried in a separate dedicated time slot, as in the cases of common-channel or channel-associated signalling in TS16 of 30-channel PCM systems and 64 kbit/s common-channel signalling in 24-channel PCM systems, the option exists for the signalling time slots to be routed through the switch block to a common pool of signalling sender/receivers. In exchanges using this option, the DLTU does not inject-extract signalling.

7.6.6 Transmission-system termination

The range of digital line systems connected to a digital exchange requires several forms of termination. To simplify connection, the DLTU normally presents the standard G703 interface, as described above. Thus, each line system (irrespective of whether it is radio, cable, optical fibre, with or without power-feeding, etc.) needs to present a standard termination to the DLTU. This interface defines the electrical characteristics of the line signal (voltage profile for pulses etc.) as well as the frame structure. The various digital line systems can thus be terminated appropriately in a transmission station close to the exchange from which the standard 2 Mbit/s interface is offered over cable to the DLTU. Normally, a digital distribution frame (DDF) is used to provide flexibility in the allocation of line systems to DLTUs and their associated switches, as shown in Fig. 7.9.

As an option, the DLTU may provide standard-line terminations for junction-type PCM links carried over copper-pair cable. This function, shown as 'termination receive' and 'termination transmit' in Fig. 7.9, comprises power feeding and test access, similar to the arrangements described for the ALTU.

The DLTU usually incorporates a frame-recovery facility within the termination receive unit which detects any sustained loss of the frame-alignment word on the incoming digital line and executes the normal recovery procedure, as described in Chapter 5. This unit is also able to monitor the mean error rate on the digital line system by detecting corruption of the frame-alignment word. Alarms are usually generated by the frame-recovery unit to indicate substained loss of frame alignment and any excessive level of mean digital error rate. Typically, alarms are raised for error rates of 1 bit in 10^5, 10^4 or 10^3 bits. Most digital exchange systems initiate the withdrawal of any line system operating at an error rate of 1 in 10^3 by commanding the DLTU to disconnect transmit and receive paths from the switch block. In such instances, the control system applies a continuous backward-busy signal via the normal signalling channel to the line system until it is manually brought back into service.

Alarms relating to a line system are normally passed to the exchange-maintenance subsystem by the DLTU. The exchange-control system or the DLTU then signals the alarm conditions to the far end of the line via the designated bits in TS0 (for 2 Mbit/s systems) in the transmit direction. Similarly, distant alarms relating to the digital line system are sent by the other exchange or multiplexer via TS0 and detected by the frame-recovery unit. These are sent to the exchange-maintenance subsystem, if appropriate.

7.6.7 Vacant time slots

There will always be more occuped time slots on the digital line system terminating at the DLTU than are passed to the digital switch block. This is because some of the non-traffic time slots are terminated within the DLTU. For a 2 Mbit/s 30-channel PCM system, TS0 is not passed to the switch block, although the frame-alignment pattern and the line-system alarms are extracted by the DLTU, as previously described. In addition, the signalling carried in TS16 of a 2 Mbit/s system may be extracted by the DLTU and passed to the signalling processing units directly, rather than by routeing through the switch block. In both cases, the DLTU injects an idle bit pattern into the TS0 or TS16 to maintain a constant 2 Mbit/s stream throughout the exchange. Standard idle bit patterns are also injected into any other time slots not carrying live traffic. This is done within the DLTU and in the time switches at the outgoing side whenever a new vacant time slot is created.

A similar situation occurs for 1.5 Mbit/s-based exchanges, where idle bit patterns need to be injected into vacant time slots and to maintain the 24-channel frame format where signalling has been extracted.

7.7 References

1 HARRIS, H. (1980): *The Line Card Gets It Together*, Telesia, **4**, pp.16-18

2 FRITZ, P. (1979): *Subscriber Line Connection Units — Present Realizations and Future Trends*, IEEE Trans. **COM-27** (7)

3 AMES, R.W. (1986): *Subscriber Line Interfaces*, British Telecommunications Engineering, **4**

4 SKAPERDA, N.J. (1979): *Some Architectural Alternatives in the Design of a Digital Switch*, IEEE Trans., **COM-27** (7), p.961

5 FLEISCHFRESSER, G.H. and KOBYLAR A.W. (1981): *Providing Digital Techniques to the Subscriber Loop*, Telephony

6 LAWRENCE, G. (1983): 'Telephones' in *Local Telecommunications*, ed. Griffiths J.M., (Peter Peregrinus Ltd), Chap. 5.

7 McDONALD, J.C. (1983): '*The Analog Termination*', in *Fundamentals of Digital Switching*, ed. McDonald J.C., (Plenum Press), Chap. 7

8 SMITH, S.F. 1978): *Telephony and Telegraphy A — An Introduction to Telephone and Telegraph Instruments and Exchanges* (Oxford University Press), 3rd ed., Chap. 1

9 SMITH, S.F. (1983): 'Exchange Interfaces', in *Local Telecommunications*, ed. Griffiths, J.M., (Peter Peregrinus Ltd), Chap. 6

10 SHARP, K.R. (1987): *Disaster Planning — Protecting Networks from Power Transients*, Data Communications, **16** (10), p.207

11 ROSENBAUM, S. (1980): *Voice Coding and Filtering. One Chip Does It*, Telesis, **4**, pp.19-23

12 WARD, R.C. (1985): *System X: Digital Subscriber Switching System*, British Telecommunications Journal, **3**

13 FLOOD, J.E. (1975): 'Telecommunications Principles', in *Telecommunications Networks*, ed. Flood, J.E., (Peter Peregrinus Ltd), Chap. 2

14 HARRISON, K.R. (1980): *Telephony Transmission Standards in the Evolving Digital Network*, Post Office Electical Engineers' Journal, **73**

15 CARNEY, D.L. et al (1985): *The 5ESS Switching System: Architectural Overview*, AT & T Technical Journal, **64** (6), Pt.2

16 SMITH, D.R. (1985): *Digital Transmission Systems*, (Van Nostrand Reinhold Co.), p.305

The handling of signalling within SPC digital exchanges

8.1 Introduction

The various ways that subscriber and trunk lines are terminated at an SPC digital exchange were considered in Chapter 7. This chapter considers how the various forms of customer and inter-exchange signalling carried on those lines are handled within the exchange. More detailed descriptions of the signalling systems themselves are given in Chapters 2 and 20.

An overview of the handling of signalling within an SPC digital exchange is shown in Fig. 8.1. There is a range of subscriber and trunk analogue and digital lines terminating on the exchange, each of which will convey some form of signalling. The Figure illustrates this variety, although not all types of line and signalling terminations will be present on every exchange. For example, trunk exchanges do not have any subscriber-line terminations. The signal-handling role within an exchange involves the routeing of signals between line terminations and appropriate signalling receivers. This routeing facility may be provided by the line-termination units, the concentration or group switch blocks, or a combination of all three of these. Thus, to accommodate the wide range of possible physical realisations, Fig. 8.1 indicates the signal-routeing function by a cloud symbol, as used in CCITT documents.

The telephone-exchange signalling functions include not only inter-exchange and subscriber-to-exchange signalling, but also conveyance of call-status information to subscribers by means of tones and recorded announcements. As the Figure shows, the latter must also be routed to the various line terminations. This chapter first considers the general routeing and handling principles of subscriber and trunk signalling. It then describes the principles of the signalling sender/receivers in a digital exchange, covering tones and recorded announcements, voice-frequency and multifrequency tone signalling, PCM/channel-associated and common-channel signalling.

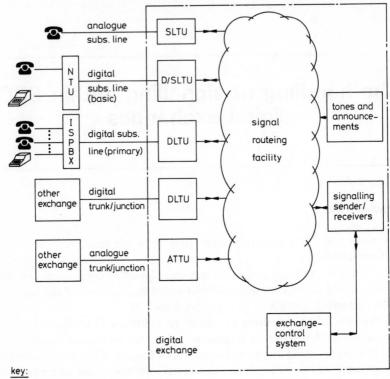

key:

DLTU digital line termination unit
SLTU subscriber line termination unit
NTU   network terminating unit
ISPBX integrated services PBX

Fig. 8.1 *An overview of the handling of signalling within an SPC digital exchange*

8.2 The routeing of signals in a digital exchange

8.2.1 Subscriber signalling

The routeing, within the exchange, of signalling from analogue subscriber lines to the appropriate signalling receivers is shown in Fig. 8.2. The routeing of the two signalling components (line and selection signals[1]) differs according to whether a dial or MF push-button telephone is used. Line (or 'supervisory') signals convey the status of the circuit. Such signals include seize and answer (when the called subscriber lifts the handset) and clear (when the subscriber replaces the handset). Selection (or 'address') signals indicate the called subscriber's number and other selection information generated by the dial or push buttons on the telephone instrument.

For a dial telephone, both the line and selection signals are conveyed using loop disconnections (LD; see Chapter 2). These signals are extracted from the

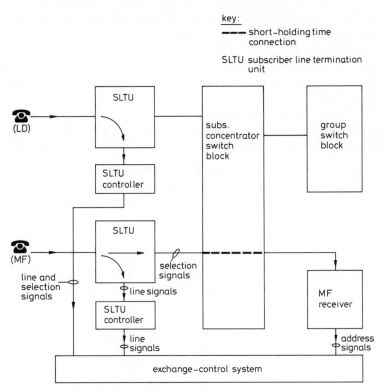

Fig. 8.2 *Routeing of analogue subscriber signalling in a digital SPC local exchange*

subscriber line by the SLTU, as described in Chapter 7. The line and selection signal outputs for a number of SLTUs are collected by an SLTU controller, where they are converted from simple loop conditions to signals indicating 'seize', digit '3', etc. These signals are sent by the SLTU controller to the exchange-control system for call processing (see Chapters 14 and 15).

With MF push-button telephones, however, the line signalling is conveyed by LD while the selection signals are carried by the MF tones. Thus, the line signals are extracted by the SLTU and passed to the exchange-control system, as for dial telephones. Access to the MF receivers is normally achieved via the subscriber-concentrator switch block (Fig. 8.2). This is because the MF receivers are required only for the set-up phase of a call. Thus, as with all short-holding-time equipment, economy is gained by sharing their use among a large number of subscriber lines. The MF receivers are, therefore, grouped in a pool of equipment which may be accessed by any of the MF subscriber lines at the start of a call. When the subscriber finishes keying the called number and the MF receiver has done its job, the MF receiver is released and made available for subsequent calls.

Therefore, two paths are established through the concentrator switch block for calls originating from an MF telephone:

(i)   The first is established from the SLTU to a free MF receiver following detection, by the SLTU, of the subscriber's 'off-hook' line signal. Fig. 8.2 indicates this path by a series of heavy dots. The path is cleared down once all selection information has been received (or if the calling subscriber should clear down prematurely) and the selection information is then passed to the exchange-control system.

(ii)   The second is established from the SLTU, through the subscriber concentrator, to the route switch in order for the speech-path connection to be made to the called subscriber after the receiver digits have been analysed.

The emergence of integrated services digital networks (ISDNs) will introduce digital subscriber lines wtih common-channel signalling. The concept of an ISDN is described in Chapter 22. Fig. 8.3 shows the routeing of the signalling

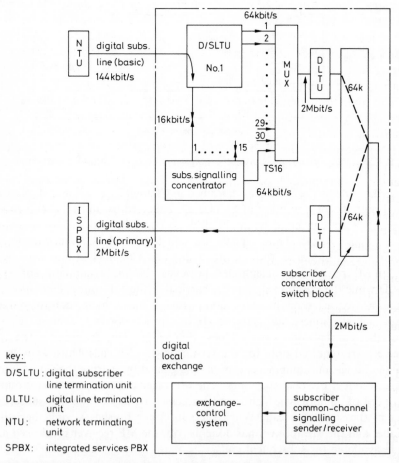

Fig. 8.3   Routeing of subscriber digital signalling in an SPC digital exchange

from two forms of ISDN exchange line terminations ('basic' and'primary') to the signalling-control system. With ISDN subscriber lines, the signalling is in both directions; however, for simplicity, only the subscriber-to-exchange direction is described below (the reverse direction follows a similar routeing).

Both the basic and primary forms of ISDN access shown in Fig. 8.3 use common-channel signalling (CSS), as described in Chapter 20. For the basic access, the CCITT standard is for a 16 kbit/s signalling channel associated with two 64 kbit/s traffic channels, making a total rate of 144 kbit/s in each direction. The 16 kbit/s channel carries the line and selection signals for both traffic channels as well as call-progress information and maintenance messages. With the primary access, which comprises a 2 Mbit/s path from an integrated services digital PBX (ISPBX), the common-channel signalling for 30 64 kbit/s traffic channels is carried in time slot 16 (TS16).

Consider first the routeing of the 16 kbit/s signalling channel in the basic access. Chapter 7 described how the 16 kbit/s signalling channel is extracted by the D/SLTU. The signals could then be passed directly over wires to the subscriber CCS sender/receivers. However, this would not represent an efficient use of the sender/receivers and would require excessive amounts of wiring in large exchanges. It is more efficient to multiplex the signalling outputs from a number of D/SLTUs.

This multiplexing may be achieved in two stages, as shown in Fig. 8.3. The first uses a signalling concentrator associated with a group of D/SLTUs. Statistical multiplexing is preferred to simple time-division mutliplexing (TDM) because of the higher multiplexing ratios obtainable.[2] Examples of such ratios are 15:1 rather than 4:1 (4×16 kbit/s) achieved with TDM. In statistical multiplexing, a number of tributaries are combined by the dynamically assigned interleaving of their contents. The system takes advantage of the 'bursty' nature of many forms of data traffic by interleaving the bursts on a common (multiplexed) bus. Unlike TDM, with its permanent assignment of regular time slots to each tributary channel irrespective of its contents, statistical multiplexing produces time slots which are irregular in duration and are assigned on a dynamic basis, according to the actual contents of the tributaries at any time. Clearly, if each tributary has a high occupancy (as with speech signals), statistical multiplexing provides no improvement in multiplexing ratio compared to TDM.

The concept of statistical multiplexing of the bursty common-channel ISDN signalling is shown in Fig. 8.4. This shows that only D/SLTUs 3, 4, 10 and 15 currently have active signalling channels; none of the other signalling channels have any messages. The messages from these four active channels are interleaved, 2 bits at a time, to form an 8-bit word. A typical signalling message of 368 bits requires 184 frames, i.e. 23 ms (184×125 μs) for its transmission. During this time, any signalling messages that arrive are either lost or held in a buffer within the signalling concentrator until a free slot became available. Thus, the higher multiplexing ratio of statistical

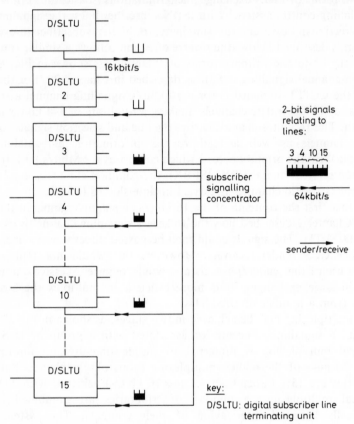

Fig. 8.4 *The concentration of subscriber line common-channel signalling*

multiplexing can only be exploited if the signalling-message traffic is low enough to avoid excessive contention for the signalling channel to the sender/receivers.

The 64 kbit/s output from the signalling concentrator is multiplexed together with the traffic-channel outputs from 15 DLTUs to form a 32-time-slot 2 Mbit/s frame, with the concentrated signalling carried in TS16, (as shown in Fig. 8.3). The subscriber-concentrator switch block terminates the 2 Mbit/s buses from many multiplexers, each serving 15 D/SLTUs or 30 analogue SLTUs. The switch block may then be used to perform a second stage of signal multiplexing by connecting the TS16s from up to 31 of the multiplexers serving D/SLTUs to a single 2 Mbit/s signalling bus.

This use of a digital switch block to perform the second stage of multiplexing requires 'semi-permanent' connections to be established between all the incoming 2 Mbit/s buses from the D/SLTU and the 2 Mbit/s bus to the signalling sender/receiver. Fig. 8.5 shows TS16s from buses A, B, C and

D are switched to time slots 1, 2, 3 and 4, respectively. These connections, although switched, are 'semi permanent' because they form a long-term link between the signalling concentrators and the sender/receiver, which is held indefinitely and not on the usual call-by-call basis. A semi-permanent connection (also known as a 'locked-up time-slot connection') may be maintained for several years. Following the failure of a switch block, the exchange-control system must re-establish the semi-permanent connections to the signalling sender/receivers. This must be the first priority, because normal call connections cannot be set up until the necessary signalling paths are provided. In the example of Fig. 8.5, all time slots other than TS16 are switched on a call-by-call basis to time slots on outgoing traffic buses in the normal way.

The 64 kbit/s CCS channel on the 2 Mbit/s primary access already contains the statistically multiplexed signalling for all of its 30 traffic channels. Fig. 8.3 shows the TS16s from all the primary accesses being switched using semi-permanent connections to the 2 Mbit/s bus to the sender/receiver. Therefore, the latter receives all the subscriber common-channel signalling, originating from both basic and primary accesses, over one or more 2 Mbit/s buses. All of the 256-bit frames on these buses, except the 8 bits of TS0, which is usually

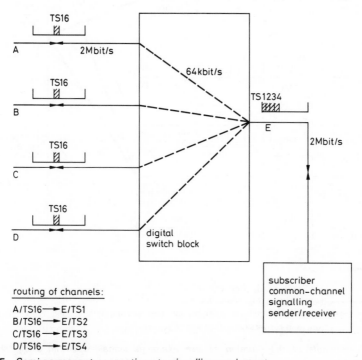

Fig. 8.5 *Semi-permanent connections to signalling equipment*

reserved for internal control messages, contain interleaved messages relating to a maximum of 30 traffic channels.

8.2.2 Inter-exchange signalling

The routeing within a local or trunk exchange of inter-exchange signalling is shown in Fig. 8.6. Although the full range of signalling systems for both analogue and digital trunks is shown, most exchanges will need to terminate only a smaller range. The number of signalling systems, and whether both

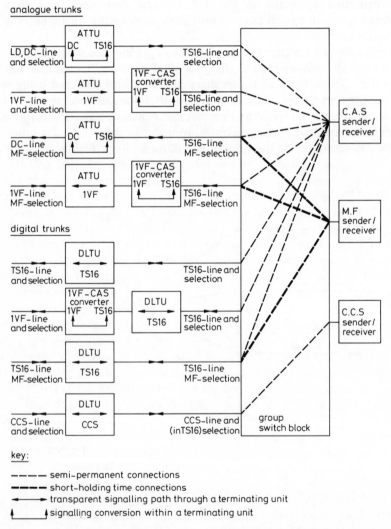

Fig. 8.6 *Example of the routeing of inter-exchange signalling within a digital trunk exchange*

analogue and digital trunks need to be terminated, depends on the network environment of the exchange. Suitable network planning can minimise the range of signalling systems that need to be terminated, as discussed in Chapter 21.

Given that a specified range of signalling systems must be terminated, the amount of signalling control equipment required within a digital exchange is minimised by first converting the variety of signalling on the lines to a basic subset. The concept is illustrated in Fig. 8.6. In that example, eight different signalling systems presented at the line termination on the left of the Figure are processed by just three types of sender/receivers: PCM/channel-associated, common-channel and MF. This is achieved by conversion within the analogue trunk terminating unit (ATTU) of all LD, DC and 1VF line and selection signals to PCM channel-associated signalling (PCM/CAS); this then enters the switch block in TS16. These TS16s may then be handled together with the TS16 channel-associated signalling directly carried on digital trunks. It is convenient to route all the individual TS16s via the digital route switch block to the CAS sender/receivers using semi-permanent connections, as described previously.

There are two types of in-band tone signalling systems: MF, which uses a burst of two simultaneous tones from a range of six tones to indicate each address digit, and '1VF', which uses bursts of a single frequency to indicate the break pulses of selection digits as well as line information (see Chapter 2). A 1VF system may be used to convey both line and selection signals for a circuit, or just the line signals of a circuit which is using MF signalling for the selection information. The routeings of 1VF and MF signalling differ; the former is handled by permanently assigned equipment, while the latter is handled by short-holding-time equipment on demand. Both methods are described below.

The conversion of 1VF in-band signalling to PCM/CAS (TS16) may be achieved using equipment associated with each analogue line input to an ATTU or by a single unit that extracts the tones from a 2 Mbit/s digital stream. The latter method is usually favoured for reasons of cost. Such a convertor deals with one 2 Mbit/s input containing up to 30 channels with 1VF tone signalling and a 2 Mbit/s output with the signalling carried in TS16. The equipment must therefore hunt for bursts of digitally encoded tones (e.g. 2280 Hz). This is achieved using digital filtering techniques.[3] In the reverse direction, the equipment converts PCM/CAS (TS16) into bursts of a tone which are injected into the appropriate channel within the 2 Mbit/s stream.

Fig. 8.6 shows the variety of ways that the ATTU and digital 1VF-to-PCM/CAS convertors are used. First, the ATTUs terminating analogue lines with LD or DC signalling perform the signalling conversion to PCM CAS (TS16). Where 1VF signalling is used, the ATTU is transparent to the signalling; the 1VF-to-PCM/CAS conversion is provided by equipment associated with the ATTU. Furthermore, 1VF may be used by channels carried over a digital trunk link. In this case, the 2 Mbit/s digital stream directly terminates on the 1VF-to-PCM/CAS conversion equipment.

MF signalling is routed on a call-by-call basis through the switch block from the calling line to the MF sender/receivers, as short-holding-time connections, (Fig. 8.6). These connections are cleared as soon as the selection signalling has ceased. The speech path is then established through the switch block between the calling line and the required outlet.

Common-channel signalling, normally carried in TS16 with 2 Mbit/s

Table 8.1 *Exchange units handling the signalling on trunk lines*

Circuit transmission	Type of signalling on trunk circuit	Exchange signalling units	
		Line signalling	Selection signalling
Analogue	LD (2 & 4 wire)	PCM/CAS S/R (conversion to PCM/CAS in ATTU)	PCM/CAS S/R (conversion to PCM/CAS in ATTU)
Analogue	DC (2 & 4 wire)	PCM/CAS S/R (conversation to PCM/CAS in ATTU)	PCM/CAS S/R (conversion to PCM/CAS in ATTU)
Analogue	1VF (4 wire)	PCM/CAS S/R (conversion to PCM/CAS in 1VF unit)	PCM/CAS S/R (conversion to PCM/CAS in 1VF unit)
Analogue	MF with DC line signalling	PCM/CAS S/R (conversion to PCM/CAS in ATTU)	MF S/R
Analogue	MF with 1VF line signalling	PCM/CAS S/R (conversion to PCM/CAS in 1VF unit)	MF S/R
Digital	1VF	PCM/CAS S/R (conversion to PCM/CAS in 1VF unit)	PCM/CAS S/R (conversion to PCM/CAS in 1VF unit)
Digital	MF	PCM/CAS S/R	MF S/R
Digital	CAS (TS16)	PCM/CAS S/R	PCM/CAS S/R
Digital	CCS (TS16)	PCM/CCS S/R	PCM/CCS S/R

Key: S/R = sender/receiver

transmission is routed to the CCS sender/receivers via semipermanent connections across the route switch block. These connections enable the CCS time slots from up to 31 input 2 Mbit/s line systems to access a single sender/receiver via a 2 Mbit/s port, using the principle illustrated in Fig. 8.5.

Table 8.1 lists the exchange units that handle the various types of signalling on trunk circuits, as described in this Section.

8.3 Tones and recorded announcements

Telephone exchanges need to advise the subscribers of the status of a call as it progresses from initiation to completion. For normal telephone instruments, the status information is passed audibly in the form of easily recognisable tones or recorded spoken announcements. Traditionally, and as a matter of convenience and economy, the conditions generally encountered during a call set-up are indicated by tones (e.g. dial tone), while the unusual and specific conditions (e.g. congestion at a particular trunk exchange) are indicated by recorded announcements. However, the CCS capability of ISDN access enables both the general and specific call-status conditions to be displayed as written messages on appropriate subscriber terminals.

8.3.1 Call-progress tones

In a digital exchange there are basically two ways of producing call-progress tones for injection into the digital speech path, namely:

(i)   The continuous generation of tones in an analogue form, which are subsequently passed through analogue-to-digital convertors.

(ii)   The continuous generation of the digital representation of tones.

The first method was used in early digital exchange systems because it exploited the existing tone-generating equipment of analogue exchanges, for which there were no practical digital substitutes. This hybrid arrangement of electromechanical analogue tone generators within a digital electronic exchange was cumbersome. However, the rapid fall in digital storage cost during the early 1980s has made the second method a practical alternative and it is now the cheaper and universally adopted solution for new exchange systems. This method is described later in this chapter.

The tone-generation equipment in a local exchange may be grouped into two sets of units: one serving the subscriber-concentrator switch block and the other serving the group switch block. The deployment of the tone generators and their position in the trunking of the exchange depends on the particular architecture of the exchange. A common method is for tones that are required during the early stage of call set-up to be provided by generators in the subscriber-concentrator unit, while the tones that cover the response of the called subscriber during call set-up are provided by generators in the group switch-block unit (Fig. 8.7).

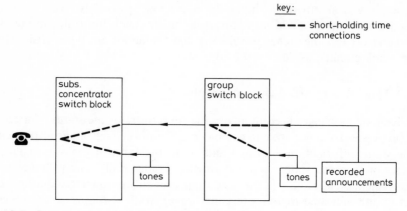

Fig. 8.7   *Routeing of tones and recorded announcements*

8.3.2 Generation of call-progress tones

The repertoire of easily distinguishable call-progress signals for public telephone exchanges is derived from a set of tones which are continuous or with regular interruptions. The format of the call-progress tones differs between administrations around the world, and these are well documented,[4] while the CCITT recommends a standard range.

The elements of a modern digital tone subsystem for use within a digital local, trunk or international exchange are shown in Fig. 8.8a. The subsystem consists of a bank of tone generators (one for each tone frequency), a selector and a control unit, which is usually microprocessor-based.

8.3.2.1 Digital Tone generators. These devices generate a digital version of a particular tone by producing the series of PCM samples that would originate from the PCM encoding of an analogue version of that tone. For a tone of 1 kHz, for example, the 8 kHz PCM encoding process produces eight different 8-bit samples representing one complete cycle of that tone. Thus, a 1 kHz digital tone generator consists of a read-only memory (ROM) containing the eight samples, which are read on to an output bus, one at a time, every 125 μs. The ROM read-address is produced by a simple cyclic counter; for example, to generate 1 kHz, the counter steps once every 125 μs from 0 to 7, repeating the sequence 1000 times per second.

In general, the number of locations in the ROM required to store the PCM samples for particular tone frequency depends on the ratio of that frequency to the 8 kHz sampling rate. For example, the ratio of the standard 1VF signalling tone of 2280 Hz to the sampling rate is 8000/2280 = 200/57. Thus, a 1VF digital tone generator requires a ROM containing 200 samples which, when read at the rate of one sample per 125 μs, generates 57 cycles of the 2280

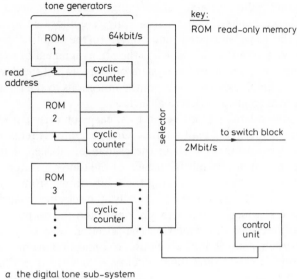

a the digital tone sub-system

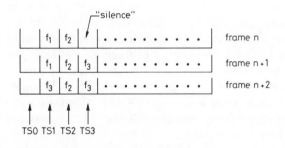

b multiplexed output from the tones subsystem

Fig. 8.8 *The digital generation of tones*

Hz tone. The same 200 samples can be read continuously for as long as the tone is required. There are techniques which reduce the ROM capacity needed for tone generation, e.g. by generating just quarter-cycles and multiplying appropriately.[5-7] In general, these methods produce tones of high accuracy because the driving frequency used to control the reading from ROM is the exchange clock.

8.3.2.2 Selection and control. As Fig. 8.7 shows, the tones subsystem is a central resource which is accessed via the subscriber-concentrator or the group switch block. The subsystem is connected to the switch blocks by a 2 Mbit/s bus, with 31 time slots available for carrying tones. There are a number of ways in which the range of call-progress tones are allocated to the 2Mbit/s output from the tones subsystem and routed to the required subscriber or trunk circuits. A common arrangement is the pre-assignment of each the tones to an individual time slot on the subsystem output bus and the distribution of these via the switch block using multipoint connections to the required traffic circuits. This arrangement is described below.

Fig. 8.8a shows a bank of tone generators, each comprising a ROM with a cyclic counter providing the read address. One generator is required for each of the call-progressing tones used in the exchange, labelled f_1 to f_3 in this example. The continuous output from each of the tone generators is switched to the output bus by a selector placing the 8-bit sample from f_1 in TS1, the sample from f_2 in TS2, and so one. Typically, the selector comprises digital-logic multiplexers, which switch between inputs and the output according to an address, as described in Chapter 6. The selection address is generated by the tones-subsystem control unit, which uses a connection memory in a similar way to the exchange space and time switches, as described in Chapter 6.

Fig. 8.8b illustrates the frame format of this example of the use of the exchange tones subsystem. Tones f_1 and f_2 are required to be continuous, and are therefore routed by the selector to time slots 1 and 2, respectively, during every frame indefinitely. Tone f_3, however, is to be applied as an intermittent signal. Thus, the selector routes to the output of the subsystem requences of f_3 and silence, according to the required cadence.

When the exchange-control system determines that a subscriber line or trunk circuit requires a call-progress tone, it instructs the switch-block control to set up a path to the time slot carrying the required tone (Chapter 6). Since any tone may be required by a number of traffic channels simultaneously, the switch block must perform a fan-out function. This form of switching from one input channel to many output channels simultaneously is known as a 'point-to-multipoint connection'.

Fig. 8.9 illustrates a point-to-multipoint connection through a T-S-T switch block between the time slots on the output bus of the tone generator subsystem and the traffic time slots. It is assumed that the input time switch A1 terminates 16 2-Mbit/s inputs, of which the output from the tone generator subsystem is one. Samples of tone f_1, f_2 and f_3 are stored in consecutive cells in the speech memory of the A1 time switch, based on a cyclic-write process (see Chapter 6). In this example, at the instant under consideration, two time slots in output time switch C1 (TS2 and TS103) require the call-progress tone f_1. This tone is also required by TS2 of time switch C2 and TS59 of time switch C3. Time slot 510 of C2 requires tone f_2, while tone f_3 is not currently required by any traffic channels. Fig. 8.9b tabulates the required connections between the group of C time switches and the tones generator, and the internal time slots through the space switch chosen by the switch-block control.

The process of point-to-multipoint connection is performed within the input

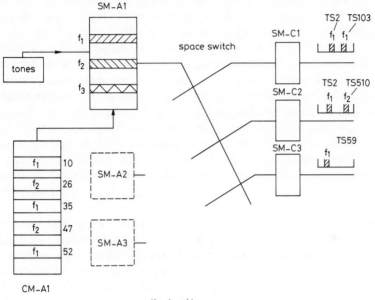

a the trunking

tone	internal time slot	outgoing bus/TS
f_1	10	C1/TS2
f_1	26	C1/TS103
f_1	35	C2/TS2
f_2	47	C2/TS510
f_1	52	C3/TS59

key:
CM connection memory
SM speech memory

note: f_3 currently not required

b required connections: example

Fig. 8.9   *Multipoint connection of tones via T-S-T switch block*

timeswitch A1. Here the same cell in A1 (containing the f_1 samples) is read during internal time slots 10, 26, 35 and 52 and identical samples are routed via the space switch to output time switches C1, C1, C2 and C3, respectively (Fig. 8.9*a* and *b*). Since f_2 is currently required only by C2/TS510, its samples are read out of A1 only once per frame (during internal TS47). The samples for f_3 are not read at all, since no traffice channels currently require that call-progress tone.

8.3.3 Recorded announcements
Recorded announcements give to the callers a verbal response indicating some particular information which was either too detailed or otherwise unsuitable

for indication by tones. Traditionally, recorded announcements have been used for messages such as advising of congestion of specific routes (e.g. 'all lines to Bristol are engaged'). However, modern SPC digital exchanges can use recorded announcements to assist in all areas of subscriber-to-exchange interaction. For example, the automatic announcement system of the System X exchanges can interact with subscribers in a conversational manner by responding to MF push-button signals.[8] This facility gives a spoken response to customers enquiring, via MF keying, about their use of the special services on the exchange, e.g. advising of the number to which calls are currently being diverted. There is currently extensive development work on speech-synthesis and speech-activated systems for telecommunications and other applications. Progress in the field is rapid, although it is likely that subscribers will still need to signal into the exchange by non-verbal keying or dialling for many years yet.[9]

There are two types of recorded-announcement subsystem which may exist within a digital exchange. The first involves the simple storage, in digitally encoded form, of the required messages (Fig. 8.10a). This requires the use of mass digital-storage devices, such as Winchester discs, because of the large volume of PCM samples involved. For example, with 64 kbit/s PCM encoding, ten messages of ten seconds each requires a total of 64 Mbit/s storage capacity. Most systems minimise the storage requirements by using one or both of the following techniques:

(i) The extraction of silent periods before storage, with automatic reinsertion during playback.
(ii) Low-bit-rate encoding, e.g. using CVSD at about 20 kbit/s or ADPCM at 32 kbit/s (see Chapter 5).

Each of the recorded announcements is allocated space on the mass-storage device and is accessed by reference to an address. Any announcement may then be read from store and be fed to any time slot on the outgoing bus to the switch block. New messages are recorded on to the system either from a local terminal or remotely from a management centre, using a dial-up connection or leased lines, as appropriate.

The second type of announcement subsystem involves the generation of sentences or phrases by reading from electrically programmable-ROM (EPROM) stores which hold a vocabulary of syllables or complete words (Fig. 8.10b). A wide range of messages can be assembled by appropriately concatenating the contents of the EPROM stores. The assembly of a particular message is created by a simple control unit which directs the reading of the EPROMs in the required sequence into a time slot on the 2 Mbit/s output bus connecting the announcement subsystem to the switch block. The great advantage of this type of system is that externally generated variables like 'time of day', 'number(s) dialled', 'names', etc. can be inserted within the sentence to make the message specific to that call. The disadvantage is that

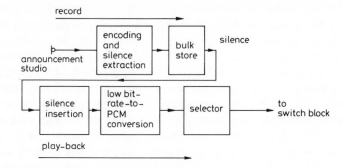

a simple storage of recorded announcements

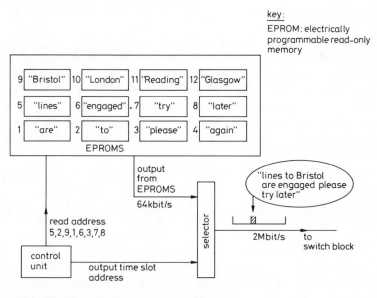

b use of EPROM stores to generate announcements

Fig. 8.10 *Recorded announcement subsystem*

new words cannot be easily added to the vocabulary because the storage of each syllable or word is the result of complex digital processing and the subsequent writing into EPROM. Further information on this fascinating subject can be found in the current literature.[10]

8.4 Multifrequency signalling sender/receivers

Despite the emergence of common-channel signalling, there is currently widespread use of multifrequency signalling systems throughout the world.

The CCITT has recommended standard systems for international and trunk route application (e.g. R2) and for subscriber signalling, (e.g. DTMF), although there are many non-standard systems in use (Chapter 2). It is therefore usually necessary for digital SPC exchanges to be able to handle an appropriate range of MF signalling systems. Although a wide range of MF signalling codes are used, there are many common features (e.g. the signal frequencies) which enable multi-standard exchange equipment to be designed.

As described in Sections 8.2.1 and 8.2.2, the MF signals from both subscribers and exchanges are routed to the sender/receivers in a 30-channel multiplexed format. Thus, each MF sender/receiver needs to serve up to 30 channels simultaneously. The number of MF signalling units required to serve a given number of subscribers or trunk circuits depends on the average signalling duration and the maximum rate of call attempts expected.

For subscriber signalling, a uni-directional path is established through the subscriber-concentrator between the calling SLTU and a free time slot on the highway to the MF receiver, while dial tone is returned to the subscriber via another uni-directional path through the concentrator. The MF unit must therefore be able to detect the first signal in the presence of the dial tone. Once the dialling (keying) is complete, the exchange-control system initiates the clear of the path through the subscriber-concentrator switch block. The time slot on the highway to the MF signalling unit is then free for another call. The procedure for inter-exchange signalling is similar, with the access being provided by the group switch block, except that signals are sent in both directions and there is no dial tone to obscure the first signal.

The conceptual arrangement of an MF signalling unit for a digital SPC exchange is shown in Fig. 8.11*a*. It consists of 30 conventional analogue-type single-channel MF sender/receivers attached to a 2048 kbit/s primary multiplexer. A standard 30-channel-frame format is used with frame alignment in TS0, and TS16 is either vacant or used as a channel to test the MF equipment (and hence not avilable for vacant subscriber use). This arrangement was used by the early versions of digital exchange systems. It was the cheapest method due to its exploitation of the well proven analogue MF sender/receivers.

More recently, developments in digital signal processing have enabled the entire MF sender/receiver function to be implemented in digital technology. A digital MF signalling unit is shown conceptually in Fig. 8.11*b*. The complex receiver circuitry is time-shared by a number of channels; four are shown in the diagram, but a typical value is eight. The received digits for each of the channels are passed by the receivers to the control unit, where they are formatted into a message and passed to the exchange-control system. The MF sender is less complex than the receivers, so a single unit is able to serve all 31 channels (30 for traffic, and one test unit carried in TS16).

A variety of systems has been developed for digital MF receivers, most based on digital filtering.[4,7,11] The requirement is for the presence of two permitted tones to be detected from a degraded received signal. This involves

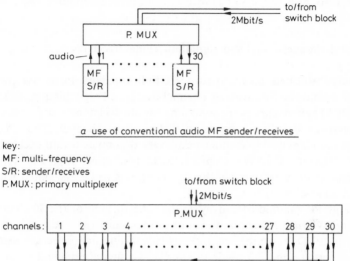

a use of conventional audio MF sender/receives

key:
MF: multi-frequency
S/R: sender/receives
P.MUX: primary multiplexer

b digital MF unit

Fig. 8.11 *MF signalling unit in a digital exchange*

several stages of filtering of the received signal, testing its level against a threshold and comparing it with the received energy in the remainder of the 4 kHz bandwidth. The relevant frequencies are detected by searching for sequence of known digital samples in the filtered input signals (i.e. the opposite process to that of digital-tone generation).

The digital MF sender has to assemble the required pair of frequencies into an MF burst which is then fed to the required output channel in the 2048 kbit/s multiplexer. This may be achieved by the digital combination of two tones, each tone being generated from ROM as described for the exchange tones unit earlier in this chapter. Alternatively, the pair of tones can be generated from ROM directly as a single signal with two frequency components.

8.5 Digital channel-associated signalling sender/receivers

Chapter 5 described how the two CCITT standard PCM systems have different methods of conveying channel-associated signalling (CAS). The 2 Mbit/s PCM system carries the signalling for all 30 channels in a separate slot, TS16, giving a signalling capacity of 2 kbit/s per channel. The 1.5 Mbit/s PCM system employs bit-stealing once every 6 samples within each of the 24 channels, giving a capacity of 1.33 kbit/s.[1] Both systems require a multiframe structure to enable the signalling bits to be associated with their traffic channels.

Although the method of conveying the CAS differs for the 30-channel and 24-channel systems, the actual signal content is similar, i.e. a simple imitation of 10 p.p.s. LD subscriber or inter-exchange line and selection signalling. Thus, the actual signal processing required by a digital exchange is similar for both systems. However, the positioning of the sender/receivers within the exchange trunking differs. The 30-channel system enables a single multiplexed 64 kbit/s signalling channel to be routed to the PCM/CAS sender/receiver as a standard channel via the switch block, whereas the T1 system needs the signals from each channel to be stripped off the 1.5 Mbit/s multiplexed bearer before it enters the switch block. The handling of CAS for the two systems is therefore described separately.

8.5.1 Channel-associated signalling carried in TS16

Earlier in this chapter the routeing of CAS in TS16 from a number of 2 Mbit/s input systems was described. The switch block routes up to 31 TS16s via semi-permanent connections on to a 2 Mbit/s output bus connected to a PCM/CAS sender/receiver, as shown in Fig. 8.6. It was also shown that the CAS may have originated from the TS16s on external digital line-transmission systems or by conversion from 1VF, or DC signalling systems terminating on the exchange. The output 2 Mbit/s highways to the sender/receivers can handle a maximum of 31 TS16s, the total signalling for 930 traffic channels (31×30). The PCM/CAS sender/receivers therefore have 2 Mbit/s ports carrying a high concentration of signalling, which can fully exploit their high-speed logic.

The PCM/CAS sender/receivers perform three main functions, namely:

(i)   Demultiplexing the input into 31 groups of TS16 contents.
(ii)   Demultiplexing the multiframe structure of each TS16 into 15 groups of signal-channel pairs.
(iii) Signal recognition and interpretation, and communication with the exchange-control system.

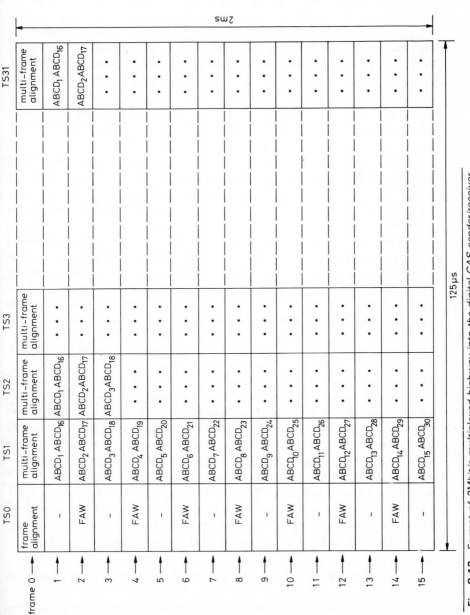

Fig. 8.12 *Format of 2Mbit/s multiplexed highway into the digital CAS sender/receiver*

These processes (i) to (iii) occur in that order in the receive direction of signalling, and in the order (iii) to (i) for the transmit direction of signalling.

The multiplexed structure of the 2 Mbit/s highway between the group switch block and the PCM/CAS sender/receivers is shown in Fig. 8.12, based on a standard 16-frame multiframe arrangement. During frame 0, the TS0 on the highway contains the normal frame-alignment word for the 2 Mbit/s system of the highway itself, while the remaining time slots carry the multiframe alignment words of the 31 TS16s semi-permanently connected via the switch block. During odd frames 1 to 15, TS0 on the highway carries the frame-alignment word (see Chapter 5). The highway time slots 1 to 31 during frames 1 to 15 carry the 4-bit signals, labelled 'ABCD', for pairs of channels from the respective TS16s.

The capacity of the 2 Mbit/s highway to the PCM/CAS sender/receiver may be reduced to 30 TS16s (i.e. signalling for 900 channels) if one of the time slots is dedicated to test equipment, so that regular routine maintenance can be carried out.[3] This will be assumed for the remainder of this Section.

8.5.1.1 Signalling in the receive direction. In the receive direction, each of the 30 2-Mbit/s streams has different origins (other exchanges, or co-sited or remote subscriber-concentrators), so each will have a different multiframe start instant in its TS16. Thus, before the format of Fig. 8.12 can be established, each received TS16 on the CAS control highway must have the multiframe start aligned to a common exchange-multiframe start. This function is achieved within the PCM/CAS sender/receiver by demultiplexing the received 2 Mbit/s stream from the highway into the consituent 64 kbit/s time slots and feeding these into multiframe re-alignment buffers (Fig. 8.13). The 64 kbit/s streams (each carrying TS16 information) are stored within these buffers, beginning the cyclic writing at the start instant of the individual TS16s frame-alignment pattern. The contents of each of these buffers are read out following the exchange multiframe start. Thus, each of the incident TS16s is delayed within the buffer by an amount equal to the mismatch between its individual multiframe start and the exchange. A maximum delay of 16 frames needs to be accommodated by the buffer, equal to 128 bits (16×8 bits).

Once aligned, the contents of the individual 31 TS16s are written once every 16 frames (16×125 μs) into each of the CAS stores, under the control of a cyclic counter, as shown in Fig. 8.13. There are 16 locations (or cells) in each CAS store. So the total contents of the 31 CAS stores will look similar to Fig. 8.12. Thus, by examining the appropriate ABCD-bit contents of the CAS stores, the CAS receivers can determine the signalling state of all the 900 traffic channels.

Before describing how the PCM/CAS receiver identifies these signalling states, it is useful to consider the example of a typical PCM/CAS (TS16) sequence shown in Fig. 8.14. The Figure shows the standard 4-bit codes that appear once every multiframe (16×125 μs = 2 ms), conveying the signalling

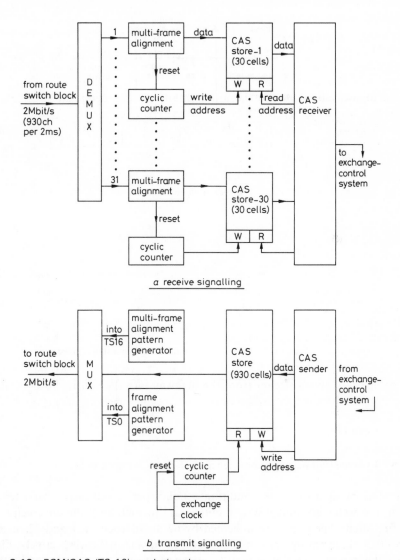

a receive signalling

b transmit signalling

Fig. 8.13 *PCM/CAS (TS 16) sender/receiver*

for one traffic channel. When the channel's line condition is in the idle state, code 1111 occurs in TS16 during the appropriate part of the multiframe. This code continues to reappear in TS16 for as long as the circuit is idle. The exchange is able to recognise that the circuit has been seized by detecting the presence of new code 0011 which persists for 10 ms, i.e. at least five repetitions of the code (one appearance every 2 ms). This checking for persistence minimises the possibility of the exchange control acting on codes erroneously

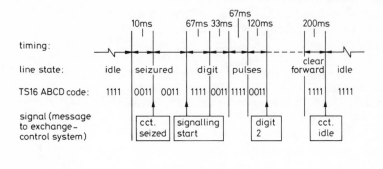

a receive signals

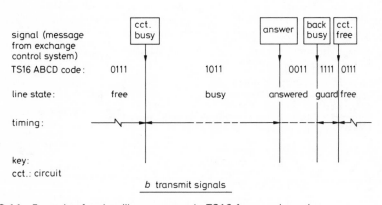

b transmit signals

Fig. 8.14 *Example of a signalling sequence in TS16 for one channel*

generated by spurious line conditions. Following a successful persistence check, the exchange-control system initiates a time-out, awaiting the receipt of the first digit. The digits are indicated by 67 ms of code 1111 and 33 ms of code 0011, representing the loop disconnect break and make, respectively. Fig. 8.14*a* illustrates the receipt of digit '2', the end-of-digit being indicated by the presence of code 0011 for 120 ms. The remainder of the example shows the 'clear forward' and 'idle' conditions appearing at the end of a successful call connection.

The required signal processing for each traffic channel therefore comprises the recognition of the received 4-bit code and the monitoring of it for the required duration of the code. The function of the CAS-store cells is to act as a form of 'call record' which holds the transient data relating to the signalling of a channel. Unlike the call record in the exchange-control system (see

Chapter 15), the CAS-store cells are permanently associated with a particular channel. Fig. 8.14 indicates the messages (shown in boxes) that are sent to the exchange-control system when each change of line state, or each digit, is recognised.

8.5.1.2 Signalling in the transmit direction. The block-schematic diagram for the transmission of TS16 signal processing is shown in Fig. 8.13*b* Unlike the receive side, all the multiframes of the transmitted 31 2-Mbit/s streams are aligned with the exchange-multiframe start, so no buffering is required. Appropriate alignment patterns are inserted into the time slots of frame 0. Fig. 8.4 shows the 4-bit codes that need to be generated for the transmit direction of signalling as a result of the output messages from the exchange-control system.

8.5.2 Channel-associated signalling with the 24-channel PCM system
With the 24-channel system, the channel-associated signalling is carried by 'bit-stealing' the least-significant bit of traffic channels every sixth frame (Chapter 5). Thus, the signalling bits must be extracted from the 1.5 Mbit/s stream before it enters the group switch block. This function is best performed by the DLTU, as shown in Fig. 8.15. The format of the 2-bit signals 'A' and

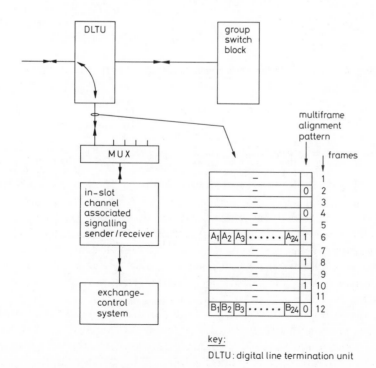

key:
DLTU : digital line termination unit

Fig. 8.15 *In-slot digital channel-associated signalling*

'B' for the 24-channel group, together with the multiframe alignment pattern during 12 successive frames, is also shown in the Figure.

The signal-processing functions are similar to the TS16 CAS system described earlier. If the CAS sender/receiver is provided on the basis of one per 1.5 Mbit/s system, the processing logic would be poorly utilised by just the 24 channels. A more efficient arrangement is for a number of DLTU-signalling outputs to be multiplexed on to a high-capacity input to a single PCM/CAS sender/receiver. The format of the multiframe, as shown in Fig. 8.12, indicates that a multiplexing ratio of $6N$ is convenient; the value of N depends on the speed of logic and on exchange-reliability aspects.

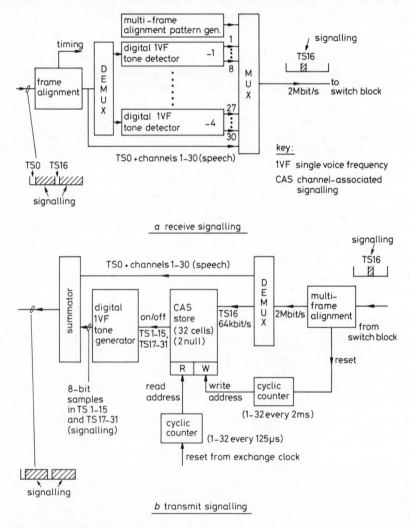

a receive signalling

b transmit signalling

Fig. 8.16 *Digital 1VF-CAS converter*

8.6 Digital 1VF-to-CAS convertor

Fig. 8.6 shows the two applications of the digital 1VF-to-PCM/CAS convertor associated with analogue and digital lines terminating on the exchange. In the case of analogue lines, the ALTU acts as a primary multiplexer forming a standard 30-channel PCM frame, with the 1VF signalling encoded in the speech channels. For digital lines, the format of the 2 Mbit/s stream at the exchange boundary is the same, the primary multiplexing being performed externally to the exchange, elsewhere in the network. In both cases, the TS16 of the 2 Mbit/s stream is devoid of signalling content.

The function of the digital 1VF-to-PCM/CAS convertor is to extract the digitally encoded single signal tone (usually 2280Hz) from each channel and to inject the corresponding 4-bit codes into TS16 for the receive direction. The reverse applies in the transmit direction. Figs. 8.16*a* and *b* illustrate the functional components required for the receive and transmit signalling directions, respectively, and these are briefly described below.

8.6.1 Signalling in the receive direction

The convertor must first be frame-aligned to the received 2 Mbit/s stream to enable each channel to be identified. A digital tone detector, similar to that described for the digital MF receivers and tuned to 2280 Hz, then examines each channel. This process requires examination of successive samples in the received channel. Tone detectors are complex devices, but system economy can be gained by time-sharing them among a number of channels, typically eight; thus, four are required for each convertor. The output from the tone detector is a simple on-off signal, following the presence or absence of tone. This is directly converted to the 0011-1111 codes for digital channel-associated signalling. These codes are then inserted into the appropriate 4 bits of TS16 corresponding to each channel. The TS16 contents relating to the 30 speech channels are inserted into the outgoing 2 Mbit/s stream by a multiplexer. As Fig. 8.16*a* shows, the result of the conversion is the transfer of all in-band signalling within each channel of the input 2 Mbit/s to TS16 on the output 2 Mbit/s stream.

8.6.2 Signalling in the transmit direction

Fig. 8.16*b* shows the block-schematic diagram of the 1VF convertor for the transmit direction. The 4-bit code relating to each channel within TS16 from the transmit 2 Mbit/s stream is written sequentially into a signal-channel store once each 16 frames (every 2 ms). Each signal-channel store is read during the appropriate time slots (i.e. TS1 for Channel 1, etc.) during each frame (once every 125 μs by the tone generator. The digital-tone generator is similar to that described in Section 8.3.2.1. The tone is set on or off directly by the 0011 or 1111 4-bit pattern read from each signal-channel cell. The digital tone is then inserted into the appropriate time slot to convey the 1VF signal for the channel. This is carried out for all time slots within a frame in turn, to provide signalling for all 30 channels.

8.7 References

1 WELCH, S. (1981): *Signalling in Telecommunications Networks* (Peter Peregrinus Ltd)
2 SMITH, D.R. (1985): *Digital Transmission Systems* (Van Nostrand Reinhold Co.), Chap. 4
3 SMITH, G.R. (1981): *System X Subsystems: The Signalling Interworking and Analogue Line Terminating Subsystem*, Post Office Electrical Engineering Journal, **74**, Pt. II
4 GRINSEC (1983): *Electronic Switching*, Vol 2 (Elsevier Science Publishers, BV Netherlands)
5 PITRODE, S.G. and LINDSAY, R.L. (1973): *Progress Tones in a PCM Switching Environment*, IEEE Trans, **COM-21**, p.143
6 TULLINS, N. (1975): *Selection of Call Progress Tones for Digial Systems*, Ibid., **COM-23**, p.301
7 MESSERSCHMITT, D.G. (1983): *'Digital terminations and digital signal processing'*, in *Fundamentals of Digital Switching*, ed. McDonald, J.C., (Plenum Press), Chap. 8
8 OLIVER, G.P. (1980): *Architecture of System X. Part 3 — Local Exchanges* Post Office Electrical Engineers' Journal, **73**, p.27
9 THANAWALA, R. *et al.* (1987): *Automatic Speech Recognition in the Public Telephone Network*, Fifth World Telecommunications Forum, Technical Symposium: Telecommunications Services For a World of Nations, Geneva, ITU, **I**, pp.235-238
10 WHEDDON, C.: *Interactive Speech Systems — Man-Machine Communications by Speech*, Ibid, **II**, pp.249-253
11 DETTMER, R. (1986): *Digital Signal Processors*, Electronics & Power, pp.124-167

Chapter 9

The design and architecture of SPC digital exchanges

9.1 Exchange-system requirements

The design of any new exchange system needs to meet a number of requirements set by the public telephone network operators (PTO). This chapter considers the range of architectures employed in SPC digital exchange designs to meet these requirements. Exchange-system requirements set by PTOs fall broadly into the categories listed below.

(i) Costs: This category covers both the capital costs and annual charges associated with an exchange system within a network.[1] The capital costs comprise the equipment costs and the cost of labour involved in installing the exchange. Annual charges cover the operational costs involved in running the exchange, including accommodation, ventilation and power, as well as the operations and maintenance activities. An important cost factor for exchange systems is the economic life of the equipment, since the capital cost must be amortised over this period.

(ii) Size range and extendability: At any time, a telephone network will comprise exchanges of various capacities according to the geographical distribution of subscribers, the dispersion of their traffic and how it is routed. Normally, the required exchange capacities increase progressively with time in response to the demand for telephone lines and the use made of them. An exchange will, therefore, either need sufficient capacity to meet the forecast demand for its life or be capable of periodic extension. In the former case, the idle capacity represents a cost burden ('burden of spare plant') because it is not earning revenue. In the latter case, the exchange system must be capable of being extended without causing disruption to service.

(iii) Facilities: An exchange system will need to provide a range of subscriber facilities, as described in Chapter 3. There may also be a requirement for an integrated services digital network (ISDN) capability within the exchange (Chapter 22). In addition, the PTO will define a range of facilities for the operations and maintenance of the system (Chapter 19).

(iv)   Performance: A public telephone exchange needs to have a high degree of reliability in order to provide a secure service. PTOs, therefore, specify strict reliability requirements (in terms of mean time between failures) for the total exchange and for conditions that would affect individual subscriber and trunk lines.

(v)   Maintainability: An exchange needs to be maintainable in order for a prescribed level of service to be offered to the subscribers. This means that an exchange system needs to incorporate effective fault-locating facilities which do not interfere with other working lines or calls. For SPC exchanges, the required level of serviceability is achieved through the use of automatic-recovery processes for software faults and the use of redundancy to reduce susceptibility to hardware faults, as described in Chapter 19.

(vi)   Environmental requirements: These requirements cover the wide range of accommodation factors, e.g. floor loading, heat dissipation, range of ambient temperatures and humidity in which the equipment must operate, etc.

9.2 Architectures of SPC digital exchanges

9.2.1 Overview

The architectural characteristics of SPC digital exchanges and the way that the above requirements are met result from both the consequences of SPC and the use of digital-switching technology. The use of SPC accounts for an exchange

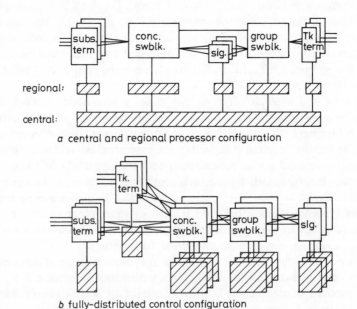

a central and regional processor configuration

b fully-distributed control configuration

Fig. 9.1 a,b *Exchange architecture overview*

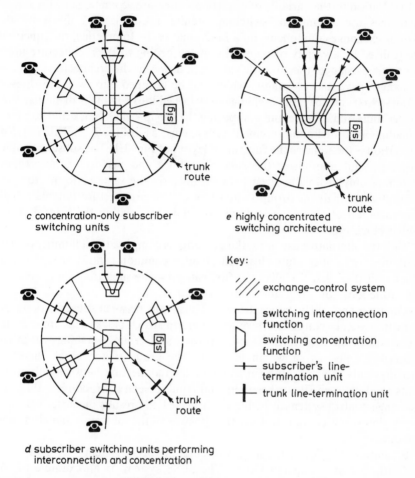

c concentration-only subscriber switching units

e highly concentrated switching architecture

d subscriber switching units performing interconnection and concentration

Key:

////, exchange-control system

☐ switching interconnection function

⬭ switching concentration function

—+— subscriber's line-termination unit

—+— trunk line-termination unit

Fig. 9.1 c,d,e *Exchange architecture overview*

structure centred around a common-control system (see Fig. 3.1 of Chapter 3). This system may be provided by a central cluster of processors or dispersed around the exchange, as described in Chapter 17. The control for most types of commercial exchange systems is provided by a set of central processors with some degree of call processing provided by 'regional processors' associated with various exchange subsystems, as in Fig. 9.1*a* (see also Fig. 11.2). An alternative form of SPC common-control architecture comprises separate processors associated with self-contained units of the exchange, as shown in Fig. 9.1*b*. The exchange-control system, whether centralised or dispersed, will employ a configuration of processors, memory and input-output systems, as described in Chapter 17.

In addition to the variants of control architecture, commercial SPC digital exchanges use a range of switching architectures. Fig. 3.1 illustrates the various exchange components on a functional basis. In practice, the functions are realised by a variety of physical units according to the architecture used. For trunk exchanges there is little variation; however, the architectures of local exchanges have many variants, which can be grouped into two categories.

In the first category, the exchange is sectioned into a number of similar subscriber units and a single unit composing the group switch block and common signalling/recorded announcement subsystems, as in Fig. 9.1c. This physical realisation is the same as the functional layout of Fig. 3.1. The subscriber units terminate and administer the subscriber lines and perform concentration switching only. All calls, even between lines on the same concentrator, are connected through the group switch block. This architecture obtains a high concentration of signalling and control equipment within the central group switch-block unit.

With the alternative category, the exchange is composed of a number of self-contained subscriber units clustered about a common central group switch-block unit (Fig. 9.1d). Unlike the first category, the subscriber units perform the functions of own-unit interconnection as well as subscriber-line termination and traffic concentration. Thus, calls between subscribers on the same unit are not connected through the group switch block. This architecture incorporates the functions of concentration and group switching within the subscriber units, as shown in Fig. 3.1. The group switch block therefore provides only the switching functions of interconnecting between subscriber units, and between subscriber units and trunks. Such a structure enables the exchange-control system to be highly dispersed and results in subscriber units being almost self-contained, since they possess all the subscriber signalling and announcement equipment.

Examples of the first category include E10B,[2] System X,[3] DMS100,[4] AXE10,[5] EWSD,[6] and NEAX61.[7] The 5ESS system[8] is an example of the second category. An interesting architectural design is used by System 12,[9] which incorporates a highly dispersed control structure while using a highly centralised switching architecture, as shown in Fig. 9.1e. This corresponds to an extreme version of category one switching architecture.

The type of technology used in a digital SPC exchange has an important influence on the system architecture. Chapter 7 describes how the subscriber line requirements (summarised by the acronym 'BORSCHT') are met using a mixture of digital semiconductor technology and electromechanical devices, e.g. relays. The latter are necessary because some of the BORSCHT functions require the handling of DC voltages and relatively high currents. Section 9.2.4 considers the relationship between the location of the BORSCHT functions, the technology employed and the architecture of the exchange system.

The technology used in the digital switch blocks (Chapter 6) has a significant influence on the hardware design and physical characteristics of the exchange system, particularly in the areas of 'equipment practice' and module size. The term 'equipment practice' is used to describe the physical characteristics of the exchange hardware. This covers the size of the equipment racks and shelves, the number of plug-in units per shelf, the pin connections between plug-in units and the shelf backplane, the methods of inter-shelf and inter-rack cabling, etc. The technology used determines the heat dissipation of the plug-in units, which limits their packing density on the shelves. The upper limit of the physical spread of the exchange hardware is set by the strict timing requirements of digital switch blocks, which results in critical cable lengths between the waveform generators and the racks of equipment (Chapter 10).

The hardware modules represent the building blocks of an exchange system. These may comprise single plug-in units or several units which form a self-contained assembly. Examples of modules used in the construction of digital space switches and time switches are given in Chapter 6. An exchange system is extended in capacity by the appropriate addition of hardware modules to the switch block and line-terminating units (e.g. SLTUs and DLTUs), together with any associated software extensions. The common (shared) exchange equipment, e.g. the exchange-control system and signalling sender/receivers, etc., will also require augmentation to match the increasing load as the switching and line-terminating hardware is extended.

A major requirement is for the exchange system to be capable of being extended without disrupting service. This condition is most easily met when the extra capacity is achieved by adding plug-in units to shelves of existing equipment. However, some extensions will require more intrusive changes to the hardware of the switch blocks or the control system. In such cases, the use of a dual- or mulitple-unit architecture (e.g. dual processor configuration or duplicated switch-block planes) enables one unit to be extended while the other continues to provide service. Each of the multiplied units may be extended 'off-line' and then returned to service. Clearly, the time for which such units are withdrawn from service needs to be minimised because the exchange will not have full protection against system failures during that period.

In practice, any one design of exchange system is incapable of covering

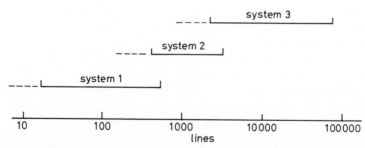

Fig. 9.2 *Family of exchange systems and their range*

economically the full range of exchange sizes required in a network. This range may typically extend from less than 100 lines to over 50 000 lines. Thus, an exchange system will require a family of designs, in which each member covers part of the required range. Fig. 9.2 illustrates the concept with a typical 3-member exchange system family. Note that there is some overlap of the top of the range of system 1 with the bottom of the range of system 2, and similarly between systems 2 and 3. This is because the physical size range of an exchange system exceeds its economical range. The latter is set by cost comparison with others in the family of exchange systems.

The security requirements for an exchange can constrain the architectural design of the system. In particular, most systems are required to meet stringent limits on the number of subscriber lines or proportion of traffic that is affected by individual hardware faults. This limits the degree of super-multiplexing within the digital switch block. Other parts of the exchange are similarly involved in the bulk commoning of circuits into pools of equipment. The solution is for digital SPC exchange designers to improve the maintainability of equipment so that, even if faults are either more frequent or have a wider effect than with analogue conventional exchanges, the faults can be traced and cleared more quickly. The result is that the outage time (product of fault incidence and time to repair) can meet the security requirements.

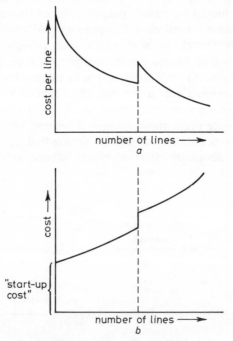

Fig. 9.3 *Capital cost characteristic of a digital local exchange*

The capital-cost characteristic of an SPC exchange is shown in Fig. 9.3. The capital cost per subscriber or trunk line decreases as the size of the exchange increases (Fig. 9.3a) due to the economy of scale.[1,10] This results from the constant cost of the common elements of the exchange (e.g. exchange control, power, etc.) which represent a decreasing overhead as the number of subscriber or trunk lines increases. The fixed level of common-control costs of an exchange are sometimes referred to as the 'start-up' cost because they are incurred even with just one line connected to the exchange, as shown in Fig. 9.3b. Most capital-cost curves have a series of steps which reflect the cost of additional common equipment required to handle the next increment of capacity, e.g. extra control processors.

9.2.2 Small SPC digital local exchanges

The start-up cost results in high costs per line for exchanges at the lowest end of the size range (Fig. 9.3). For example, the economical smallest local exchange unit of a system with maximum extendable capacity of 40 000 lines is around 2000 lines; however, this varies according to the actual system design, facilities offered and manufacturers' prices.

There are two approaches to the provision of economical small SPC local exchanges. The first is the specific design of switching systems optimised for operation with small capacities, with a relatively low maximum size. This cost optimisation is achieved by restricting the switch block to a small-growth range and it may also include restriction of the customer and administration facilities provided by the exchange-control system. A number of successful commercial digital SPC exchange systems designed for simple basic telephony service in remote rural areas are available.[11] These use small-capacity switch blocks, typically employing a single T-stage, which is cheap and requires minimal switch-block control. The exchange-control systems are usually only microprocessor-based, which is adequate for the relatively low volume of call processing and administrative-system processing requirements.

The alternative approach, made possible by PCM transmission and common-channel signalling, is to share one large common-control system and a number of support functions among several remotely located subscriber switching units, as described below.

9.2.3 Remote subscriber switching units

The concept of a remote subscriber switching unit (RSSU) is shown in Fig. 9.4. In the example shown, four RSSUs depend on a centrally located parent local exchange for common-control and routeing functions. Each of these RSSUs is relatively small, with between 400 and 1500 subscriber lines. However, when added to the 20 000 subscriber lines of the parent exchange the common equipment is shared by a total of 23 700 lines, which constitutes an economical load for the digital SPC system of this example. In addition to sharing exchange-

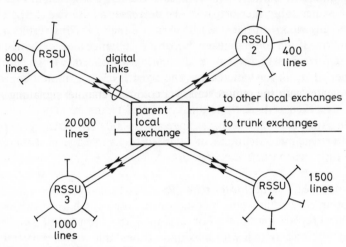

Fig. 9.4 *Remote subscriber switching unit concept*

control functions, the RSSU configuration enables other resources, such as
maintenance and operations support, to be centralised (see Chapter 19).

A typical architecture for the combination of the RSSU and parent local
exchange is shown in Fig. 9.5. The principle is to locate remotely the subscriber-concentrator switch block, which in a normal digital SPC local exchange
would form part of the concentrator switch block and would be collocated
with the group switch block. Thus, the digital link between the switch blocks
is extended over external PCM line systems. Under normal conditions, originating traffic is concentrated at the RSSU and then carried over the PCM link

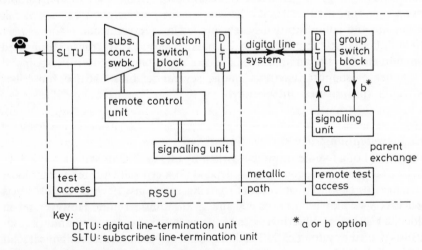

Fig. 9.5 *Remote subscriber switching unit architecture*

to the group switch block for interconnection with other RSSUs or trunk and junction routes. The reverse applies for traffic terminating at the RSSU. Calls between subscribers on the same RSSU are either switched within the RSSU or 'tromboned' via the parent-exchange group switch block by being carried back and forth over the PCM link. Tromboning is necessary where the RSSU provides traffic concentration only. In addition, the exchange-control system communicates with the remote units via common-channel signalling (not necessarily CCITT standard) carried in the PCM links.

Most commercial digital SPC exchange systems offer the use of remote subscriber switching units. For example, E10B[2] and System X[3] employ concentration-only type RSSUs. (Own-RSSU calls are tromboned via parent exchanges.) The DMS100,[4] AXE10,[5] 5ESS[8] and Systsem 12[9] are examples of systems in which the RSSU provides local interconnection of subscribers on the same RSSU.

As Fig. 9.5 shows, remote subscriber switching units contain equipment in addition to the standard subscriber-concentrator unit. The roles of these additional items, i.e. isolation switch block, remote control system, common-channel signalling system and test network, are described below.

9.2.3.1 Isolation switch block. Because the RSSU is largely dependent on the control system of the parent exchange, the availability of service for its subscribers depends not only on the RSSU reliability, but also the reliability of the PCM link and the parent exchange. The required availability of the RSSU and parent local exchange will be provided by the appropriate system of replication and protection, as described in Chapter 18. Similarly, the link to the parent exchange must be protected by suitable replication. In practice, a minimum of two PCM line systems provide the link; each carries the common-channel signalling, so the latter is at least duplicated. Most administrations attempt to route the PCM systems over diverse physical paths in order to minimise the effect of any external damage to the transmission plant between the RSSU and parent exchange. However, despite these precautions there remains a probability of the RSSU becoming isolated.

There are two problems caused by the isolation of an RSSU. The first is loss of the control-processing power provided by the parent exchange. The second is loss of the transmission path which will prevent calls being routed via the parent exchange. This will include calls between subscribers on the RSSU if trombone working is employed. For the latter case, the disruption to service is minimised by the inclusion of an isolation switch block within the RSSU, the use of which is only invoked during isolation from the parent exchange.

An isolation switch block usually comprises a single space switch which provides a loop on the output buses from the subscriber-concentrator switch block. This loop enables calls to be connected between subscribers on the same RSSU. Under normal conditions, i.e. when the RSSU is in communication with the parent exchange, the isolation switch block either provides a permanent through-connection to the transmission link or it is switched out of the connection path. Since the isolation switch block provides a simple loop between time slots on the buses of the concentrator switch-block output, the blocking probability in isolation conditions is far greater than for the normal routeing via the parent switch block.

9.2.3.2 Remote control and common-channel signalling equipment. As previously discussed, a typical exchange-control system is physically implemented as a central unit with a number of units distributed around the exchange. Thus, some element of the exchange-control function may be included in the subscriber-concentrator unit in addition to the switch and switch-block control units (Chapter 6). Under normal conditions, the remote control unit of the RSSU needs only to provide the extra control functions necessary for remote operation of the concentrator switch block. These functions comprise the control of the termination of the common-channel signalling channels to the parent exchange and the management of RSSU system alarms. It is important for failures in the basic function of the RSSU (e.g. loss of power supply, faults in the switch blocks, etc.) to raise alarms which are routed to the parent exchange either over the common-channel signalling link or via separate cable links.

Under link-failure conditions, when the RSSU is isolated, the remote-control equipment has the added role of controlling the connections through the concentrator and the isolation switch block. The required call-control processing capacity is minimised by limiting these connections to basic telephone calls without any additional facilities. There is usually no charging or traffic-monitoring processing associated with such calls. Switch-block capacity and call-processing limitations may also require the restriction of service during isolation to a small range of designated subscribers. The extent to which the RSSU can provide call connections during exchange isolation is the result of the design compromise between the need for continuity of basic service and minimisation of the cost and complexity of the remote-control unit. Clearly, costs need to be minimised in order to obtain the maximum advantage from centralising the control system.

9.2.3.3 Remote test facility. Whether remote or not, the subscriber-concentrator unit provides the capability for simple electrical testing of each subscriber line. Access between the testing equipment and the line is gained either via a relay in the subscriber line-terminating unit (SLTU), as described in Chapter 7, or through the special switch matrix.[8] For an RSSU, the testing equipment is usually provided as a common pool at the parent exchange. Test access to subscriber lines on the RSSU is then provided from the SLTU relay or special switch matrix over a metallic pair of wires to the test equipment at the parent exchange. The use of metallic test wires is not possible for long RSSU-exchange links because of the excessive transmission loss. In such cases, local test equipment is required in the RSSU remotely controlled from the parent exchange using the common-channel signalling link. This creates an additional processing load on the RSSU remote-control system.

9.2.3.4 RSSU dimensions. The traffic capacity of an RSSU is determined by the structure of the subscriber-concentration unit. However, practical considerations covering the cost of providing PCM links to the parent exchange and the relative economics of providing adequate security for large units limits the effective size range of an RSSU design. Commercial RSSU systems have maximum link capacities of between 8 and 16 PCM systems. For link security, each RSSU requires a minimum of two PCM systems to the parent exchange. Two 30-channel PCM systems provide a link capacity of about 48 erlangs, assuming a loading of 0.8 erlang per circuit. The number of subscribers that can be supported by an RSSU is limited by the sum of their originating and terminating traffic. For example, with an RSSU switch block capacity of 170 erlangs, a maximum of 850 subscribers with an average traffic of 0.2 erlangs per line can be supported, compared to 2125 subscribers with an average traffic of 0.08 erlangs per line.

The use of PCM transmission should impose almost no restrictions on the distance between RSSU and its parent. In practice, plant costs and reliability constraints as well as network topology tend to limit distances to around 30 km. Exceptionally, RSSUs may be many hundreds of kilometres away from the parent exchange.[12]

9.2.4 The location of the BORSCHT functions

Chapter 7 described how the subscriber-line exchange functions (summarised by the acronym 'BORSCHT') can be provided by a single module, the SLTU, permanently associated with each line. Although this represents the current approach used in most commercial systems, there are now some important variations in the location of the BORSCHT functions. These variations result from the introduction of digital transmission in the local network, the support of new services and improvements in technology. In considering the architectural options for locating the BORSCHT functions, the codec will be considered first.

9.2.4.1 Location of the codec function. The range of possible locations for the analogue-to-digital convertor (codec) for subscriber and trunk terminations on an SPC digital exchange is shown in Fig. 9.6. For the conventional analogue subscriber line, the codec is normally located within the exchange, along with other BORSCHT functions (fig. 9.6a). An option is to provide analogue-to-digital conversion at a primary multiplexer in the local network (Fig.9.6b). This offers a pair-gain advantage, because a number of subscriber lines are multiplexed on to a single 4-wire bearer, as well as enabling entry into the exchange at the 1.5 Mbit/s or 2 Mbit/s level. Although this gives

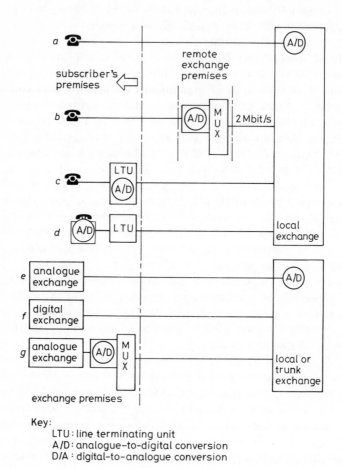

Fig. 9.6 *Location of the codec function*

a saving in SLTUs, the BORSCHT functions must still be provided for each circuit by the multiplexer.

A number of techniques are now available for digital transmission over local cable, as described in Chapter 22. These enable one or two 64 kbit/s duplex channels to be extended digitally from the local exchange to the subscriber's premises to form an integrated services digital network (ISDN). This form of access enables the codec to be located either within the network terminating unit (NTU, see Fig. 9.6c) or within the subscriber's telephone set (fig. 9.6d). With the codec at either of these locations, the subscriber is able to gain direct access to the digital channels for the conveyance of non-voice services, such as digital data communications. The progressive introduction of digital transmission into the local network will decrease the number of subscriber lines that require analogue-to-digital conversions within the digital exchange, with a corresponding increase in codecs provided at the subscriber's premises.

Analogue trunk transmission routes require a codec at the exchange (Fig. 9.6*e*), which is usually provided within the analogue trunk termination unit (ATTU) on a group rather than per-circuit basis (Chapter 7). Careful planning by an administration can, however, eliminate the need for such codecs within the exchange by ensuring that only digital trunk routes are terminated. Thus, analogue-to-digital conversion should be provided at the distant analogue exchange in a PCM multiplexer (Fig. 9.6*g*). The ideal situation is for codecs to be eliminated on all trunk and junction routes; this is achieved when all exchanges in the network are digital (Fig. 9.6*f*).

9.2.4.2 Location of all the BORSCHT functions. Fig. 9.7 shows three possible layouts of the BORSCHT functions within the trunking of a digital local exchange. All three architectures have been used by commercial systems in order to provide minimum-cost solutions with currently available switching technology.

Since the circuitry to provide the BORSCHT is both expensive and physically bulky, the cheapest solution is to locate the functions between the concentrator and group switch block. With a 10:1 concentration ratio, the solution should be about 10% of the cost of providing the BORSCHT functions permanently for each subscriber line. However, it is not possible to concentrate the off-hook supervision of subscriber lines; this must be on a per-line basis so that detection of the off-hook signal can initiate connection through the concentrator. The secondary fuses must also be placed on the line side of the switch to ensure constant overvoltage protection. The configuration with the maximum concentration of BORSCHT functions, all but 'O' and 'S' (over-voltage and off-hook supervision), is shown in Fig. 9.7*a*). It should be noted that the address signalling elements of supervision (S*) may be placed after the concentrator.

The concentrator switch-block technology required for the configuration of Fig. 9.7*a* must allow the passage of the high ringing currents (R) and the exchange-battery voltage (B) as well as a metallic path for test access (T). The location of the codec and hybrid transformer requires the switch block to provide an analogue 2-wire path of high crosstalk immunity. The first exchange to use this architecture was the first commercial digital public exchange, the E10A launched in 1970.[13] It used analogue reed relays in the concentrator switch block.

Fig. 9.7*b* shows all but the codec part of the BORSCHT functions located on the subscriber's side of the concentrator. Although this configuration requires more equipment per line, it has less stringent requirements for the subscriber-concentrator switch block, because high voltages and DC paths need not be transmitted.

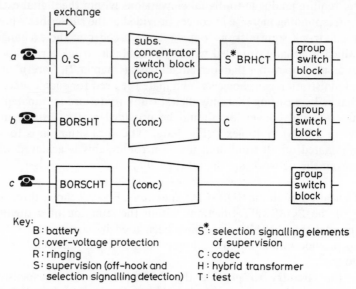

Key:

B : battery
O : over-voltage protection
R : ringing
S : supervision (off-hook and selection signalling detection)

S^*: selection signalling elements of supervision
C : codec
H : hybrid transformer
T : test

Fig. 9.7 *Possible locations of the BORSCHT functions*

The configuration shown in Fig. 9.7c, with all the BORSCHT functions on the line side of the subscriber-concentrator, is necessary if digital switch blocks only are to be used. Using current technology, digital switch blocks cannot provide DC loops on metallic paths or accept high voltages; thus, a line terminating unit (SLTU) must provide all the BORSCHT functions, with the possible exception of the codec, as previously discussed.

To summarise (refering to Fig. 9.7), the earliest digital local exchanges used analogue electromagnetic (e.g. reed relay) concentrator switch blocks using configurations (*a*) or (*b*). In order to achieve all-digital exchanges, configuration (*c*) has been widely adopted since the early 1980s. Recent advances in semiconductor technology have enabled configurations (*a*) and (*b*) to be realised using analogue electronic switch blocks.[8] The choice of BORSCHT location is influenced by cost, reliability and available technology in the context of the overall architecture of the switching system.

9.2.5 The routeing of internal messages

An important architectural feature of digital SPC exchange design is the method of routeing internal messages between control systems. These messages pass between central exchange-control processors and the dispersely located processors associated with specific parts of the exchange. The format of the information within such messages is specific to particular system designs; the use of international standards is inappropriate and so proprietry interfaces and formats are applied. However, the transport mechanism for

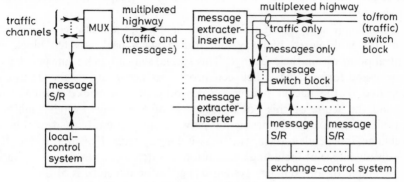

a use of a separate message switch block

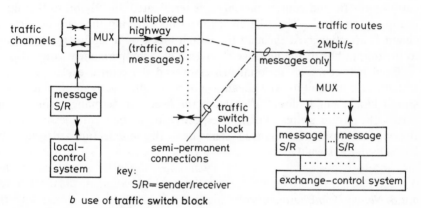

b use of traffic switch block

Fig. 9.8 *Message routeing within an exchange*

conveying the messages may use standard message protocols, such as CCITT X25.

There are two basic architectures for the routeing of internal messages: one uses a message switch and the other uses the digital group switch block. A simple representation of the two options is shown in Fig. 9.8. Both options use message sender/receivers to originate and terminate the transmission of messages at all participating processors. Control information between processors is transported by a packet-message-transmission system. The output message from a processor, which is in the form of a binary word, is converted into a data packet by the message sender. This packet comprises the original binary word enveloped by an address and reference bits (header) and some error-correcting bits (tail). The start and finish of the packet it indicated by a binary pattern. A stream of message packets is inserted into one of the 64 kbit/s time slots of a multiplexed highway carrying traffic channels. This may be within a standard 2 Mbit/s frame, in which case TS16 is used to carry the messages, or it may be within a higher-order (super-multiplexed) frame structure.

With the message-switch architecture (Fig. 9.8a), the messages are extracted from the multiplexed highway and routed via a packet-switch block at a central point within the exchange. This special switch block routes packets on a store-and-forward basis between processors according to the destination indicated in the packet header.[14] (The principles of packet switching are beyond the scope of this book.) The use of a message switch enables an optimised control and messaging network to be established within the exchange, independent of the traffic-switching systems. However, it does have the disadvantage of including an additional switch block within the exchange. An example of an exchange system using this architecture is 5ESS.[8]

With the alternative architecture (Fig. 9.8b), the multiplexed highway, carrying traffic and control messages, is terminated directly on to the digital group switch block. The 64 kbit/s time slots carrying the various message channels are connected through the switch block on a semi-permanent basis to the output bus of the message receiver. One semi-permanent connection is required between each regional processor and the central exchange-control system. This method has the advantage of not requiring an additional packet-switch block, but it does suffer from the need for permanent connections across the group switch block, irrespective of amount of message traffic flowing. Examples of exchange systems using this architecture are System X,[3] EWSD[6] and System 12.[9]

9.2.6 Network-database control

The exchange architectures described so far in this chapter comprise a control system incorporated in the exchange, which, in the case of RSSU working, also extends control to its dependent exchange units. However, there is now a new architectural feature emerging in which some elements of an exchange's control function reside at special network centres, known as service control points (SCPs). The SCP contains a database system supporting call-control software accessible from a number of exchanges for a range of specialised services. Calls are initially handled in the normal way by the local exchange. Where a specialised service is requested, i.e. when indicated by the number dialled, the originating local exchange or its trunk exchanges sends a request for control information using common-channel signalling to an SCP. Control information is returned by the SCP over the CCS link, according to the records within its database and the control software for the invoked service. The SCP may request more information from the subscriber, and the local or trunk exchange will need to generate an appropriate recorded announcement.

This architecture, known as intelligent network (IN), centralised intelligence (CI)[15] or a network-database system,[16] has a number of advantages, as described below.

(i)   It enables new services to be implemented relatively quickly because the software of only a few SCPs requires changing, rather than the software at all the local exchanges in the network. This shorter time between specifying the requirements of a new service and its implementation means that the telephone administration is able to be more responsive to the rapidly changing business service market.

(ii)   Once new services have been proven and the demand reaches an economic level, the control-system software of the relevant local and trunk exchanges can be upgraded to include the service facility, thus freeing the SCP for more new specialised services.

(iii)   Where the traffic levels of a specialised service remain low, the control can remain with the SCP, since it would be uneconomical to incur the cost of modifying the software of the local exchanges in the network.

(iv)   The SCP may also provide services in which the call-routeing information varies according to the contents of the database. This means that, for example, variations according to staffing levels at bureau answering centres can be incorporated.

(v)   The SCP has an overall view of the way that the specialised traffic is originating on the network and can decide on network management action to improve traffic flow or, by redistributing calls to different answering bureaux, decrease the time to answer.

An example of a specialised service is wide-area telephone services (WATS), known as the '800' service in the USA and '0800' service in the UK. With this service, vendors are able to advertise nationally using just one telephone number (beginning with '800' or '0800'); subscribers calling the number are routed to the designated vendor's reception centre free of charge. A variant is the 0345 service, which enables such calls to be charged at local rate, irrespective of distance. Different reception centres can be used for various geographical areas according to the service provider's instructions to the telephone administration.[17]

The network database architecture will also provide a convenient method of implementing networked centrex (see Chapter 3) and virtual private networking (VPN). Both networked centrex and VPN exploit the fast connection capabilities of digital SPC exchanges in a public switched network to give the same level of private communication for a company with dispersed offices as a private system using PABXs at each site linked by leased-line circuits.

With VPN, where virtual private links are established between PABXs via the switched network, the PABXs are connected to local exchanges in the normal way. However, the lines have a special class of service which ensures that, as soon as calling is detected, the network database is accessed for routeing and call-control information. The database holds the details of extension numbering at all participating sites and any incoming or outgoing call restrictions imposed by the company on its private network. The database will also act as the billing and private network management centre.[18]

9.3 Digital cross-connection systems

Digital SPC exchange systems are also used in public telecommunications networks to provide interconnection of leased-line circuits. Such circuits, which provide a semi-permanent link between two locations in a company's private network, are charged only an annual rental and so do not incur call (or usage) charges. The exchanges that provide the necessary semi-permanent switching are known as digital cross-connection (DCC) units. With the DCC units located at local exchange sites, digital 64 kbit/s leased-line connections can be quickly established between subscribers in the exchange catchment area, using the local copper-cable pair network. As with the PSTN, wide-scale leased circuits can be provided by the linking of DCC units by 2 Mbit/s (or 1.5 Mbit/s) trunk routes and, where appropriate, the use of trunk DCCs for transit connections between distant 'local' DCC units.

DCC units differ from digital SPC telephone exchanges in two ways: local-line termination and control. DCC systems do not directly terminate local lines because they have only 2 Mbit/s (or 1.5 Mbit/s) ports. Thus, subscriber local lines must terminate on digital primary multiplexers, either collocated with the DCC units in local exchange buildings or at remote local exchange sites.

The form of subscriber line-termination unit (SLTU) at the multiplexer differs according to whether the local line is analogue or digital. In the case of analogue lines, the SLTU must provide all the BORSCHT functions except supervision (S); the latter is not required because no call charging or call routeing is involved. Signalling on analogue leased lines must be passed transparently by the network. MF or 1VF in-band signalling is encoded into PCM by the SLTU and carried in the 64 kbit/s channel. However, in order for the signalling to remain in-band throughout the network, the SLTU must convert the DC signalling on the subscriber's line to MF or 1VF rather than digital channel-associated signalling (CAS) in TS16. The subscriber line-termination units for digital lines (D/SLTU) provide the BORSCHT functions in conjunction with a network termination unit (NTU) located on the subscribers premises, as described in Chapter 7.

The connections through the digital switch block of the DCC unit (typically a TST configuration) are made on a semi-permanent basis, as described in Chapter 6. Switch and swtich-block control systems are similar for both DCC and telephony digital SPC exchanges. However, the exchange-control system for the DCC unit is much simpler than for the telephony exchange because of the lack of subscriber-initiated call-control processing. The DCC control system is required only to establish new connections, clear-down unwanted connections and provide the necessary maintenance support. Therefore, DCC control systems are relatively simple. They may be provided by small computer systems (e.g. personal computer) located with each DCC unit. Access to the control unit is from a local or remote operator's console using an appropriate man-machine language. The system may alse be extended to allow subscribers to gain access to the control system, via a dial-up modem connection, which, using a set of instructions, allows them to reconfigure their designated circuits directly.

For a full level of network management and speed of provision of circuits connected via several DCC units, the exchange-control system should be centralised. The digital cross-connection network of British Telecom, which provides the KiloStream digital leased-line service, comprises about 80 DCC units and over 700 remote multiplexer sites controlled by a duplicated centralised system which contains a database covering the routeing of every circuit. Control links between the centre and each DCC unit are provided over the packet-switched network. This arrangement also allows alarm signals from all DCC units to be handled centrally.[19]

9.4 Trends in exchange-system architecture

Digital SPC telephone exchanges have achieved a state of maturity. They are now operating in most of the world's networks, and in many places they have totally replaced the analogue switching systems. Their introduction is following, to various extents, a corresponding conversion of transission links from analogue to digital. This will result in the progressive establishment of integrated digital networks, i.e., digital exchanges linked by digital transmission (Chapter 21). Already, there are indications of the directions that digital exchange architectures will take to reflect the increasingly digital transmission environment as well as the new facilities required by both subscribers and telecommunications administrations.

The predominant exchange-architectural trend is the move of many of the BORSCHT functions out of the exchange into the local line or to equipment located at the subscriber's premises. As described earlier in this chapter, this trend results from the increasing use of digital transmission over subscriber lines, where the BORSCHT functions reside in either the NTU or the subscriber's apparatus. The provision of digital transmission over subscriber lines is usually incorporated in an ISDN system. This enables the subscriber to exploit a digital interface for attachments, as described in Chapter 22. However, where the BORSCHT functions are provided by a multiplexer in the local network, the subscriber still has a conventional telephone with an analogue copper pair connection. In either case, the exchange needs to provide fewer BORSCHT functions.

This trend of removing the BORSCHT functions from the exchange system can be regarded in two ways. One is that a digital SPC local exchange system need not then provide the BORSCHT functions. This results in a cheaper exchange due to the elimination of the costly SLTUs. However, it does require well defined and enduring boundaries to the exchange system so that network multiplexers, network terminations and subscriber terminals (apparatus) can be developed independently by a variety of manufacturers, while providing compatible BORSCHT functions.

The alternative view is that the BORSCHT functions remain part of the local exchange system, irrespective of their location. If the functions are dispersed between subscribers' premises and an exchange building, then the digital SPC exchange system must be considered to encompass these locations. The system design may therefore have functions and electrical signal levels apportioned between the various exchange elements in an optimum way, without the need to adhere to rigorous interfaces at the boundaries of the exchange, multliplexer or NTUs. The choice of alternative is determined by the commercial and regulatory climate surrounding telecommunications networks and the supply of equipment within a country. Where the climate is liberalised, the first alternative will apply.

Another consequence of the progressive digitalisation of the local network is that the entry level to the exchange ceases to be a single circuit. Rather, it is at 144 kbit/s (ISDN CCITT 'basic level') giving two 64 kbit/s channels, or 1.5 Mbit/s or 2 Mbit/s giving 24 or 30 64-kbit/s channel capacity, respectively (Chapter 22). With current designs of exchange systems, these digital lines terminate directly on the subscriber-concentrator switch block. However, if the traffic loading of such links is high, the concentrator switch block can become congested. Increasingly, therefore, future designs of digital exchanges will allow the direct termination of highly loaded digital links from subscribers on to the digital group switch block. Such an arrangement would require subscriber signalling receivers to be accessible from the group switch block.

The influence of network databases is likely to increase in the future. Telecommunication administrations will probably concentrate on the storage of an ever-growing range of subscriber-specific-service data within network databases, leaving the exchange-based control systems to hold the call-routeing data for the basic telephony and ISDN services.

Finally, there is evidence that future digital SPC exchange systems will embrace a range of network functions other than those of the PSTN. Thus, a single exchange unit might comprise (telephony and ISDN) circuit and packet switching as well as leased-line cross-connection. The first example of such an architecture has been developed by Northern Telecom for their 'SuperNode' system.[20,21] This enables their exchanges to incorporate the functions of telephony local and trunk switching, signal transfer for CCITT No. 7 signalling (STP), packet switching, digital cross-connection and network service database. The architecture is based on a central control system('DMS-core'), providing basic control functions, which is linked to a variety of peripheral processors to provide the required additional functions. Messages between the core and peripheral control systems are routed via a collocated message switch ('DMS-bus'). The facilities of a SuperNode may be extended to other SuperNode sites via CCITT No. 7 common-channel signalling.

A view of the long-term trends in SPC digital exchanges is given at the end of Chapter 22.

9.5 References

1 LITTLECHILD, S.C. (1979): *Elements of Telecommunications Economics*, (Peter Peregrinus Ltd), p.47

2 GRINSEC (1983): *Studies in Telecommunications*, Vol. 2 (Electronic Switching), Pt. VII (Elsevier Science Publishers BV)

3 TIPPLER, J. (1979): *Architecture of System X. Part 1. An Introduction to the System X Family*, Post Office Electrical Engineers' Journal, **72**

4 SWAN, R. (1983): *DMS Family Evolution*, Telesis, **10** (3), pp.2-5

5 EKLUND, M., LARSON, C. and SORME, K. (1976): *AXE10 — System Description*, Ericcson Review, No.2, pp.70-89

6 *Special Issue: EWSD Digital Switching System*, Telecom Report 4, 1981, (Siemens)

7 SUEYOSHI, H., SHIMASAKI, N. and KITAMURA, A. (1982): '*A Versatile Digital Switching System for Central Offices — NEAX61*', in *Electronic Switching: Digital Central Office Systems of the World*, Ed. Joel, A.E., Jnr., IEEE Press, (John Wiley & Sons Inc.), p.81

8 CARNEY, D.L. et al. (1985): *The 5ESS Switching System: Architectural Overview*, AT&T Tech Journal, **64** (6), Pt.2

9 STC (1983): *System 12. ITT 1240 Digital Exchange, A Technical Description*, STC Telecommunications Ltd)

10 REDMILL, F.J. and VALDAR, A.R. (1985): *Selecting a System*, African Technical Review, pp.31-35

11 GIBBS, J.W. and TRUDGETT, P.A. (1983): *UXD 5B: A 600 Line Digital Local Exchange*, British Telecom Eng., **2**

12 SKEY, P. (1983): 'Overseas Practices' in *Local Telecommunications* ed. Griffiths, J.M. (Peter Peregrinus Ltd)

13 JOEL, A.E. (1982): *Electronic Switching: Digital Central Office Systems of the World*, IEEE Press (John Wiley & Sons Inc), Chap.1

14 ROBERTS, L. (1978): *The Evolution of Packet Switching*, Proc. IEEE, **66**, pp.1307-1313

15 WALKO, J. (1988): *Switching on the Network*, Communications Systems Worldwide, pp.44-47

16 MANTERFIELD, R.J. (1987): *Migration of Intelligence in Evolving Networks*, Fifth World Telecom Forum, Technical Symposium: 'Telecommunications Services For a World of Nations', ITU, Geneva, Vol.II, pp.33-37

17 KETTLER, H.W. and PIGGOTT, S. (1986): *Advanced 800 Service Goes Internationl*, AT&T Technology-Products, Systems and Services, **1**,(1)

18 SPRING, P.G. (1986): *SDN — A New Approach to Business Networks*, AT&T Technology-Products, Systems and Services, **1**,(1)

19 MARSHALL, J.F., ADAMSON, J. and COLE, R.V. (1985): *Introducing Automatic Cross-Connection into the KiloStream Network*, British Telecommunications Engineering, **4**, p.124

20 TSUI, S. (1987): *The Technology Behind the DMS SuperNode*, Telephony

21 VOSS, B. and WOOD, R.G. (1987): *A Messaged Based, Distributed Computing Engine For DMS-100*, Paper B3.4.1, presented to the International Switching Symposium, Phoenix, Arizona, 15-20 March 1987.

Chapter 10

Digital-network synchronisation

In describing the process of digital switching in the earlier chapters of Part II, it was assumed that all the digital transmission systems connected to the exchange are operating at the same speed (i.e. bit rate) as the digital logic within the exchange. As will be explained, this condition, known as synchronism, is necessary for any digital exchange connected to digital transmission links. This chapter considers the requirements for exchange-timing sources and how the timings within a digital network are kept adequately synchronised.

10.1 The timing aspects of digital networks

10.1.1 Timing-distribution methods
The electronic components that constitute digital transmission and switching systems are based upon synchronous digital logic and so require a timing signal for their operation. Within a transmission or a switching system the timing signal is provided by a central clock source. The signal from the clock, in the form of a binary waveform may be distributed over wires directly to the various digital logic devices (Fig. 10.1a) where the distances are relatively short, as within a digital exchange. This distribution is by cables, usually replicated for security, radiating from the exchange clock to every rack or shelf of switching equipment; distribution within the rack (or shelf) is usually by a bus accessible by each plug-in module.

However, where there are long distances between the clock source and the components of the system, as with digital transmission systems, the timing needs to be carried along with the transmitted digital signal, as shown in Fig. 10.1b. This is achieved using a line code (see Chapter 5). At the distant end of the digital transmission link, the timing is extracted from the attenuated and distorted signal by the terminal regenerator and used to produce a new

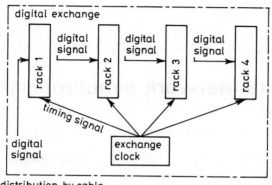

a direct distribution by cable

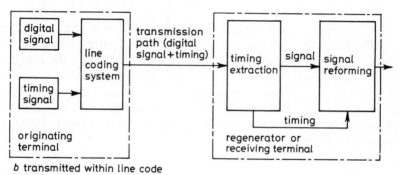

b transmitted within line code

Fig. 10.1 *Timing distribution*

undistorted replica of the digital signal. This may then be fed to a primary multiplexer, high-order multiplexer, digital distribution frame or digital switch block. Alternatively, in the case of long-distance routes, the regenerated signal may be retransmitted to line for onward transmission.

10.1.2 Timing arrangements for digital networks

The timing arrangement for a digital (e.g. PCM) line system which has analogue interfaces at both ends is independent. The transmit and receive directions of transmission can operate at their own bit rates, although they would both be nominally the same. For example, with 2 Mbit/s systems, both clocks ought to be within a range of 200 bits (+ 50 p.p.m. of the nominal line rate).[1] However, for operational convenience, the timing may be taken from a clock in one of the primary multiplexers, with the distant end using the extracted timing to drive the receive direction of transmission (Fig. 10.2a). Thus, the frequency of the master multiplexer, f_a bit/s, drives both directions of transmission due to the loop at the far end.

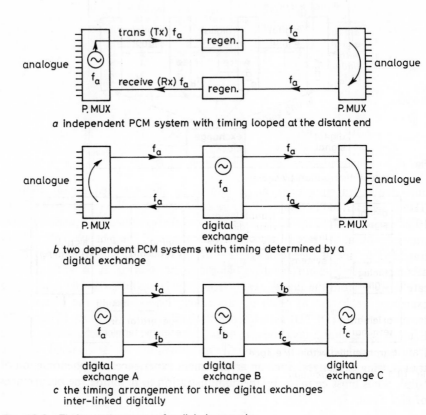

a independent PCM system with timing looped at the distant end

b two dependent PCM systems with timing determined by a
digital exchange

c the timing arrangement for three digital exchanges
inter-linked digitally

Fig. 10.2   *Timing arrangements for digital networks*

Similarly, a digital exchange which interfaces only at the analogue level to external circuits may operate with independent timing. In theory, the exchange clock may then have any frequency which gives adequate PCM sampling quality (see Chapter 5). However, in practice, a frequency approximating to the nominal 2048 kbit/s or 1544 kbit/s is used.

When digital-line systems are connected via a digital interface to a digital exchange, the latter assumes control of the timing. This is shown in Fig. 10.2b, in which the digital exchange clock (f_a) drives both the transmit directions of the two dependent PCM systems and the timing is then looped at the distant primary multiplexers. The network is thus operating at f_a, the frequency of the single exchange clock.

The situation becomes more complex once two or more digital exchanges are interconnected digitally, as shown in Fig. 10.2c. Most administrations aim to establish a network in which all the digital exchanges are interconnected by

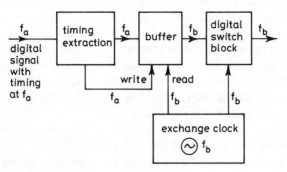

Fig. 10.3   *The timing arrangement at exchange B*

only digital transmission. This is known as an integrated digital network (IDN), as described in Chapter 21. The arrangement within exchange B of Fig. 10.2c is shown in Fig. 10.3. This shows how exchange B receives digital transmission at the rate of the distant exchange, f_a (or f_c), but switches and transmits to line at the exchange rate of f_b. The mismatch in frequencies is accommodated by a buffer into which the received bit stream is written, at the rate derived from the line (f_a), and from which the bit stream is read at the exchange clock rate f_b. Such a buffer may be located in the digital line-termination unit (DLTU) associated with the digital switch block (Chapter 7). Each of the digital links shown in Fig. 10.2c has different frequencies in the transmit and receive directions of transmission, as determined by the transmitting exchanges. Such an arrangement experiences the phenomenon of 'slip'. Steps must be taken by the administration to minimise the occurrences of slip within their IDNs.

10.1.3 The concept of slip

Slip occurs periodically in a digital system in which there is a mismatch between input and output line frequencies, the rate of slip depending on the degree of mismatch. Each slip comprises a digital error resulting from the insertion or loss of one or more bits. A simple explanation of how slips occur is given with reference to Fig. 10.3. Assume that the buffer shown in the Figure is initially half full. If the rate of writing into the buffer, f_a, is greater than the rate of reading from the buffer, f_b, the buffer contents will gradually increase. When full, the buffer contents will need to be dumped. Alternatively, with f_a less than f_b, the buffer contents will gradually decrease. When empty, some additional content will be required. Both the dumping and replenishment of the buffer contents give rise to slips.

There are two classes of slip: uncontrolled and controlled. The former are unpredictable in both timing of occurrence and their extent. The occurrences of the latter can be predicted and may be controlled so as to cause the minimum disruption. Controlled slips may involve a bit, time slot, frame or multiframe. It is usual for controlled slips in digital SPC exchanges to be in units of one frame, known as 'frame slip'.

The mechanism of digital frame slip is shown in Fig. 10.4. When the incoming-line digital rate is faster than that of the exchange, the frame duration of the line system is correspondingly shorter than the exchange frame (Fig. 10.4a). Since the exchange is reading out at a slower rate, the aligner buffer progressively fills up. When full, the contents of the aligner (frame F) are dumped and the aligner fill reduced to zero.

Fig. 10.4b shows the alternative situation where the aligner buffer is progressively emptied as a result of the exchange reading out faster than the input line rate. The contents of frame D must be repeated to enable the aligner to be refilled. In both cases the rate of slip, frame repetition, or dumping, is determined by the magnitude of the discrepancy between the two frequencies and hence the rate at which the fill of the aligner buffer changes. When the two frequencies are equal, the aligner fill is constant, and its level depends on the constant phase difference between the frame start of the incoming line system and that of the exchange. The relationship between aligner fill and the

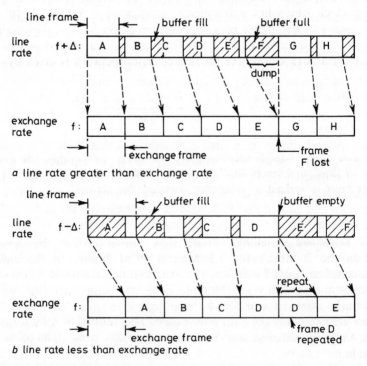

Fig. 10.4 *Examples of frame slip*

timing discrepancy forms the basis of network-synchronisation systems, as described later in this chapter.

10.1.4 The effects of frame slip

The perceived effect of a frame slip incurred at one or more digital exchanges in a call connection depends on the type of service or application involved. Four main cases are considered below.

10.1.4.1 Digitally encoded speech. The high degree of redundancy in speech means that, with standard 8-bit digital encoding (i.e. PCM), the effect of one sample being in error as a result of a frame slip is small. The human perception of an error in a PCM sample depends on the degree of correlation between the erroneous and the true samples. It has been estimated that only about 5% of frame slips are perceived by telephone users, in the form of a brief click.[2] However, the use of low-bit-rate speech encoding, such as 32 kbit/s CVSD or ADPCM (see Chapter 5), make the effects of slip more noticeable because of the lower levels of encoding redundancy.

As an illustration of the levels of slip that can occur, consider a typical exchange clock which has a frequency accuracy of $\pm$ 3 parts in 10^7. The average frequency difference between two such clocks is $\frac{2}{3}$ of the maximum range.[3] For a connection involving n exchanges, the number of frames slips per hour for a 2048 kibt/s (256 bit frame) digital network is given by

$$(2n-1) \times \frac{2}{3} \times \frac{3 \times 10^{-7} \times 2048 \times 10^3 \times 3600}{256}$$

For a value of $n = 4$ (equivalent to a typical trunk connection), the average number of slips per hour is 40. Thus, with PCM encoding, only about 1 to 2 slips per hour is noticed by telephone users of this network.

10.1.4.2 Telephone signalling. Frame slips may also affect the signalling carried in the 2 Mbit/s (or 1.5 Mbit/s) PCM frame. In the case of PCM/channel-associated signalling (PCM/CAS), the isolated slip of just one frame does not cause loss of the multiframe alignment since that requires several frame-alignment words to be received in error (see Chapter 5). Thus, a single frame slip only injects a minor corruption into the CAS words. The PCM/CAS system incorporates sufficient redundancy to be immune to such corruption.

Clearly, if the slips are uncontrolled, or are controlled but occur very frequently, multiframe alignment will become lost. Recovery of this alignment takes several milliseconds, during which time no signalling can be transported. It is possible that this break will create disruption to any current signalling sequences, which may result in call set-ups being in error or abandoned.

Common-channel signalling (CCS) systems (Chapter 20) have error-detecting mechanisms which invoke a retransmission of any CCS messages corrupted by any disruption to transmission, such as frame slip.

10.1.4.3 Data transmission over telephone channels. Where telephone calls are used to convey data by the use of voice-frequency modems,[4] frame slip produces an isolated disruption to the data signal. However, because this form of data transmission is analogue, the slip induces a slight impairment to the signal, which is unlikely to result in an error in the received data.[2]

10.1.4.4 Digital data. The effect of slip on a 64 kbit/s channel containing digital data is more serious than for telephony because of the higher information content involved. The accepted criterion for digital-data transmission performance is the percentage of error-free one-second intervals (EFS). Each frame slip causes a burst of up to eight errors in both directions for each 64 kbit/s channel on the digital line system concerned. In the case of 40 slips per hour, calculated above, only 3560 seconds in each hour are error-free. The percentage of error-free seconds is thus given by

$$\text{EFS} = \frac{3560 \times 100}{3600} = 98.88\%$$

This figure is unacceptable for digital data transmission. The CCITT[5] recommended level is 99.9% EFS and is intended to cover all network-induced impairment, of which slip should form only a small contribution.[2,6] Where necessary, data customers use error-detection techniques which request the retransmission of corrupted blocks of data. This enables the effects of slip and other errors to be minimised. However, excessively high slip rates cause frequent retransmissions, with a consequent reduction in data throughput.

10.1.4.5 Digital transmission links. Notwithstanding the various effects of slips described above, it should be noted that persistently high levels of slip cause digital line systems to be automatically withdrawn from service. This is undertaken by the digital trunk termination units (DTTU, see Chapter 7) at digital exchanges. Persistent high levels of slip cause continuous loss of frame and multiframe alignment which, on detection by the DTTU, result in that unit being 'busied out'. On receipt of a maintenance message from the DTTU, the exchange-control system withdraws the offending 2 Mbit/s line termination from operation by changing its status within the exchange-data store. This ensures that all existing call connections to channels on that 2

Mbit/s path are cleared (and possibly re-routed) and no new connections can be established until the link is returned to service.

10.1.5 Causes of slip

The filling or emptying of aligner buffers, resulting in slips, can be caused by several factors, individually or collectively, as described below.

10.1.5.1 Imperfect clocks. In practice, any two clocks set to the same nominal frequency and left to run freely always have some degree of timing mismatch. This is due to the two main characteristics of practial clocks. The first is the degree of accuracy with which a clock can be tuned. This is determined by the short-term stability of the clock. If a type of clock can only be tuned to within x Hz of a set frequency, any two such clocks tuned simultaneously are initially within the range of $2x$ Hz of each other.

The second characteristic is that of frequency drift or 'stability'. This is the degree to which a clock will change its natural frequency over a period of time as a result of ageing. Practical clocks exhibit both short-term variations in frequency (limiting their tuning accuracy) and long-term 'drift' away from the original set frequency.

10.1.5.2 Transmission-delay variation. The propagation time of a digital signal over a transmission system may vary and cause aligners to fill or empty, even if the clock frequencies at both ends of the line are identical. Temperature changes significantly affect the propagation time of transmission over cables. For example, a 10 deg C change in temperature over a 50 km length of paper-insulated cable carrying a 2 Mbit/s system causes a change of 6 bits in the aligner fill. The propagation-delay variation over coaxial cable is about 1/10th of that for paper-insulated cables.[2]

The other cause of propagation-delay variation is that of a physical change to the path length via an earth satellite. Even geostationary satellites depart from their orbit position; a dither of up to 0.5° may occur without correction by the on-board rocket motors, causing delay variations of over 1000 bits duration.[2]

Both these causes of transmission-delay variation exhibit slow daily cycles, as well as seasonal cycles in the case of temperature variation. This slow change to the frequency of the received digital signal is often referred to as 'wander'.[7]

10.1.5.3 Jitter. Jitter is the name given to the alternating variation of the instantaneous frequency of a digital signal from the long-term nominal value. It results in the displacement of the individual signal elements (bits or ternary digits) from their ideal positions in time, which can be viewed as phase modulation. All digital transmission and switching networks experience jitter as a result of the accumulation of minor timing inaccuracies introduced by the gating logic of regenerators and multiplexers in the network. Jitter is cumulative and thus worsens with the length of digital transmission systems.[7]

10.1.5.4 Network rearrangements. For operational reasons, for example in the event of a cable breakdown or planned works on the network, the circuits between two exchanges may be re-routed over other cables. When the new routeing incurs a different propagation time, there is a resulting change in the fills of the aligners at both ends of the link. This leads to the possibility of one or more slips, some of which may be uncontrolled. This is a particular problem where automatic re-routeing of transmission links is used for service protection (see Chapter 21).

10.1.6 Timing sources

Three types of timing sources are currently available for use in digital SPC exchanges: quartz crystal, rubidium and caesium. The choice is governed by the role that the timing source is to take in the network and by the following parameters:

(i)   Cost.
(ii) Reliability.
(iii) Short-term stability (which determines the *accuracy* to which the timing source can be tuned)
(iv) Long-term stability (i.e., *drift*)

The caesium beam is an atomic timing source that is internationally accepted as a standard. Although these sources have a worse short-term stability than the other form of atomic clock, rubidium, they do have infinitely high long-term stability, i.e., zero drift. Such sources are, however, expensive and unreliable.

The rubidium-vapour-cell timing source is cheaper than the caesium type, but it has a long-term stability of about 5 parts in 10^{11} per month. The short-term stability is better than caesium, allowing more accurate tuning.

Quartz-crystal timing sources are used extensively for exchange and multiplexer clocks. as well as for many other applications. They are cheap and reliable and have good short-term stability. They have a relatively poor long-term stability, with drift rates ranging from parts in 10^6 to parts in 10^8 per month, depending on the quality of the crystals and the level of temperature compensation used. However, they do age in a predictable way. Their frequency can be changed by varying the voltage across the crystal to provide voltage-controlled crystal oscillators (VCXO). This enables these sources to be controlled externally by more stable timing sources, as described later. For best results, quartz crystals are usually housed in an electrical oven to minimise the effects of ambient temperature variation on the output frequency.

The characteristics of the various types of timing sources are summarised in

Table 10.1. This also gives a guide to their prices and reliability normalised against the values for temperature-compensated quartz-crystal oscillators.

Table 10.1   *Timing sources*

Type	Long-term stability (drift)	Short-term stability (tuning accuracy)	Price (normalised)	Reliability (normalised)
Temperature-Compensated Quartz Crystal	± 2 parts in 10^6 p.a.	± 2 parts in 10^6	1	1
Oven-controlled Quartz Crystal	± 2.4 parts in 10^7 p.a. (± 2 parts in 10^8 p.m.)	± 1 part in 10^8	10	1
Rubidium	1 part in 10^{10} p.a.	1 part in 10^{12}	$100-200$	0.2
Caesium	nil	1 part in 10^{11}	500	0.1

10.2 Methods of controlling the timing of digital networks

The effects of jitter and propagation-delay variations can be adequately absorbed by a suitably sized buffer between each digital line system and the exchange. The frame-alignment buffer within the digital line-termination unit (DLTU, see Chapter 7) with its full-frame capacity is usually more than adequate for this. However, as explained earlier in this chapter, simple buffering will not eliminate slips due to frequency mismatch between different exchange clocks. There are two approaches to restricting the occurrence of frame slip within an IDN, namely:

(i) Plesiochronous operation: With plesiochronous operation, each exchange clock operates independently. Slips are kept to an acceptably low level by using clocks of high stability which are periodically manually re-tuned so that they all operate within close limits of a nominal network frequency.

(ii) Synchronous operation: A synchronous network has all the component clocks controlled by an automatic mechanism so that they operate at a single network frequency. In practice, the clocks may be maintained at the same mean frequency, but with short-term variations, as described later in this chapter.

In general, for acceptably low levels of slip, plesiochronous operation requires clocks of at least moderate quality which must be manually retuned. Synchronous operation enables cheaper lower-quality clocks to be controlled by just one high-quality master clock in the network. However, the latter does require a control system. There is, thus, a complexity and cost trade-off

Table 10.2 *Performance objective for overall slip rate on a 64 kbit/s international connection*[8]

Performance	Mean slip rate	Portion of time
Satisfactory for all types of service	Max. 5 slips in 24 hours	Greater than 98.9%
Satisfactory for speech but other services degraded	Between 5 slips in 24 hours and 30 slips in 1 hour	Less than 1.0%
Unacceptable for all services	Worse than 30 slips in 1 hour	Less than 0.1%

between plesiochronous and synchronous modes of slip control. The choice is primarily decided by the slip-rate objectives for the network.

Before describing these two modes in more detail, the relevant CCITT recommendations governing the acceptable levels of slip are considered. The CCITT has recommended slip-rate objectives for international digital connections,[8] i.e., for calls established over two national digital networks linked by an international digital circuit. The provisional performance objectives for slip rates on a 64 kbit/s international connection are summarised in Table 10.2. These objectives, which relate to the total connection, have been apportioned to the component networks as shown in Table 10.3. As with most performance apportionments, the local network is allocated the majority of the impairment. This is because the high traffic loading and capacity of the international and trunk networks warrant higher expenditure to ensure high performance than the large, widespread but lower-loaded, local network.

The CCITT has also recommended that digital interconnection between national digital networks should be on a plesiochronous basis with a maximum rate of slip in any 64 kbit/s channel of not greater than 1 in every 70 days per digital international link.[9] Consequently, this recommendation requires all

Table 10.3 *Apportionment of slip performance*[8]

Network portion		Unacceptable	Satisfactory
		Less than:	Greater than:
International	(2%)	0.6 slips/h	4 slips/1000 h
Trunk	(9%)	2.7 slips/h	18 slips/1000 h
Local	(40%)	12 slips/h	80 slips/1000 h

Note: Between unacceptable and satisfactory is the range 'acceptable but degraded'.

clocks controlling network nodes with international links (i.e., international exchanges) to have a long-term frequency stability not worse than 1 in 10^{11}. Consideration is also being given to synchronous international digital networks.

10.3 Plesiochronous networks

The quality of the timing source within an exchange in a plesiochronous network is set by the objective slip rate. For example, for national digital telephony a slip rate of 500 per hour can be tolerated. This figure means that for an average call of 3 min, 25 slips occur, of which only one is heard as a click by both users. For oven-controlled quartz-crystal oscillators,this level of operation can be sustained by retuning each timing source about once or twice a year. However, if the CCITT standards for digital international connections apply, which are essentially based on data-network requirements, then similar clocks will require retuning up to 300 times per year! Alternatively, the standards can be met by using atomic clocks at each main exchange. This is not only an expensive option, but the short life of such timing sources creates serious operational difficulties.

Plesiochronous networks also require a mechanism to enable the exchange clocks to be retuned periodically to a reference. This means providing access at each exchange to an accurate reference sourse. There are three options for this, namely:

(i)   Provision of a mobile reference clock.
(ii)   Provision of a central reference clock in the network and arranging for the clock in each exchange to be returned progressively down the routeing hierarchy, working away from the reference. Clearly, the adjusting errors and drifts at intervening nodes limit the accuracy of this method.
(iii)   Provision of a reference signal to be distributed throughout the country, similar to the pilot frequencies provided for FDM transmission networks.

Plesiochronous operation may be favoured by administrations during the early implementation of digital switching in national networks, when there are few digital links, a sporadic disposition of exchanges and no integration with digital data services. However, as the network implementation progresses, the cost of synchronising tends to be comparable with plesiochronous operation for telephony-only applications. Where integration with digital data services or adherence to CCITT recommendations is required, then synchronous operation is the cheaper and more pracital solution.

10.4 Synchronous networks

With synchronous operation, just one or two high-quality (and cost) atomic reference clocks can control the frequency of the cheaper lower-quality clocks in the remainder of the network. The synchronisation mechanism ensures that the whole national network operates at a single frequency. This gives slip-free operation under normal conditions, unlike a plesiochronous network which will always experience some slips. The essential element of a synchronous network is a timing source, in each exchange, whose frequency can be altered electrically, e.g., voltage-controlled crystal oscillators within a phase-locked system.

Timing information is disseminated from the reference source to each exchange clock via a synchronisation-control network. This network comprises 'synchronisation (sync.) nodes' located at each exchange and 'sync. links' joining the sync. nodes. Where possibile, the sync. links are also used to carry normal traffic; for example, a 2 Mbit/s line system may operate as a sync. link and carry 30 speech channels. In some circumstances, 2 Mbit/s paths may have to be provided just for sync. links.

A sync. node determines its drift from the reference frequency by measuring the increase or decrease in phase difference between the exchange-frame start and that of a sync. link terminating on the exchange. As described at the beginning of this chapter, this phase difference is given by the aligner-fill level, and the change in the fill indicates how fast or slow the exchange is in relation to the reference frequency. For reliability, a sync. node may have several sync. links providing timing information. If all sync. links at a node are lost due to transmission or synchronisation-system failure, the timing source is left to run at its natural frequency. Clearly, there will be increasing incidence of slip at the exchange until the sync. links are restored and the exchange is brought back into synchronism with the network.

There are two broad categories of synchronous network operation: despotic and mutual.

10.4.1 Despotic synchronisation systems

The term 'despotic synchronisation' covers a number of techniques where a single reference clock dictates the frequency of all other clocks in the network. With despotic systems, at each sync. node changes in the aligner fill are used in a phase-lock method directly to control the frequency of the local timing source. Thus, the frequency of the exchange is constantly kept in synchronism with the reference. Where a number of sync. links are used, only one is operative at any time, all others being in standby mode. The direct-control nature of despotic synchronisation results in the nodes changing their frequency due to temperature variation of propagation time, as well as reacting to genuine timing differences between clocks. Despotic sync. networks may be of the master-slave or hierarchical master-slave type.

10.4.1.1 Master-slave synchronisation. With a master-slave configuration, just one reference clock directly controls the frequency of all nodes in the network via individual sync. links. This form of operation is relatively simple to implement compared to the mutual-synchronisation systems. However, the need for direct links from the reference to all nodes can be onerous in a large network. Clearly, the reference node must be made highly reliable because all nodes directly rely on it for their timing.

10.4.1.2 Hierarchical master-slave synchronisation. The hierarchical master-slave system is a more economical solution for large networks and is also more robust from a reliability point of view. The synchronisation control from the master node is disseminated via a hierarchical routeing through the network. Thus, the master node is linked only to a set of principal nodes which, in turn, feed timing information to lower-level nodes, and so on. This can take advantage of the star and mesh characteristics of a network, and the sync. links follow the traffic network routeing. In the event of a failure of the master (reference) node, one of the principal nodes can automatically assume command. However, this form of master-slave synchronisation method does suffer from the possibility of a 'rogue' node adversely controlling the frequency of all dependent nodes.

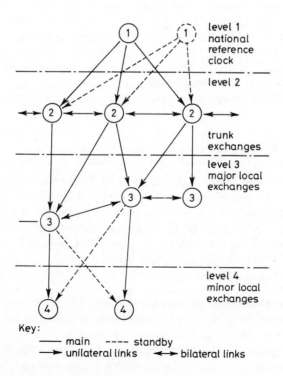

Fig. 10.5 *Typical four-level synchronisation hierarchy*

10.4.2 Mutual-synchronisation systems

In mutual-synchronisation systems, groups of sync. nodes compare their timing with each other and each determines the necessary corrections to its clock so that all nodes operate at the same frequency. This means that each sync. node has several sync. links, each providing timing information. An appropriate algorithm must be used by the sync. node's equipment to determine whether the rate of its clock should be advanced, retarded or left as it is in order to remain in synchronism. This information is used to generate periodic adjustments to the clock frequency in the form of a single quantum increase or decrease. Thus, mutual-synchronisation systems ensure that exchanges attain the same average frequency in the long term, although short-term deviations may occur. This is known as mesochronous working.

The sync. network is arranged as a hierarchy with one or more reference nodes at the top level. Sync. links between hierarchical levels are unilateral with control downwards, i.e. 'effective' at the lower-level node. Sync. links between nodes in the same hierarchical level are bilateral with control exerted ('effective') at both ends. Fig. 10.5 shows a typical four-level synchronisation hierarchy with unilateral and bilateral links. Mutual synchronisation systems may be single-ended or double-ended, as described below.

10.4.2.1 Single-ended mutual synchronisation. With single-ended control, changes to the clock frequency are governed by the average fill variations of the aligner buffers on all effective sync. links at the node. Fig. 10.6a shows a single-ended unilateral sync. link from exchange A to exchange B, where the change in aligner fill is detected by a phase comparator. Exchange B determines the necessary frequency shift, as a simple step increase or decrease (or neither) lasting for a few milliseconds, based on the change to the aligner fill. If there are several links into B, the required correction of B's clock must be based on a majority decision. In the case of a single-ended bilateral link, two such decision processes are undertaken, one at each end of the link, as shown in Fig. 10.6b.

The net result of this majority decision activity is that a mesh of sync. nodes mutually agree on a common-network frequency. Where the network-master node is operative, this will determine the network frequency of the mesh layer below. If there is no master node (as a result of failure or an incomplete network) the mesh layer will determine its own frequency. The disadvantage of single-ended working is its vulnerability to transmission-delay variation due to temperature changes on the links.

10.4.2.2 Double-ended mutual synchronisation. The double-ended mutual method overcomes the influence of temperature variations by subtracting the change of phase as determined at one end of the sync. link from the change of phase as determined at the other end of the link. The required interchange of phase information is achieved by signalling within the sync. link between the two sync. nodes, as described later. Fig. 10.6c shows the arrangements for

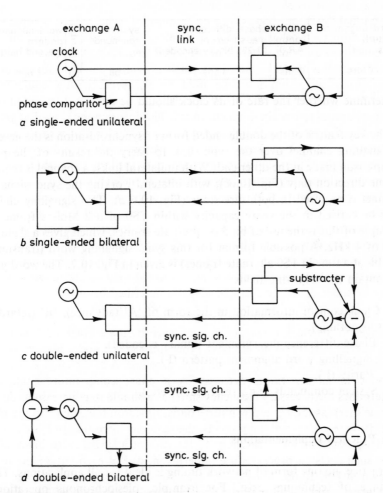

Fig. 10.6 *Mutual-synchronisation systems*

a unilateral link and Fig. 10.6*d* shows the arrangements for a bilateral link. The elimination of the effects of temperature variation is explained below.

The difference in phase bewteen exchange B and exchange A (as given by the link) determined at exchange B is $\Delta(\emptyset_B - \emptyset_A) + T$, where $(\emptyset_B - \emptyset_A)$ is the phase change due to the relative discrepancy in the two clocks and T is the change in phase due to temparature variation. The net phase change after substracting the results of a similar comparison made at exchange A is given by

$$\emptyset_{net} \Delta (\emptyset_B - \emptyset_A) + T - [\Delta (\emptyset_A - \emptyset_B) + T]$$
$$= 2 \Delta (\emptyset_B - \emptyset_A), \text{ since } \Delta (\emptyset_B - \emptyset_A) = - \Delta (\emptyset_A - \emptyset_B)$$

Thus, the temperature effect T is cancelled.

word–alignment pattern (first half)	timing control (A,R,O)	phase–difference measurement (0–255 binary encoded)	system commands	1–0 parity	word–alignment pattern (second half)
6 bits	2 bits	8 bits	4 bits	1 bit	6 bits

Fig. 10.7 *Synchronisation-signalling word format*

The key feature of the double-ended form of synchronisation is the need for a signalling channel over the sync. link to carry the results of the phase comparison made at the other end. With unilateral links, a channel is required in one direction only (Fig. 10.6c); with bilateral working the sync. signalling channel is required in both directions (Fig. 10.6 d). This signalling channel may be carried in the spare capacity within TS0 of a 2 Mbit/s frame. An example of this is the use of bit 5 of alternate frames, which gives a signalling rate of 4 kHz. A possible format for this sync. signalling word (transmitted one bit at a time in TS0 alternate frames) is given in Fig. 10.7. The word needs to contain the following fields:

(i)   Clock control information in the form of 'A' (advance), 'R' (retard) or 'O' (do nothing).
(ii)   Phase-difference measurement (binary encoded).
(iii)   Signalling word alignment pattern (F).
(iv)   Parity (P).
(v)   System commands (e.g., reset, fault, etc.).

10.5 Practical implementations

In practice, no one form of network timing is ideal for all applications. Thus, a range of techniques exist. For example, plesiochronous operation is employed on all digital international links, because this enables each national network to operate independently. Countries may thus develop their national networks without undue constraints. Plesiochronous operation between national networks requires all international exchanges to be timed from atomic-standard clocks.[9] Many networks, including those in the USA, Italy, Australia and the UK have a single operational national reference clock (NRC) at the top of the national synchronisation hierarchy. Alternatively, several countries, notably Canada, Japan, France and Switzerland, employ simple master-slave control from the NRC and between levels of the hierarchy.[1,10,11] Such networks have a simple topology with worker links to the parent node and standby links to other nodes. Master-slave networks do not require signalling between clocks, the timing being inherent in the digital streams themselves. On the other hand, mutual synchronisation systems, which have the complexity of inter-node signalling, overcome the vulnerability of the purely master-slave networks. The UK has implemented a double-ended mutual synchronisation system within its trunk network and the FRG has installed a single-ended mutual synchronisation system.[12]

However, the UK is using simple master-slave synchronisation with worker-standby links between the trunk or main local exchanges and their dependent minor local exchanges. This hybrid arrangement offers economy at the bottom end of the hierarchy, where the cost and complexity of mutual synchronisation cannot be warranted.

10.6 The exchange-clock unit

Now that the elements of digital-network timing have been described, the components of a clock unit within a digital SPC exchange can be considered.

As shown in Fig. 10.8, the exchange-clock unit contains a secured assembly of timing sources and waveform-generation equipment. In addition, the

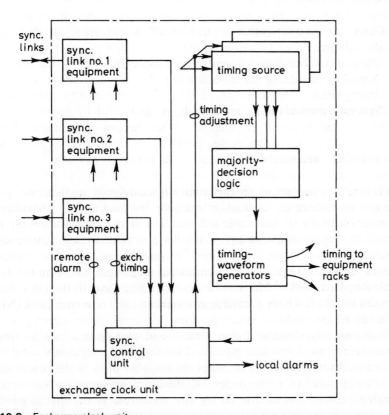

Fig. 10.8 *Exchange clock unit*

exchange-clock unit incorporates either network synchronisation equipment or, in the case of plesiochronous operation, appropriate slip-control and manual retuning equipment. Since failure of the clock unit will completely disable the exchange, the unit is designed to be highly secure, usually employing redundancy in all its key components.

For most national exchanges, temperature-controlled (oven) quartz oscillators are used as the timing sources. Usually, only the national-reference clocks use atomic-standard source.[13] In either case, the timing sources are duplicated or triplicated with their outputs processed by decision logic on a worker/standby or majority basis, respectively. The output from the decision logic is then used by the timing-waveform generators to derive the required timing waveforms for all the components of the exchange. These waveforms are typically at the sampling frequency of 8 kHz and the basic bit rate of 2048 kHz; there may also be a requirement for waveforms at 64 kHz and 4 or 8 kHz. The timing-source oscillator usually operates at the highest required frequency, or the highest common-multiple frequency, so that the individual timing waveforms may be derived by a process of frequency division from the single source.

Distribution of the various waveforms from the timing-waveform generators to the exchange equipment is by cable, usually triplicated for security. These cables take the timing to each rack of equipment, where backplane buses distribute the timing internally. The length of the cable runs are critical at the high frequencies used in digital SPC exchanges, since the electrical propagation times are of the same order as the digital-element timings. Distribution-timing errors may be eliminated by ensuring that all distribution cables are of the same length irrespective of the actual distance from the clock to the various equipment racks. This practice, of course, places an upper limit on the physical spread of an exchange unit (see Chapter 9).

With synchronous operation, a number of sync. links (2 Mbit/s or 1.5 Mbit/s paths) terminate on the exchange. Each link has terminating equipment that monitors the rate of change of the aligner fill resulting from the frequency mismatch between the exchange and that link (Fig. 10.8). The results from each sync. link terminating equipment are processed by a synchronisation-control unit.

In the case of master-slave configurations, only the output from one link is operative; all others are in standby mode. The output directly produces a correcting voltage which is applied to the electrically-alterable timing source of the exchange clock.

With mutual synchronous operation, the outputs from all sync. links are monitored by the sync. control unit. The necessary corrections are derived using a decision algorithm that takes the majority of unopposed indications from the outputs of the sync. links. The double-ended mutual synchronisation signalling channel is controlled by the respective sync. link termination equipment. Both this equipment and the sync. control unit need to monitor the activity on each of the sync. links terminating at the exchange. Excessive activity on any link indicates that the distant clock is free-running or else there is a fault in the system. In either case, the exchange-control system must be advised so that an alarm or fault report may be produced.

The implementation of a digital-network synchronisation system involves the planning of an auxiliary network of links (bilateral and unilateral) and nodes that must overlay the emerging traffic network of digital exchanges and routes. The principles involved in the introduction of such a network are described in Chapter 21.

10.7 References

1 CCITT (1984): Red Book, VIIIth Plenary Assembly, Malaga-Torremolinos, October 1984, Recommendations of the Series G: Volume III — Fascicle III.3, Rec. G703

2 SMITH, R. and MILLOTT, L.J. (1984): *Synchronization and Slip Performance in a Digital Network*, British Telecommunications Engineering, **3**

3 BOULTER, R.A. and BUNN, W. (1977): *Network Synchronization*, Post Office Electrical Engineers' Journal, **70**, p.21

4 ADAM, T.W. (1977): 'Data Services', in *Telecommunication Networks*, ed. Flood, J.E. (Peter Peregrinus), Chap.10

5 CCITT Rec. G821, op. cit.

6 McLINTOCK, R.W. (1984): *Error Performance Objectives for Digital Networks*, British Telecommunications Engineering, **3**

7 KEARSEY, B.N. and McLINTOCK, R.W. (1984): *Jitter in Ditigal Telecommunications Networks*, Ibid., **3**

8 CCITT Rec. G822, op. cit.

9 CCITT Rec. G811, op. cit.

10 ABATE, J.E., et al. (1980): *Switched Digital Network Synchronisation*, Telephony, **10**, pp.33-41

11 McDONALD, J. (1983): 'Digital Networks', in *Fundamentals of Digital Switching*, (Plenum Press), Chap.10

12 MITCHELL, P.A. and BOULTER, R.A. (1979): *Synchronization of the Digital Network in the United Kingdom*, Proc. IEEE International Conference on Communications, Boston MA, June 1979

13 BUNN, W. (1978): *National Synchronization Reference Clock*, Post Office Electrical Engineers' Journal, **71**, p.132

Part III
Exchange control by stored program

Chapter 11

Introduction to SPC

11.1 Overview

There are now a large number of SPC telephone exchange systems on the market and in operation throughout the world. They all possess individual design characteristics, both in switching equipment and in the manner of providing control. Although all modern systems are digital, early SPC exchange systems were based on electromechanical switching and analogue transmission. It should therefore be noted that SPC refers only to the method of control and does not imply anything about the type of switching, signalling or transmission equipment being controlled.

The control of a telephone exchange was introduced in Chapter 2. Now, Part III of the book:

(i)   Establishes how control is achieved by software.

(ii)   Expands on the earlier brief description of exchange control.

(iii)   Explains in detail the software techniques used in the processing of calls.

(iv)   Describes the data structures used to facilitate the software techniques.

(v)   Provides an introduction to the CCITT Specification and Description Language (SDL)[1-3] which is used in the specification and design of the call-processing aspects of SPC.

(vi)   Discusses the organisation of software within SPC systems, showing how this has evolved.

(vii)   Gives a briefing on maintenance functions, and how the integrity of an SPC system is maintained by hardware redundancy and software fault-tolerance routines.

In studying these principles and techniques, it should be remembered that there is never a single correct way of achieving a desired goal in software. Each designer is likely to devise some innovation and each programmer to introduce something of his own character in the implementation. The text is, therefore, not a definitive design manual. It is intended to introduce the reader to the functions carried out within the control system of a modern telephone exchange. It establishes why the functions must be performed; in many cases, particularly in call processing, it shows how they may be performed in software. Design is a discipline which demands pragmatism and inspiration, as well as knowledge. It is hoped that the following chapters will not only provide the knowledge required for readers to embark on SPC design, but also provoke thought to allow designers to bring independence and innovation to their work.

It is unlikely that any existing system will be found to duplicate everything stated in these chapters. The principles, however, will be found in all systems. While all modern SPC systems are modular and, in the main, adhere to accredited design principles, the scope of individual functions is defined differently in each. Not only is this the case in software modules but, where distributed processing is used, in hardware. Since a processor can only handle one instruction at a time, and good software is written as a collection of subroutines or task programs, it is a matter of design choice as to how functions are defined and distributed, particularly in call processing.

Traditionally, exchange control was almost exclusively concerned with call handling. One great advantage of SPC is that it allows the easy provision of other functions, such as checks on the integrity of the system, maintenance facilities for the switching, transmission, terminal and control equipment, and the handling of exchange and call data. Further significant advantages of SPC are

- space-saving,
- compatibility with digital switching and transmission equipment,
- it allows common-channel signalling,
- it facilitates the easy provision of customer facilities,
- electronic equipment is labour-saving,
- miniaturisation,
- greater traffic-handling ability and, thus, increased capacity,
- flexibility to expand systems,
- flexibility to introduce new features,
- flexibility to replace individual functions with equivalents or improvements in newer technology,
- flexibility to load and modify exchange data,
- the ability to implement dynamic network management,
- flexibility in the collection of data for management purposes, such as marketing, status monitoring, accounting and observing the effects of recent actions.

11.2 Call processing

Call processing is the main purpose of the control system of an exchange and, in all cases, the ability to handle calls is an important criterion in judging an exchange. A substantial proportion of the processing invested in a call is in setting it up. Thus, one test of an exchange is how many call attempts it is capable of handling in a given time. Once a call has been set up, the amount of further processing required to monitor its state and meter its duration is small in comparison. Not all call attempts result in a successful set-up, but there is very little difference in the amount of processing required between a call which receives an answer signal and one which receives a busy tone.

For successful processing of any call, the data concerned with its progress must be available within the control system. An important aspect of SPC is, therefore, the storage of information relating to exchange equipment, subscribers and calls, and the access to these data. The data structures used in SPC to increase efficiency of the storage and data access are described in Chapter 12 and are referred to throughout Part III. Readers who are not familiar with the principles of data structures may either refer to Chapter 12, as required, or use the chapter as a tutorial prior to studying the remainder of Part III.

Chapters 14 and 15 deal with the principles of call processing. Not only do they rely on data structures, but they also use the CCITT Specification and Description Language (SDL), as a tool in their explanations. A description of SDL and its use in call processing is therefore presented in Chapter 13. A study of this chapter will show that a call is not set up in a single process or by a single program. The call-processing software is designed to advance the call, at any given time, as far as the current information allows. The call is then left until further information, in the form of signals, arrives. Meanwhile, the processing power of the control system is used to handle other calls. In this way, the control system is time-shared by a number of calls. The maximum number of simultaneous call arrivals that can be handled depends on the processing power and, when this number is exceeded, congestion results. The call-processing software, therefore, consists of a number of programs, which are activated appropriately to control the transit of a call from one state to another.

In early SPC systems, a single processor, duplicated for security, provided all the processing power and contained all the software required for the control system. With the miniaturisation of computers, the development of communications software and the advance of software-engineering techniques, all modern SPC systems use distributed processors. Usually there is also a main central processor, consisting of a super-minicomputer (again, replicated for security). In this configuration, the distributed microprocessors are often referred to as 'remote processors' (RP).

The principles of call-processing software, described in Chapters 14 and 15, are general and do not depend on whether processing is centralised or distributed. This issue is discussed in Chapter 17.

11.3 Distribution of control

In the first SPC system, the No.1 ESS, all processing was carried out in a duplicated central processor.[4] Line conditions were detected and interpreted by the central processor. With the advent of microelectronics, microprocessors became available to assume some of the processing load, in the form of RPs. Whereas RPs may be used to execute any chosen functions, remotely from the central processor (CP), one of their earliest applications in SPC was to form the interface between line terminations and the CP, thus performing an initial analysis of received signals. In carrying out this function, their common task is to receive line signals, but the amount of processing which they carry out varies from system to system. They may count dial pulses and, having determined the decimal digits which they represent, send these, one by one, to the CP. On the other hand, they may store the digits and, when there are sufficient of these to make routeing possible, send them as a single message to the CP.

A further refinement in the use of RPs is illustrated in Fig. 11.1. Here the interfaces between the CP and other equipment are more standardised. They are all message-based. The CP is not called on to perform functions preliminary to routeing. RPs detect line conditions and receive all signalling information, and also preprocess all signals before passing them to the CP in message-based format.

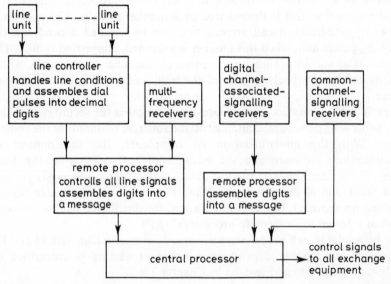

Fig. 11.1 *A system based on the principle of centralisation but with some distribution of processing*

The most recent SPC systems make use of distributed processing on a different basis. Although RPs still perform line scanning and digit reception, as in Fig. 11.1, distribution is more often on a functional than a hardware basis. RPs therefore need to interact with each other, as well as with the CP. Indeed, the use of communications software allows the interconnection of RPs, using local-area-network (LAN) principles. With every communication between software processes being in standardised message format, the processes themselves may be distributed over a number of RPs. This is described at greater length in Chapter 16.

One advantage of having RPs at the input is that, with many RPs functioning at the same time, parallel processing is achieved, and therefore greater throughput for the exchange. Although every call may eventually require the attention of the CP, a great deal of preliminary work is carried out by the RPs. A second advantage is improved reliability. With all processing carried out centrally, a processor failure may result in an exchange shut-down. In a distributed system, failure of certain processing functions results only in the loss of a number of terminations. Moreover, with evolution towards local autonomy within remote concentrator units, even a CP failure would not bring down the whole exchange. Reliability increases with equipment redundancy, and it is usual to build redundancy into RPs as well as the CP, so as to improve hardware fault tolerance.

A third advantage of distributed processing is that, with the processing load distributed, the complexity of the central processor system's software can be reduced. A reduction in complexity is reflected in improved reliability and maintainability. Further, with storage of data at the RPs which require it, rather than in a single central store, it can be arranged that no processor need access storage beyond its direct addressing capability. All data can be held in random-access memory (RAM). This eliminates the need for frequent disc access and increases the efficiency and speed of call processing. However, in case of memory corruption, copies of the software still need to be held in backing store for reloading to RAM storage. The backing store may consist of discs, or of bubble memory,[5] the access to which is considerably faster.

11.4 Software functions

In addition to call processing, which has already been discussed, a number of functions are essential. SPC software may, therefore, be considered in three further categories.

11.4.1 System software
The system software of an SPC system performs the usual operating-system functions. These include

- control of timing, to ensure that processes are executed, either at specified times (such as alarm calls) or periodically (such as scanning),

- control of the flow of processing, to ensure that high-priority processes are handled speedily, and that all processes are carried out,
- scheduling, to ensure that processes are executed according to predetermined order and timing,
- interrupt handling, to permit high-priority events to gain precedence under certain conditions,
- storage management, to control the storage and access of exchange, subscriber and call data,
- interprocess communications, to standardise and facilitate communication between software processes as well as between processors,
- input/output control, to allow communication between the system and the outside world,
- man-machine language interpretation, to define and provide the protocols of communication between terminals and the system.

In addition, the control of overload within the exchange demands an overload controller. In fact, every stage of call processing must contain checks for the symptoms of overload. When any of these are detected, the remedies taken may involve actions such as the reorganisation of processes by the executive program, or rescheduling of tasks by the scheduler. Whereas overload detection may occur within the application software, initiation of curative action may take place within the system software.

A discussion of the system software in a typical SPC system is given in Chapter 16.

11.4.2 Maintenance software
The control system must perform maintenance functions on both itself and the exchange switching equipment. Tasks include:

- Detection of faults. This is achieved by tests which are included in all types of applications software.
- Diagnostics to isolate faults. Diagnostic programs are activated as a result of the detection of faults.
- Disconnection or busying of faulty equipment. This is carried out after diagnostics have identified the faulty equipment.
- Correcting faults, when this is possible. This may take the form of reloading or reconfiguring software, or reconfiguring hardware so as to bring redundant equipment into service.
- Output of messages. Messages include error messages and information on actions taken.

Typically, maintenance routines account for a large proportion of the total software, although they demand only a fraction of the processing resource (see Fig. 17.7). However, if the availability of the system is to be guaranteed, the overhead of maintenance software cannot be avoided. Maintenance functions and the integrity of the system are discussed in some detail in Chapter 18.

11.4.3 Administrative software

For an administration to plan and manage its networks, data must be collected from exchanges. Uses of these data include planning, marketing, charging, accounting, performance monitoring, maintenance and network management.

In an SPC exchange, a call record is created in software for every call (see Chapter 14) and, when the call has been cleared, all or some of the data in the call record may be stored for later use. The data's collection, storage, processing and output require dedicated administrative software.

11.5 Presentation of information

The most onerous exchange functions are those of a local exchange, in which originating and terminting calls must be processed. For local-exchange call processing, a great deal of data relating to each local termination must be stored and accessed. The principles of control of transit exchanges, such as trunk and international exchanges, are implicit in those of a local exchange; however, they are simpler, because those exchanges do not (in most cases) have to cope directly with subscribers. There are differences in the requirements of the various types of exchange, but the basic principles are common, except that a local exchange has wider needs. For this reason, the following chapters mostly treat call processing from a local exchange's point of view, although it is also explained how transit calls are handled.

11.6 SPC bibliography

The purpose of this part of the book is to introduce principles, rather than to describe particular systems. In drafting the text, however, reference is made to a considerable number of sources, which present information on a number of systems, and the publication of which has spanned many years. Some of these papers are referenced in the following chapters.

While the more recent papers describe the trends in SPC and the architectures of current systems, they do not offer the detail which was provided in early works. Serious students of SPC are advised to refer to the special issues of the Bell System Technical Journal given in References 6 and 7. In studying these, allowance must be made for the fact that the switching systems being controlled were not digital but analogue electromechanical systems. This having been said, the control system software principles are well described. The two main conferences at which manufacturers present up-to-date information on their systems are the International Conference on Communications, held annually in the USA, and the International Switching Symposium (ISS), held usually every three years, in different countries. The proceedings of both these conferences are always informative. The last two

ISSs to be held are given in References 8 and 9. Moreover, many manufacturers and administrations produce journals, and, from time to time, these contain useful articles on SPC.

11.7 References

1 CCITT (1977): Orange Book, Vol. V1.4 — Recommendations Z101-Z103. ITU, Geneva

2 CCITT (1981): Yellow Book, Fascicle V1.7 — Recommendations Z101-Z104. ITU, Geneva

3 CCITT (1985): Red Book, Fascicle V1.10 — Recommendations Z100-Z104. ITU, Geneva

4 HARR, J.A., TAYLOR, F.F. and ULRICH, W. (1964): *Organisation of No.1 ESS Central Processor*, Bell System Technical Journal, **XLIII**

5 TROUGHTON, D.J. et al. (1985): *System X: The Processor Utility*, British Telecommunications Engineering, **3**

6 Bell System Technical Journal, Vol. XLIII, September 1964. Special issue on the No.1 Electronic Switching System (in 2 parts)

7 Bell System Technical Journal, Vol.48, October 1969. Special issue on No.2 ESS

8 ISS'84, Florence, May 1984

9 ISS'87, Phoenix, March 1987

Chapter 12

Data structures for SPC software

12.1 Introduction

There are many ways in which data are stored and accessed within a computer, and the method chosen for any application is determined by the type of data and the use to which it will be put. The configuration of data within the store is known as the data structure. If judiciously chosen, it can achieve economies of storage or run-time efficiency. Usually there is a tradeoff between these. The use of one data structure for a certain type of data may result in a saving in storage but the need to process more instructions to access and use it. Another data structure may achieve processing economies at run-time, at the expense of the need for increased storage.

Data structures, therefore, play a significant role in SPC design, and this book would be incomplete without a description of some of the more important ones. It should be noted, however, that this chapter does not present an exhaustive description of data structures, bur rather a discussion of those which are important in SPC. A very full description of data structures is provided in Reference 1, and the use of translation tables in SPC is described in Reference 2.

12.2 Pointers

Each word in computer memory has a unique number, or address, by which it can be accessed. In 'direct addressing', the central processing unit (CPU) uses this address to access the required location, either to insert information into it or to extract the information stored there. Often, however, it is advantageous to use 'indirect addressing', in which the CPU is given the address of an intermediate location. In this location is stored the address in which the information may be found, or from which the search for it may proceed. When a location contains the address of another location, it is said to contain a 'pointer' and to 'point' to that location. Fig. 12.1 shows location A containing a pointer to location B.

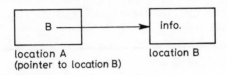

location A   location B
(pointer to location B)

Fig. 12.1   *Location A contains the address of location B*

12.3 Packing of data

Frequently an item of data requires storage of less than a word or a number of bits not equal to an integer number of words. Storage economies can therefore be achieved by 'packing' more than one item of data into a word, by allocating 'fields' within words to individual items of data. Care must then be taken that the field size accommodates the maximum size of a data item. A field may be as small as a single bit, and, in SPC, it often is; but it may also be larger than a single word. A number of fields, stored in consecutive words, may be considered to form a 'node' in storage, for example, as in Fig. 12.2.

It is also possible to design subfields within fields. For example, a circuit number may consist of the concatenation of a PCM-system number and a time-slot number. Each may need to be accessed individually, and be allocated to its own field. However, locating these two fields consecutively also allows the complete circuit number to be accessed as a single field. Fig. 12.2 shows the principle: fields 7 and 8 are accessible entities in their own right, but are also subfields of field 9.

The fields stored in the same node are normally related in some way. For instance, they may all contain items of data pertaining to a single subscriber, or an item of equipment. In an exchange, a data entity necessary for one subscriber or item of equipment must be replicated for every subscriber or similar item of equipment. The saving of a word in the design of a node could thus result in an economy of many thousands of words over a whole exchange. If this saving is repeated for all the various types of data in use in SPC, and

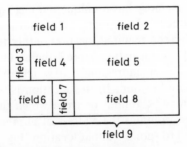

Fig. 12.2   *A node consisting of 3 words and 9 fields*

is multiplied by the number of exchanges in use in a network, considerable financial savings result. SPC therefore makes extensive use of data packing.

Packing and unpacking of data require more machine instructions than simply using a single word for each data item; hence they require more processing time. This is one instance of the tradeoff between storage and run-time efficiency. However, languages used for telecommunications programming are designed to handle fields of any length, including single bits, and this has lessened the effect. In addition, the distributed processors in modern control systems achieve a considerable degree of parallel processing and so offset the effects of run-time inefficiencies.

12.4 Sequential linear lists

A number of similar nodes stored sequentially form a sequential linear list. The list is 'linear' because each node has not more than one predecessor and not more than one successor. Sequential location implies that nodes which are adjacent in the list are adjacent in the store, and this provides a direct means of addressing any node. The address of a node is taken to be the address of the first word of that node. Given that all nodes of a particular type are of equal length (say, n words), the address of node $M = A + (M - 1)n$, where A

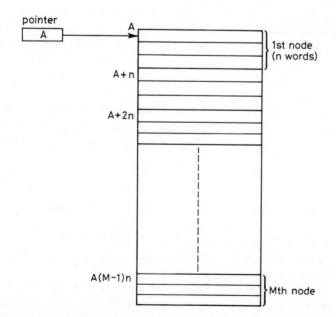

Fig. 12.3 *A sequential linear list of M nodes, each of N words, with the address (A) of the first node being held in a pointer*

is the address of the first word of the first node (see Fig. 12.3). A program written to process the list would, therefore, not need to contain information on the list itself, other than the structure of a node and a means of accessing the first node. Such access can be achieved by use of a pointer, as in Fig. 12.3, and it can be seen that the programmer need only make permanent the address of the pointer; the location of the list can be determined when the data is loaded, at which time the address of the first node is inserted into the pointer. By this use of indirect addressing, programs can be written independently of their data.

In general, lists and other data structures may be formed of nodes of variable length. However, as these are not prevalent in SPC, the techniques for handling them will not be discussed here.

When a sequential linear list is created, a number of sequential locations, representing a chosen number of nodes, is set aside for its use. Data can be inserted into a node when required, and the empty nodes comprise the space which is available for the storage of further data. When the list is full, 'overflow' occurs, and further data cannot be accommodated. In many instances in SPC, lists of various types are used during call processing. The dimensioning of such lists is crucial to ensure that the call-handling design objective is met. Overflow of a list in this context implies overload in the system and congestion in the exchange.

12.4.1 Using the sequential linear list as a queue

A queue implies that items are inserted at one end and removed from the other, i.e. it operates on a first-in-first-out principle. Such a mechanism is sometimes required in SPC and must therefore be provided in software. One data structure which can provide it is the sequential linear list. Programs must be written to perform the necessary functions and it must be decided how many nodes are required to comprise the list. Then, an area of storage consisting of the equivalent number of sequential locations is set aside. Dimensioning of a queue is important; too many nodes waste storage, but too few means that the queue overflows and data cannot be inserted. In SPC, this may mean that a call cannot be set up. Fig. 12.4a shows (horizontally) the sequential list of nodes available for use in the queue. The limits of the queue are marked by the pointers FIRST and LAST, which point to the first and last nodes, respectively. In order to access the queue to insert or remove items, there are pointers F and R to the front and rear of the queue of filled nodes. In Fig. 12.4a, in which all nodes are empty, R points to the first free node (i.e. FIRST), but, since there is no entry in the queue, F is given a null value and therefore does not point to any part of the queue.

Figs. 12.4b, c and d show the queue after one and four insertions and then two deletions. Fig. 12.4e shows that, as the process progresses, insertions can reach LAST while there is available space in the queue, and f shows how using the queue circularly ensures that this space is not wasted. Then, with further

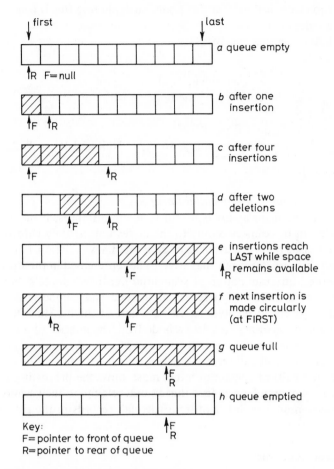

Fig. 12.4 *Using a sequential linear list as a queue*

insertions and deletions, F and R chase each other around the circle. If insertions consistently exceed deletions, R may catch up with F. In this case, the queue is full, as in Fig. 12.4*g*. If the queue is then emptied, F and R remain pointing to the same node, as in Fig. 12.4*h*. F is then given a null value and R is made to point to the first node, i.e. FIRST.

The algorithms to insert and extract nodes can be summarised as follows. (The symbol ': =' is used, as in Algol, to mean 'becomes equal to'.)

Insert

If R = F, then queue full. Cannot insert.

Else R points to first free node. So, insert item into R.

Make R point to next node. Thus, if each node contains N words, R : = R + N.

(No need to check if R = F at this point, since, before this is necessary, there may be a deletion.)

If R = LAST, then R := FIRST.

Delete

   If F = Null value, then queue is already empty.

   Else, extract the contents of the node to which F points.

   F := F + N.

   If F = LAST, then F := FIRST.

   If F = R, then queue is empty, so F := Null and R := FIRST.

12.4.2 *Limitations of a sequential linear list*

While a sequential linear list is useful in many instances, it has limitations. Each node can be connected only to the nodes immediately before and after it. This has disadvantages, as items of data often have relationships with other items, and these relationships may need to be represented in the storage of the data. Further, an examination of sequential linear lists reveals that insertion or deletion of a node anywhere but at the end is a cumbersome and time-consuming process. It involves rewriting a part of the list. In SPC, there are times when it is necessary to delete a node from the interior of a list; then the overheads, if a sequential linear list were used, would be unacceptably high. For example, a list of call records (each of which contains all the information pertinent to a call) of calls in progress must allow the processing of any call at any time: it cannot be arranged that signals relating to the various calls will arrive in accordance with the order of the call records in the list.

12.5 Linear linked lists

To overcome the inadequacies of a sequential linear list, a method of storage known as 'linked lists' can be used. In this, an extra field, known as a 'Link Field', is included in each node. This link, or pointer, is made to point to the next node in the list. The removal of a node, or the reordering of the list, then only involves the changing of the value of one or more pointers, and not the rewriting of the remainder of the list.

   As with a sequential linear list, the nodes of a linked list must be preallocated to the use of the list, although the initial allocation may later be incremented. The available space is usually provided in the form of sequential storage, but this is not essential. Indeed, an advantage of linking is that, if the list's total available space needs to be increased, achieving this does not depend on the availability of sequential storage at the end of the existing list. The unused nodes are linked together and, when the list of nodes in use is empty, as at the initialisation of the storage, the resulting list of available nodes

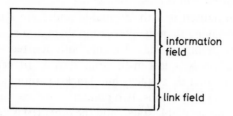

Fig. 12.5 *A node of a linked list seen as two fields*

consists of all the nodes. Again, a pointer to the top of the list is necessary if the list is to be found.

A node may contain a number of fields, each storing an individual item of data. However, for the purpose of describing linked-list operations, a node will be considered to consist of two fields: one composite 'information field' and one 'link field' (see Fig. 12.5). In the descriptions of operations on linked lists which follow, the symbol 'LINK (I)' is used. This refers to the contents of the link field of a node designated, or pointed to, by I. Thus, A := LINK (A) means that the pointer A acquires the value which is stored in the link field of the node which had been pointed to by A (see, for example, Fig. 12.6). This is an example of indirect addressing, and the following descriptions will expand on it.

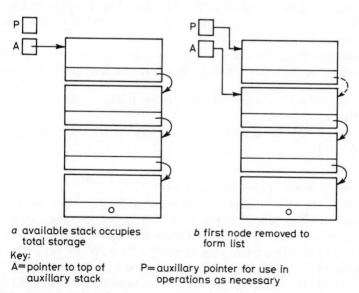

a available stack occupies   *b* first node removed to
   total storage   form list

Key:
A = pointer to top of   P = auxillary pointer for use in
    auxillary stack   operations as necessary

Fig. 12.6 *A stack of linked nodes*

12.5.1 Stack operations on a linear linked list

When storage is initialised and the available nodes are linked together, the simplest way of managing them is by treating them as a stack (The 'available stack'). A stack, by definition, has elements both inserted and deleted at one end only. It therefore is a last-in-first-out device. In general, the list of available nodes is treated as a stack, and stack operations are important in SPC. To perform stack operations, two pointers are necessary: one (A) always points to the top of the available stack, and the other (P) is kept as an auxiliary pointer for use when required. These pointers are used in the programs written to access the available stack.

Removing a node from this stack is carried out by a number of stack operations, many of which are pointer manipulations. The stack, before and after the operations, is shown in Fig. 12.6*a* and *b* respectively. The stack operations, in pseudocode, are given below.

> Find whether there is a free node by checking the value of the pointer of the node to which A points:
> If A has a null value, there is no free node: the list has overflowed.
> Else, there is a free node. Then
> Make P point to top (i.e. P := A)
> Make A point to second node (i.e. A := LINK (A)).

Notice that, if the pointer manipulations are carried out in the wrong order, the address of the first node is lost forever. However, if the operations are done correctly, the first node still points to the second; but it has been removed from the stack and can be added to the main list of nodes in use by using the value of the pointer P. If it is attached at the rear of the list, its link field will be given the null value (i.e. LINK (P) := Null).

Returning a node to the stack assumes that there is a pointer (P) pointing to the node to be returned. The process is then the reverse of the above:

> Make the link of the node point to the top of the stack
> (i.e. LINK (P) := A).
> This node is now the top of the stack, so A must point to it
> (i.e. A := P).

12.5.2 Using a linear linked list as a queue

Recall that the queuing principle demands that nodes are processed in a strict order, such that the first in is the first to be processed. Processing must therefore always commence at one end while nodes are added at the other. It is clear that, to facilitate processing, there must be a pointer (F) to the front of the queue and link pointers must point backwards, as shown in Fig. 12.7*a*, which represents the queue horizontally. Also, to avoid traversing the queue each time a node is to be added, there must be another pointer (R) at the rear.

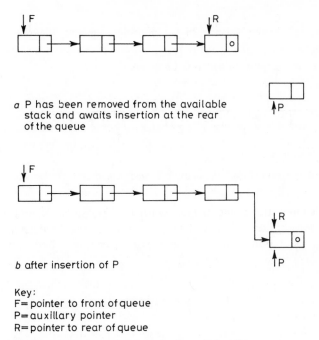

a P has been removed from the available
stack and awaits insertion at the rear
of the queue

b after insertion of P

Key:
F= pointer to front of queue
P= auxillary pointer
R= pointer to rear of queue

Fig. 12.7 *Insertion of a node at the rear of a linear linked list*

Assuming that the node to be inserted has already been removed from the available stack (as above) and is pointed to by P, insertion consists of:

The link field of the node, pointed to by R, must point to the node pointed to by P (i.e. LINK (R) : = P)

And R must now point to the new rear of the queue (i.e. R : = P).

If, during processing, the first node is always processed and then removed from the queue, deletion consists of simple stack operations, again using P as an auxiliary pointer:

P := F
F := LINK (F)

P now points to the node removed from the queue, which is returned to the available stack as described earlier.

12.5.3 Removing a node from the middle of a linear linked list
In SPC, there are occasions when the first node in a queue may be examined but not processed. Then, the first node must remain in its place while another node, somewhere inside the queue, is processed and removed. Clearly, to relink the list, it must be possible to back-up in the queue, and, to achieve this, there must be a pointer to the node preceding the one to be deleted. Processing

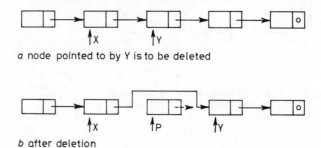

a node pointed to by Y is to be deleted

b after deletion

Fig. 12.8 *Deleting a node from the body of a linear linked list*

therefore includes traversing the list with two pointers (X and Y) always pointing to two adjacent nodes. Then, when a node to be deleted is found, there is always a pointer to its predecessor (see Fig. 12.8). The deletion operations then take the following form:

LINK (X) : = LINK (Y)
Return Y to available stack, as already described.
Y : = LINK (X)

12.5.4 Insertion in the middle of a linked linear list

It is also sometimes necessary to insert a node in the body of a linked list. An example of this is the placing of timers in chronological order (see the end of this chapter). Once again it is necessary to traverse the list, using two pointers, X and Y, as in Fig. 12.9. This ensures that when the appropriate slot is found for the new node (after X and before Y) there is a pointer to the node which

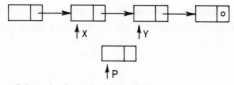

a P is to be inserted into list

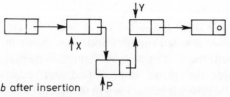

b after insertion

Fig. 12.9 *Inserting a node into the body of a linear linked list*

will precede the new node. This allows insertion, as seen in the following operations:

Get the new node, pointed to by P, from the available stack.
LINK (P) : = LINK (X)
LINK (X) : = P.

12.6 Translation tables

The need to translate information from one form to another is common. In software, it is achieved by storing the information in translation or 'look-up' tables and using known information as a reference or 'key' to the location in which the rquired information is stored. In some cases, the key must also be stored. The implementation of tables in software is usually by one of the methods of storage already described (i.e. sequential linear storage or linked lists), depending on the flexibility required. The word 'table' is used because the data being stored are usually presented on paper in tabular form. The searching techniques used for accessing the tables take two main forms: scanning and key transformations.

12.6.1 Scanning
The key is stored as well as the information. When a search is made, each stored key is compared in turn with the search key until a match is found. The required information is then extracted. For example, a table may contain subscribers' class-of-service (COS) records, each consisting of (say) four words, one each for directory number, equipment number, semipermanent data and transient data (see Fig. 12.10). A program to scan the table would contain a pointer, P, to the first word of the table. If the directory number were used as the key, then the directory number being sought would be compared in turn with the contents of addresses P, $P+4$, $P+8$,.. $P+4i$,.. $P+4N$, until a match were found. Then, if the equipment number were required, it would be found in location $P+4i+1$, the semipermanent data in location $P+4i+2$, and the transient data in location $P+4i+3$ (where the value of i is between zero and N).

Because a comparison is made in every node until a match is found, the data do not need to be stored in any particular order. New data may be added at the end of the table and no gaps need be left in the table. A table to be accessed by scanning is therefore very economical on storage, but the scanning process is expensive in time. However, although storage is economical, in that the minimum number of nodes can be stored, the keys must be stored as well as the other information, at the expense of an extra word per node. This can be used to advantage. In the example given, if the equipment number, instead of the directory number, is known, scanning can still be used to achieve a reverse

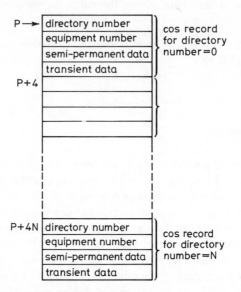

Fig. 12.10   *A table in which the key is stored*

translation. In this case (again using Fig. 12.10), comparisons would be made with locations $P+1$, $P+5$, $P+9$,.. $P+1+4i$,.. $P+1+4N$.

12.6.2 Key transformations

If a direct relationship exists between the key and the location in the table of the information to be retrieved (the object parameter), the latter can be accessed by direct addressing. Then scanning is unnecessary and it is also unnecessary to store the key. For example, a table of COS records, stored in directory-number order with no gaps (see Fig. 12.11), may be addressed directly, using the directory number as a key. With the first location of the table having address P, the equipment number corresponding to directory number N must be stored in location $P+3N$, given that directory numbers start at zero. Similarly, the corresponding semipermanent and transient data are stored in locations $P+3N+1$ and $P+3N+2$, respectively. These formulae are 'key transformations', by which the basic key (directory number) is transformed into an address at which the object of the search is to be found.

The advantages of key transformation (also referred to as 'indexing') are that each node occupies a minimum of storage, because keys do not need to be stored, and accessing the table is direct and fast. The penalty, however, is that nodes must be in their proper places; therefore, if data are sparse, there are gaps in the table and thus a waste of storage. For instance, if an exchange had the capacity for ten thousand subscribers (0000–9999), there would have to be 10 000 nodes (or 30 000 words) in the table for the example given. If only

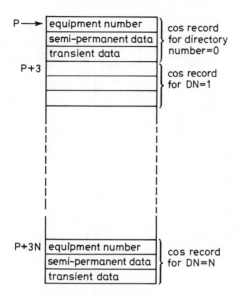

Fig. 12.11   *A table in which the key is not stored*

1000 numbers were allocated, a table of 10 000 nodes would still be required, but only 3 000 words would be used, with 27 000 being empty.

When almost the total possible data population is in use, it is worthwhile to store the data in a single table (or list) and access it by key transformations. The few gaps in the table do not constitute a major wastage, and there is a significant run-time advantage. On the other hand, when there are very few data, scanning is worthwhile, for it allows compact storage with no concern for order. Then, the time penalty in scanning is not great.

12.6.3 Two-stage tables

It was seen above that linear storage is satisfactory for two conditions: (*a*) nearly the full complement of nodes in use; (*b*) very few nodes in use. However, in many cases neither of these extreme criteria apply. Then, it is advantageous to arrange storage in a tree structure, in which a node may have any number of successors but no more than one predecessor. In SPC it is usually implemented in the form of two- or even three-stage tables. In two-stage tables, the information (COS records, in these examples) is stored in the secondary tables, while the primary tables contain pointers to the appropriate secondary tables.

There are many ways of realising this, depending on the spread of data, and searching the tables can be by combinations of scanning and key transformations. One example is shown in Fig. 12.12. If the directory numbers in use in a 10 000 line exchange only fall within (say) three A-digit ranges, time

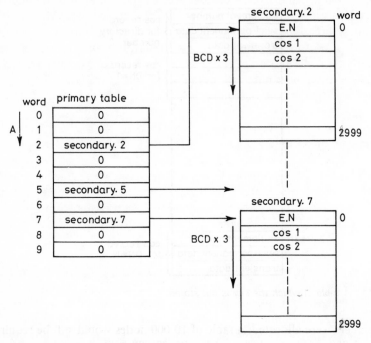

Fig. 12.12 *Two-stage table access: primary by A; secondary by BCD $\times$ 3*

and storage can be saved by locating the COS records in tables corresponding to the BCD-digits. These form the secondary tables and are directly addressed by a key transformation consisting of BCD $\times$ 3. The primary table is addressed by the A-digit and its entries consist of pointers to the appropriate secondary tables. Storage is thus reduced from a minimum of 10 000 nodes to 3 000. The time penalty is merely accessing one extra address. The method offers considerable flexibility: when the exchange is expanded, the new COS records can be added simply by allocating a secondary table of storage for each new A-digit, and inserting pointers to them in the appropriate locations in the primary table.

12.6.4 Class-of-service records

The above examples have introduced the data structures and accessing techniques used for translation tables in SPC. However, it should be said that, although key transformations do not require the key to be stored, there are times when it is necessary to store it for other purposes.

When a subscriber lifts the handset, the resulting signal is detected at the terminating equipment in the exchange. The subscriber's class-of-service (COS) record must, therefore, be accessed with the equipment number as the key, and the record must contain the subscriber's directory number. However, when an incoming call arrives, addressed to that subscriber, it is the directory number which must be the key, and the equipment number which must be looked up in order for the call to be connected to the correct terminating equipment. Therefore, if there is to be a single COS record for each subscriber's termination on the exchange, it must contain both the directory and exchange-equipment numbers, as in Fig. 12.10, regardless of the methods of storage and search being used.

12.7 Presentation of data

This chapter has shown methods of storing data which facilitate either efficiency of access or economy of storage. The means of storage are also based on the relationships between data entities, as was shown in the use of pointers which lead from chosen entities to related ones. Such representations in memory of data relationships form the principles of databases and, in particular, relational databases.

However sophisticated the actual storage of exchange, subscriber and call data, there must be some means of representing them in a two-dimensional form, which nonsoftware engineers can easily comprehend. It is, therefore, usually presented in tabular form, which allows the layman to examine the data independently of their method of storage. Certain terms are used in connection with such presentation of data.[3] These are explained with reference to Fig. 12.13.

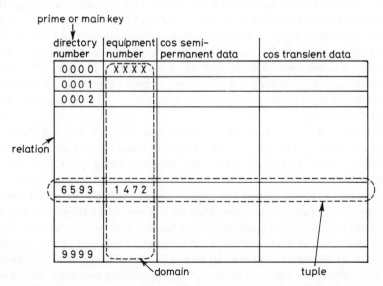

Fig. 12.13 *Normalised or two-dimensional representation of a relational database*

The Figure shows a tabular presentation of the sequential storage illustrated in Fig. 12.10. In the terms of this 'normalised' (i.e. two-dimensional) representation, the table itself is referred to as a 'relation'. The implication is that a relational database consists of a number of relations. The key most frequently used in searching the relation is shown in the table, whether or not it is actually in memory. It is known as the 'prime key' or 'main key' and defines the data entity in question. All the attributes (which are themselves data items) relating to an entity are represented horizontally as a row of the table, or a 'tuple' of the relation. N items in the row comprise an 'N-tuple'. Finally, all the items which fall under the heading of any one attribute make up a column of the table, and this is known as a 'domain'.

This explanation will not aid the understanding of the storage or handling of data in software, but it explains terms which are in use both in computer science and, increasingly, by exchange manufacturers.

12.8 Software timers

In call processing, whenever a call is left in a steady state to await an input signal from a subscriber, a timer is set. Therefore, a time-out is a possible input signal at that steady state. If the time-out occurs before the subscriber's action, the call is put into the 'parked' state; this avoids exchange equipment and common storage being allocated indefinitely to a line on which there is no call activity. It should be mentioned that the terms 'watchdog' and 'watchdog timer' are often used in the same sense that 'timer' is here.

Timers are also used for internal processes. An example is timing the durations, or defining the limits, of the 'make' and 'break' conditions during dial-pulse signalling, so as to distinguish between pulse signals and interdigital pauses. In this case, timing is required to an accuracy of about 10 ms, whereas, in the case of subscribers' actions, timing is usually in the order of seconds. The same principles of timing are applicable in both cases, and the accuracy required simply determines the frequency with which the timers are checked within the software.

A timer may consist of a node of two fields, one for the timing and the other to contain a pointer to the call record of the call being timed (see Fig. 12.14). When timing must be initiated, the task program being processed calls a subroutine which exists to set a timer of the appropriate accuracy. A parameter, passed to the subroutine, defines the time which must elapse before a time-out should occur. The subroutine then finds a free timer and sets its pointer field to address the appropriate call record. It then sets the timer field in one of two ways, as described below, in Sections 12.8.2. and 12.8.3. Finally, it inserts, in a field reserved in the call record for the purpose, a pointer to the timer. This allows the timer to be found and the timing cancelled if another input signal occurs within the specified time.

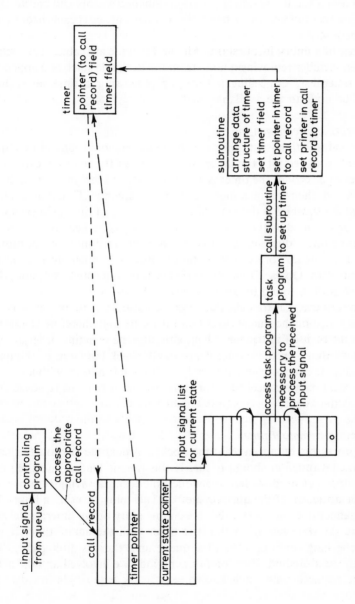

Fig. 12.14 *Setting a software timer*

12.8.1 Implementation

Timers may be provided as either a sequential linear list or a linked list. In the former case, each timer needs to have a flag bit which is set when the timer is put into use. Then the scanning program only decrements and checks those timers whose flag bits are set. When a timer times-out, or is cancelled, its flag bit must be reset.

In the case of a linked list, timers not in use form the available stack. Setting a timer then includes removing a node from the available stack and appending it to the timer list. Cancelling a timer or processing a time-out includes returning the node to the available stack.

12.8.2 Timing by decrementing

In this method, the time-span, over which the timer should run before a time-out is given, is loaded into the timer. If scanning of the timers is carried out in multiples of a second, the timer is set in seconds; if scanning is in multiples of a millisecond, then the timer must be set in milliseconds. Timers are scanned at a set frequency, which is determined by the category of timer. On each scan, a timer is decremented by the time between scans, and checked. When it equals zero, the time-out is due, and an input signal is inserted into the appropriate call-processing queue. For example, a timer which is scanned once per second may be set to 10 s. On each scan, this number is decremented by 1, and, after the tenth scan, it is found to equal zero.

A useful refinement is to make the timer list a queue. A timer, on being set, is given its place in the queue according to its setting; therefore it must be inserted in the body of the queue. Then, although the scanning program must decrement all timers in the queue, it need only check the front of the queue for zero value. It continues to check only until it finds a timer with a nonzero value. When a timer is cancelled before it times-out, it must be removed from the body of the queue.

12.8.3 Timing by actual time

Since the processor contains a real-time clock, it is possible to know the actual time at which a time-out should occur. If this time is loaded into the timer, a scanner does not need to decrement the timer, and a saving in processing time can be achieved. If the queuing method is employed, using a linked list, then the scanner does not even need to examine every active timer; it only has to compare the time setting of the first timer with the actual time. If they coincide, checking continues down the queue until there is discrepancy. This is a very efficient method when long time periods are involved and the actual time of the required time-out is known, e.g. in the case of alarm calls.

12.8.4 Timing by subtraction

In the above descriptions, a timer carries out a process which terminates in a time-out signal. The signal is then used within the software to initiate a process. It should also be mentioned, for completeness, that the timing of some events, like the duration of a call, may be carried out by subtracting the actual start time from the actual termination time of the event. In such cases, the start and termination times are stored in the call record.

12.9 References

1 KNUTH, D.E. (1973): *The Art of Computer Programming*, Vol.1, 'Fundamental Algorithms', (Addison-Wesley), 2nd ed., Chap.2

2 HILLS, M.T. and KANO, S. (1976):*Programming Electronic Switching Systems*, (Peter Peregrinus Ltd)

3 MARTIN, J. (1976): *Principles of Database Management*, (Prentice Hall)

Chapter 13

The specification and description language

13.1 Steady states of a call

System design depends on a specification of requirements which should, ideally, be correct and complete. Software, being notoriously difficult to correct, or even to understand, needs to be designed with particular care, and, thus, accurately specified. So, what are the software requirements for call processing?

An SPC system must process a large number of calls simultaneously, and is therefore time-shared between them. Although no SPC system relies on a single processor, the principles of call processing are best described with respect to a single call and a single processor. Since it can only handle one call at a time, a processor can only give the impression of simultaneous processing by its speed. All calls awaiting attention must be left dormant until the processor returns to them. This is facilitated by the fact that a call is a steady-state process: it remains in a steady state until an event occurs to initiate its transition to another steady state. Thus, the processor does not need to leave a call in an arbitrary condition determined only by the time that an interrupt occurs — as is the case in, say, batch processing. It can leave the call in a logically determined steady state while processing other calls.

Specification of call processing can therefore take the form of the specification of all the possible steady states of a call and the transitions between them. Although there is a large number of states, the task is eased by the fact that, while a call is in any one state, there is a limited number of events that can occur and, therefore, a limited number of other states to which the call can progress. For example, a call in the state of 'waiting for the first digit' from subscriber A, whose handset has been lifted, cannot jump arbitrarily to the 'conversational' state. The only possible events that can occur are:

(i)   Subscriber A replaces the handset, returning the call to the idle state.
(ii)   Subscriber A dials the first digit, initiating progress of the call to the state of waiting for the second digit.

(iii)   Subscriber A does nothing and the control system times-out and places A's line in the 'parked' state. This is a means of releasing exchange equipment for use with other calls when a subscriber has not progressed a call within a given time. The line is left with no tone, because no equipment is connected to it. Eventually, replacing the handset initiates the return of the line from the parked to the idle state. Another call may then be initiated by lifting the handset once more.

Listing the states and the events which lead from one state to another is a logical process and straightforward for a telecommunications engineer, but the total specification becomes extremely complex. It requires a graphical language, rather than a purely textual one, if it is to be unambiguous. Further advantages of a graphical language are:

(i)   It is easy to read and, therefore, can readily be checked for completeness and correctness.
(ii)   It is easily expandable.
(iii)   It can be implemented directly in software, so that, if the specification is correct, there is a high probability of the software being correct.

The Japanese were the first to use the state-transition diagram as a call-processing specification language.[1] The principle was quickly adopted by other manufacturers of SPC systems and, in the early 1970s, the CCITT started work on the development of a standard specification language. This resulted in recommendations for the CCITT Specification and Description Language (SDL) in the Orange Book,[2] and further explanations and recommendations for enhancements, four years later, in the Yellow Book.[3] Most recently it has been further defined in the Blue Book.[4] It is also summarised in Reference 5. SDL is now widely used in specifying call-processing requirements for SPC systems. Because it will be used in subsequent chapters, to show the requirements and implementation of call processing, its principles are outlined in the following Sections.

13.2 The recommended symbols and their definitions

The recommended symbols are shown in Fig.13.1. The CCITT has specified drawing conventions[3] and issued a stencil to facilitate their creation. The following definitions, given by the CCITT, govern the use of the symbols:

State: a condition in which the action of a process is suspended while awaiting an input.
Input: an incoming signal which is recognised by a process.
Output: an action within a transition which generates a signal which, in turn, acts as an input elsewhere.
Decision: an action within a transition which asks a question whose answer can

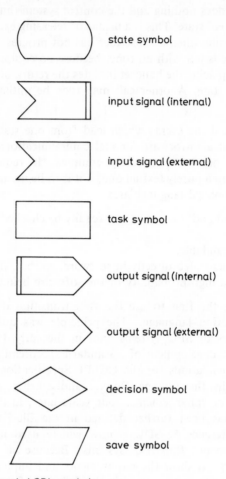

state symbol

input signal (internal)

input signal (external)

task symbol

output signal (internal)

output signal (external)

decision symbol

save symbol

Fig. 13.1 *Recommended SDL symbols*

be obtained at that instant and which, as a result of the answer, chooses one of several paths to continue the transition.

Task: any action within a transition which is neither a decision nor an output.

Save: the postponement of recognition of a signal when a process is in a state in which recognition of that signal does not occur. The 'save' symbol has been useful in the specification of signalling systems (e.g. CCITT No.7) but is not usually applicable in call processing.

It is also important, especially in order to understand the distinction between internal and external signals, to put these definitions into the context of modular software design. The CCITT defines a 'functional block' as 'an object of manageable size and relevant internal relationship, containing one or more processes' (see Fig. 13.2).

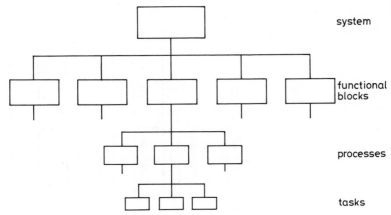

Fig. 13.2 *Functional decomposition of software*

A process is defined as performing a logic function that requires a series of information items to proceed, where these items become available at different times. It is either in a state awaiting an input, or in a transition.

A signal is defined as a flow of data conveying information to a process. If the information flow is from a process in one functional block to a process in another functional block, it is an external signal. If the flow is between processes within the same functional block, it is an internal signal.

From the latter definitions, it is dependent on the choice of functional block boundaries whether signals are internal or external. The same basic signals may be given different designations in different designs. It should be noted, however, that 'external' does not necessarily mean external to the exchange, although subscribers' signals are external both to the functional block which handles them and to the exchange.

13.3 The use of SDL

As an example of the use of SDL, Fig. 13.3 shows a high-level specification of a local call. In the main, the processes are seen from the A-subscriber's point of view and the exchange operations are not specified. Further examples of the use of SDL are given in the descriptions of various aspects of call processing in Chapter 15 and in Fig. 14.8. This Figure specifies a local call, but only as far as the state of 'awaiting the first digit'. The specification is at a lower level than that of the example of Fig. 13.3. It specifies the exchange operations as well as the subscriber's signals. It is closer to a design which an implementer would use. For example, it recognises that, when the 'A off-hook' signal is received, checks on the A-subscriber's line have to be carried

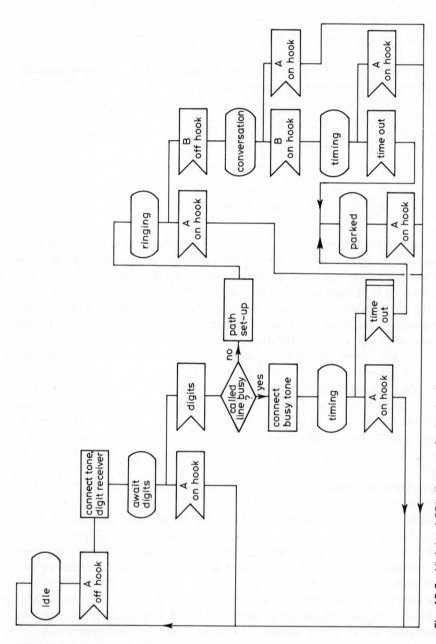

Fig. 13.3 *High-level SDL diagram of a local call*

out; also that, in attempting to connect a signal receiver to the line, there is a possibility of blocking.

13.4 Advantages of SDL

SDL is a pictorial language which is easily understood. It is, therefore, suitable as a high-level overview specification for management, as well as a detailed specification. SDL represents the paths that a call can take during its existence. Each telephonic steady state is unambiguously defined, and the input signals which are possible while a call is in a given state can be determined accurately.

The functional specification is based on the logical progress of a call, and is thus stated in terms of telephonic events. It is independent of the controlling processor or of any specific computer language.

An SDL specification can be drawn by considering a single call. SDL is not explicitly concerned with the other calls in progress, although, implicitly, it makes allowance for them. The design of complex software is therefore simplified.

SDL is easily extendable. New facilities, and thus new states and signals, can be introduced without major redesign of the existing diagram. Errors found can also be modified easily.

The understandability and unambiguous brevity of the representation make SDL suitable both as a design document and for direct translation into a computer implementation of call processing.

13.5 Direct interpretation of SDL into software

Whereas the majority of SPC software is written in procedural programs in a high-level language, the call-processing software is usually an exception. As described, a call remains in a steady state until an input signal causes a transition to another steady state. The SDL defines this exactly and specifies the tasks which must be carried out by the control system in order to facilitate the transition. A direct interpretation of the SDL can be achieved by the use of the data structures of linked lists and translation tables, which were described in Chapter 12. In effect, the SDL is stored directly, in memory, in the form of software data structures. This carries the advantages of SDL into the implementation. In particular, the facility to effect change or expansion without disturbing the design structure makes it less likely that faults will be introduced during correction or extension.

When the call-processing software is written in the form of a direct translation of the SDL diagram, a single program, or 'interpreter' may be written to access the tables and lists, and to progress a call by interpreting the data within them. In this sense, the call-processing software is said to be 'data-

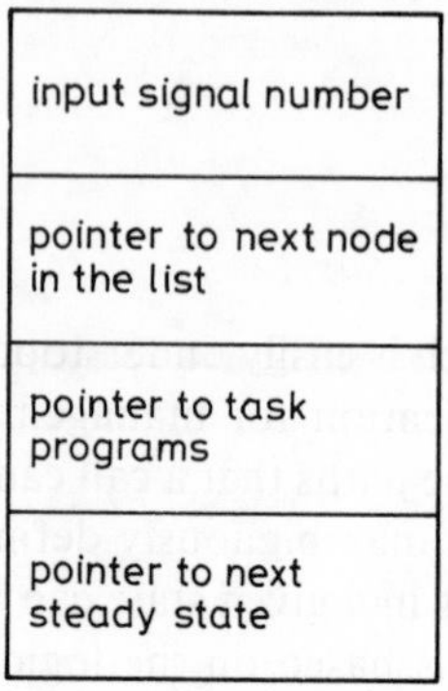

input signal number
pointer to next node in the list
pointer to task programs
pointer to next steady state

Fig. 13.4 *A node in an input-signal list*

driven'. Call processing is also said to be 'event-driven', since processing of a call is initiated by an event, in the form of the arrival of an input signal. Becuse an event, in the present context, is merely data to the software, the two are equivalent here.

It should be mentioned that, whereas the flow of processing is directed by the data structures, the task programs shown in SDL diagrams are routines written to perform specific functions, and cannot be generalised. They are written as procedural programs in executable code, which may be assembly language but is usually a high-level language.

In the implementation of SDL in software, all the input signals which could possibly occur when a call is in a given steady state are represented as the nodes of a linked list. Each node consists esssentially of four fields (see Fig. 13.4):

(i)   The input signal number, to be used as the key in searching the list.
(ii)   The pointer to the task programs which must be activated if that input signal occurs.
(iii)   An indication of the call's next steady state.
(iv)   A pointer to the next node in the list.

Thus, when a given input signal occurs, its node contains all the information necessary for the transition of the call to the next steady state, which itself follows logically from the input signal.

13.5.1 Practical implementation
The use of data structures to achieve call processing is illustrated in Fig. 13.5, and the following description is based on this Figure. A single program, the interpreter, is written to process the transition of any call, whatever its type, from any state to a succeeding state. This program, in effect, controls call processing. First, the interpreter extracts an input signal node from a queue of arriving input signals; then, using the equipment number as a key, it

accesses the appropriate call record. (The contents of the nodes in the arrival queue which make this possible are described in the following chapter.) Then, the interpreter accesses the current-state field, whose content is a pointer to the top of that state's input-signal list. It follows this pointer and compares the input signal just received with that in the first node of the list. If they do not match, it follows the list pointers to successive nodes, scanning as it goes, until it finds a match. If there is a task program associated with that input signal, the interpreter's next duty is to execute it, so it checks the contents of the task-program pointer field. If this is zero, or some other chosen null character, there is no task program to be executed. Otherwise, the interpreter calls the task program as a subroutine. This initiates the telephonic events necessary for the transition of the call to its next state. It is then necessary to update the call record so that, when the next input signal arrives, it contains a pointer to the correct list. For this purpose, the next field in the node contains a pointer to a translation table (the steady-state table), from which the appropriate pointer is extracted and loaded into the current-state field of the call record. This ensures that the input-signal list corresponding to the current steady state is always stored in the call record and is thus accessible for processing when the next input signal arrives. The interpreter then seeks the next input signal (for the next call) from the input queue.

It is possible for the node itself to contain the pointer ready for loading into the call record, thus obviating the need for the steady-state table. However, for frequently-referenced states, such as 'park' or 'idle', there would then be a large number of occurrences of the address of the equivalent input-signal list. A rearrangement of storage would require all these to be updated, and would incur the possibility of leaving one or more incorrect pointers. Having the state references in a permanent table means that the data in the nodes of the lists need never change, and the rearrangement of storage would require the alteration only of pointers in the table.

13.5.2 Branching

As can be seen in Fig. 14.8, in the following chapter, the transition from one steady state to another is not always determined only by the input signal, but sometimes depends on a decision made within a task program. For example, when an 'off-hook' signal is received, it is a check of the class-of-service record which reveals if the subscriber is permitted to make outgoing calls. Depending on this, the call goes either into the 'awaiting first digit' or the 'parked' state. Clearly, the node of Fig. 13.4 is not suitable for such branching, as it directs the call to a steady state determined only by the input event.

In a list whose task programs contain branches to different steady states, there must be a variation on the procedure described in Section 13.5.1 above. For example, when the task program returns control to the interpreter, it passes a parameter (usually 1,2,3 etc.) which defines the branch to be taken. The nodes in the list contain a field for each possible branch (see Fig. 13.6)

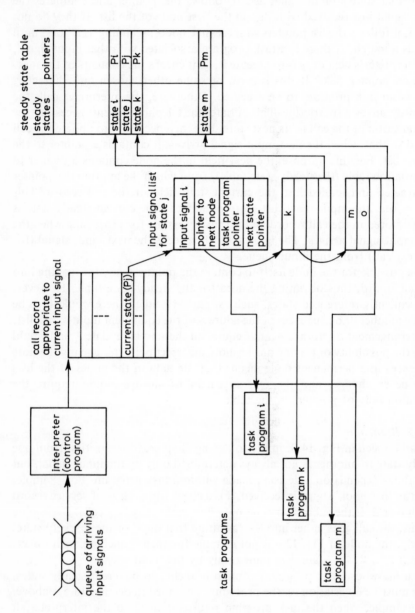

Fig. 13.5 *Interpretation of SDL in software*

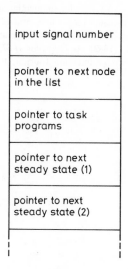

| input signal number |
| pointer to next node in the list |
| pointer to task programs |
| pointer to next steady state (1) |
| pointer to next steady state (2) |

Fig. 13.6 *A node which allows branching to be controlled interpretively*

and in each field there is a pointer to the appropriate steady-state table address. The interpreter is written to follow the pointer which is appropriate to the value of the parameter received from the task program. Because the same interpreter is used for all lists, a parameter must be returned from all task programs, whether or not they make decisions which result in branching. Nodes must be designed to contain the number of 'next-steady-state' pointer fields demanded by task programs addressed by their task-program pointers.

13.5.3 Run-time efficiency

Direct translation of the SDL into software increases the probability of error-free call processing. Moreover, the interpreter which interprets the data structures can be relatively short. However, the price of this simplicity and increased chance of correctness is run-time inefficiency.

Since the task consists mainly of searching, intepreting addresses and carrying out comparisons, execution requires the processing of a large number of instructions. This is unavoidable, but it is mitigated in two ways. First, the interpreter may be written in assembly language to increase its speed. Secondly, list processing can be optimised by judiciously arranging the nodes of a list in order of likelihood of occurrence of the input signals.

13.6 References

1 KAWASHIMA, H., FUTAMI, K. and KANO, S. (1971): *Functional Specification of Call Processing by State Transition Diagram*, IEEE Transactions on Communications Technology

2 CCITT (1977): Orange Book. Vol V1.4 — Recommendations Z101 — Z103

3 CCITT (1981): Yellow Book, Fascicle V1.7 — Recommendations Z101 — Z104

4 CCITT (1989): Blue Book, Fascicle X.1 — Recommendations Z100

5 GERARD, P. and BIERMAN, E. (1982): *An Overview of SDL, The CCITT Specification and Description Language*, ITU Telecommunication Journal, **49**

Chapter 14

Elements of control

The Specification and Description Language (SDL), described in the preceding chapter, is used in the specification and design of call-processing software. Also, by using a software interpreter, SDL can be translated directly into the call-processing software itself. However, the transition of a call from one steady state to another depends on input signals, and these must be detected. Thus, to introduce call processing, this chapter presents a discussion of the way in which signals arrive at the control system and of the data necessary for handling them.

14.1 Receipt of external signals

Signals from sources external to the exchange arrive in four main forms. These are common-channel, channel-associated time-slot 16, multifrequency and loop-disconnect signalling, and each has its own type of signal receiver. Common-channel signalling is used on digital routes from other SPC exchanges; time-slot-16 signalling is used on digital routes from analogue sources, which may be exchanges, or any other sources which can justify a PCM link; multifrequency and loop-disconnect signalling are both from analogue sources, which may be other exchanges or local customers. The handling of calls incoming from other exchanges will be discussed in the following chapter; this chapter is concerned with originating calls.

As was described in Part II, the method of termination of PCM systems with channel-associated time-slot-16 signalling usually includes the permanent connection of time-slot 16 to a signal receiver. This receives and interprets the incoming signals and passes them to the control system.

14.1.1 Subscriber's line-termination unit (SLTU)

Originating calls from an analogue subscriber's line must be recognised at the subscriber's line-termination unit (SLTU). Detection of new calls requires constant scanning of the SLTUs in order that all changes of state on the lines

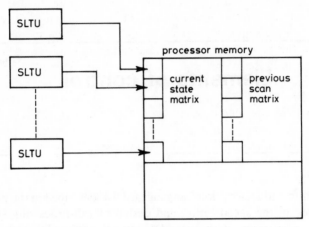

Fig. 14.1 *Scanning of SLTUs by microprocessor*

are recognised. These changes are intepreted into telephonic input signals which are used to progress the call.

Scanning is usually carried out by a microprocessor,[1,2] sometimes taking the form of a 'line controller'[3] to which the SLTUs are permanently connected. Line signals arriving at the SLTU, if in analogue form, must be converted to a digital form suitable for direct input to the microprocessor. The digital-signal codes are stored in a dedicated location in the microprocessor's memory, i.e. within a 'current-state matrix', as in Fig. 14.1. At a regular frequency, the microprocessor scans this matrix, comparing each location with its counterpart in a 'previous-scan matrix', which stores the states as they existed on the previous scan. If the scan reveals a difference in the contents of the two corresponding locations, the line state has changed. The input signals derived from these changes are then queued to be processed by the call-processing programs (see Fig. 14.2).

14.1.2 Off-hook and on-hook signals
Representation of the basic states of a line (on-hook and off-hook) requires only one bit. The matrices of Fig. 14.1 must, for this purpose, consist of one

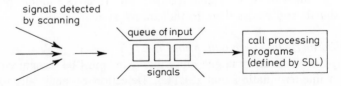

Fig. 14.2 *Scan results queued for processing*

bit per SLTU. A change in a bit, when comparing the current-state and the previous-scan matrices, indicates a change in the state of the line. A concatenation of the current and previous bits thus provides an adequate definition of the input signal. '01' might, by convention, indicate the off-hook signal, i.e. that the state of the line had changed from on-hook to off-hook; '10' would then indicate the reverse, i.e. the on-hook signal.

Frequency of scanning for on- or off-hook signals need not be high. If a subscriber receives the dial tone within a second of lifting the receiver, the service is likely to be considered satisfactory. A scan rate of 100ms is therefore quite adequate; however, with distributed processing, a much faster rate is often provided. As an example with which to explain the method of scanning, the figure of 480 SLTUs per remote processor will be used. Assuming a 32-bit word in the processor and a requirement of one bit per SLTU, 15 words are required for each of the current-state and previous-scan matrices. The SLTUs must be connected in a known numerical order, so that the scanning microprocessor can equate the bit being examined with a particular item of equipment. Ascending order of SLTUs is usual (see Fig. 14.3), and the previous-scan matrix is designed to match this. It should be noted that this correspondence between storage and equipment numbers can be considered as permanent. It is the translation between equipment number and subscriber's directory number which is likely to change. This is provided by translation tables.

In scanning, the examination of each bit individually is not efficient. The probability of a change in state of any one line within a 100ms period is exceedingly small, and that of a change in one or more of 32 lines is still not significant. It is therefore economical to compare a word, or even several

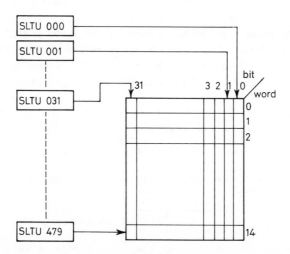

Fig. 14.3 *Mapping 480 SLTUs onto 15 32-bit words*

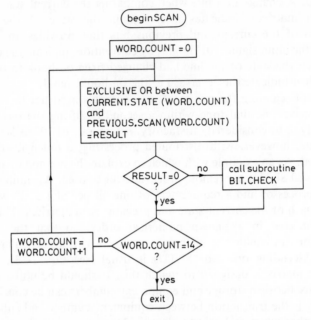

Fig. 14.4 *Scanning for on-hook and off-hook signals*

words, at a time rather than one bit at a time. If the words are different, only then is it necessary to examine them further to detect which bit or bits have changed.

The method of scanning[4] (see the flowchart of the program, SCAN, in Fig. 14.4) is to compare in turn each word in the current-state matrix with its corresponding word in the previous-scan matrix, using the exclusive OR function. The result is stored in location RESULT and is zero for equality. Thus, a test of RESULT for a value of zero is sufficient to determine whether there is a change of state of one or more of the 32 lines represented by the bits of the word being tested. If there is no change, the scan proceeds to the next

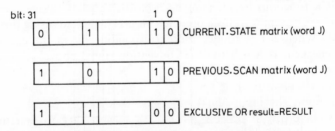

Fig. 14.5 *Exclusive OR of corresponding words from CURRENT.STATE and PREVIOUS.SCAN matrices*

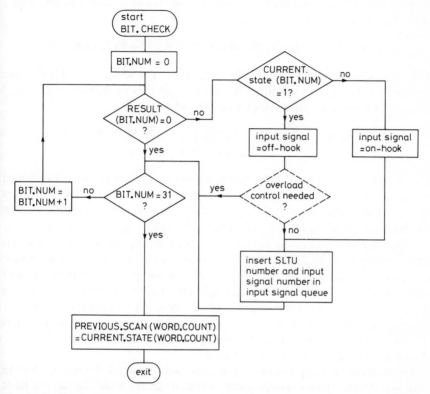

Fig. 14.6 *Routine bit.check, to examine each bit of result*

word, using a counter WORD.COUNT to control the testing of the words. A nonzero RESULT indicates that at least one line has changed state and the position(s) of the bit(s) in RESULT which equal 1 defines which lines they are (see Fig. 14.5). Subroutine BIT.CHECK (see Fig. 14.6) is then called to identify them.

As the storage in most SPC computers is bit-addressable, subroutine BIT.CHECK functions by accessing each bit of RESULT in turn and testing it for zero, while keeping a check of which bit is being tested, using the counter BIT.NUM. When a 1 is found, the corresponding bit in the current-state matrix is checked. If it is 1 (by the convention stated above), the off-hook event has occurred (as at bit 17 in Fig. 14.5). If it is zero, then the on-hook event has taken place (as in bit 31).

At this point, the input-signal number is known (01 or 10, as already explained). In order to process the call, it is necessary to determine the equipment number of the SLTU on which it has arrived. With the equipment connected and scanned in ascending numerical order (see Fig. 14.3), this can be achieved by knowing which bit of which word is under consideration. Thus, from Fig. 14.6,

$$\text{SLTU Number} = \text{WORD.COUNT} \times 32 + \text{BIT.NUM}$$

This identifies the equipment within its scanner. The absolute SLTU number within the exchange, given 480 SLTUs per microprocessor scanner and scanner numbers commencing at zero, can be derived from

$$\text{SLTU Number} = \text{WORD.COUNT} \times 32 + \text{BIT.NUM} + \text{Processor Number} \times 480$$

Owing to the possibility of noise on the line giving the appearance of a change of state, a persistency check is usually made before off- or on-hook conditions are accepted as genuine. For this, a record of new inputs is kept so that a check can be made of whether they persist for more than a given number of scans.

Following this, processing of the call proceeds by inputting the events which have occurred to the call-processing software, as in Fig. 14.2. With a single processor carrying out all the tasks of call processing, queuing is essential to ensure that scanning of all the exchange terminations is completed within a given time. With scanning distributed to remote processors, each with a limited number of lines under its control, the call-processing software is concentrated in one or more processors, to which the events must be sent. Again a queuing mechanism is required.

The methods of implementing a queue in software are described in detail in Chapter 12. For the current purpose, a linked list could be used as the basis of the queue. In this case, each node would consist of three fields: SLTU equipment number (to identify the line), input signal number (01 or 10, as already explained, to identify the event which has occurred) and the link to the next node (see Fig. 14.7). However, because there is no requirement to extract nodes out of sequence, using a linked list offers no advantage over the sequential linear list, which would save on the link field.

The SLTU number and the input-signal number are then inserted into the node at the rear of the event queue, and the scan proceeds. When all the 32 bits of the RESULT word have been checked, it is necessary to update the previous-scan matrix by replacing (overwriting) the word defined by the value of WORD.COUNT with its equivalent from the current-state matrix. Once this has been done, control is returned to subroutine SCAN. When all 15 words of the current-state matrix have been checked, the processor can

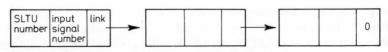

Fig. 14.7 *Nodes in a linked list of line-state input signals*

recommence scanning or proceed with other tasks, according to its programmed schedule. The method of scheduling tasks will be discussed in Chapter 16.

At any time in the operation of an exchange and for a number of reasons, overload may occur. The subject is discussed more fully in Chapter 16, but it is appropriate here to mention that one method of overload control is to suppress new call attempts. This can easily be achieved (see Fig. 14.6) by including code in subroutine BIT.CHECK which, under overload conditions, suppresses off-hook events and thus the processing of the calls.

14.2 Class of service

While the lines are being scanned, all on-hook and off-hook events which have occurred since the previous scan are detected and queued. The nodes of the queue must then be examined in order to progress each of the calls. An on-hook signal requires that the switch path through the exchange is cleared down, and any data that need to be retained (e.g. for purposes of billing, maintenance or planning) is stored, output, etc. according to the exchange procedure.

After an off-hook signal, preparations must be made to receive the address digits (see Fig. 14.8). The subscriber is awaiting dial tone, but, before this is provided, it must be confirmed that the calling line is legally in service, that outgoing calls are not barred and if a multifrequency signalling receiver must be connected to the line. These tasks are accomplished by the interrogation of the calling subscriber's class-of-service (COS) record. A COS record exists for each termination on the exchange and consists of a number of data fields containing 'semipermanent' and 'transient' data which define the termination and its current status.

14.2.1 Semipermanent data

Semipermanent data define the line and what is allowed to occur on it. These data are not directly alterable by the subscriber. However, if the line type and/or status are changed by arrangement with the administration, they must be updated.

Examples of semipermanent data are line type, barring level and whether or not customer facilities are permitted:

(*a*)   Examples of line type are:

000   unassigned line
001   domestic subscriber's line
010   business subscriber's line
011   announcement within exchange (a recorded announcement which may be connected to a subscriber's line in certain conditions)

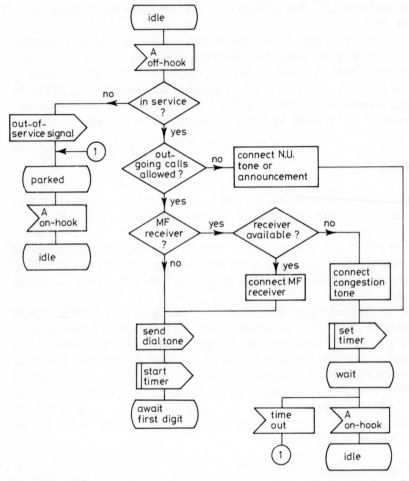

Fig. 14.8 *SDL representation of the first stage of processing of an originating call*

100   coin box
101   data terminal
110   PBX
111   ISDN line.

In general, line types are mutually exclusive; thus, the above eight options could all be represented uniquely by three bits (see Fig. 14.9), as per the binary representations which precede the line types above.

(*b*)   Examples of barring level are:

00   no outgoing call restrictions
01   all outgoing calls barred

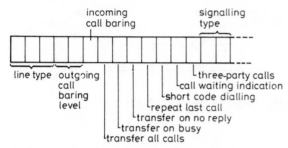

Fig. 14.9 *Examples of semipermanent data fields in the COS record*

10   non local calls barred
11   international calls barred
      incoming calls barred.
The first four of the above examples are mutually exclusive, so they can be represented uniquely by two bits, as suggested above. The fifth needs a bit of its own (again see Fig. 14.9).

(*c*)   Customer facilities offered differ from administration to administration and are increasing in number. The list below does not purport to be exhaustive and only offers examples to illustrate the principle of the COS record:

transfer all calls
transfer on busy
transfer on no reply
repeat last call
short-code dialling
call-waiting indication
three-party calls.

If these customer facilites are offered by an administration, the subscriber can obtain the use of any of them by arrangement with the administration and by paying the appropriate fee. They are therefore not mutually exclusive and require one bit each to define their availability to the subscriber (see Fig. 14.9). A binary 1 in the appropriate bit may, by convention, indicate that the facility is permitted to the subscriber, in which case a binary zero would mean that it is prohibited.

The semipermanent data of the above examples of line type, barring level and customer facilities are seen, in Fig. 14.9, to require ten data fields, totalling 13 bits. A further requirement is an indication of the type of signalling (multifrequency, loop-disconnect or common-channel) on that line, and a further field of two bits is required for this.

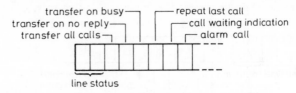

Fig. 14.10 *Examples of transient data fields in the COS record*

14.2.2 Transient data

Transient data define the current status of the line and, therefore, alter in accordance with the subscriber's actions. Typically, they embrace the line status (or busy level) and the facilities enabled.

(*a*) Examples of line status are:

line out of service
line free
line engaged
line parked.

These four states are mutually exclusive and can be represented within a two-bit field (see Fig. 14.10).

(*b*) Facilities enabled refer to those facilities which a subscriber is permitted to use but which may be enabled or disabled in accordance with the subscriber's choosing. Examples, from the earlier list of facilities are:

transfer all calls enabled
transfer on busy enabled
transfer on no reply enabled
repeat last call enabled
call-waiting indication enabled
alarm call enabled.

One bit is needed for each facility (see Fig.14.10) to indicate whether it is enabled or not. In addition, when a transfer facility or the repeat-last-call facility is enabled, there is a need to store the directory number involved. The COS record may be designed to accommodate such numbers, or they may be stored elsewhere. This point is discussed in the following chapter.

14.2.3 Use of class-of-service data

When a subscriber is initiating a call and the off-hook signal is received by the control system, the line-status and barring-level fields must be checked to ensure that the line is in service and that outgoing calls are not barred (see Fig. 14.8). If both are true, it must be determined what type of signalling is in use on the line, so that the appropriate type of signalling receiver may be connected. If dialling is commenced with the trunk or international prefix, the barring level must again be checked to ensure that such calls are permitted on that line. If the subscriber attempts to use or enable a special facility, a check of whether or not this is allowed must also be made.

Before a call is connected, within the exchange, to a subscriber's line, the control system must check the subscriber's COS semipermanent data to ensure that the line is of a type to which calls may be directed, and that incoming calls are allowed. It must then interrogate the transient-data fields to determine if the line is busy or if a transfer facility is enabled. If the latter is the case, the call must be directed to the transfer number, which will be located in a designated area of storage within the COS record or elsewhere.

14.2.4 Access to the class-of-service records

There must be a COS record in store for each line termination, and there must also be a means of finding it. There must be some key, or reference, by which it can be accessed. Key transformations (see Chapter 12) are possible if the records are stored in strict numerical order; otherwise scanning must be used. When a new call is detected, it is the subscriber's line-termination unit which provides the initial indication, and the number of this equipment may be deduced as described earlier in this chapter. Thus, if the COS records are stored in terminating-equipment order, they can be accessed directly by a key transformation, using the equipment number as the key.

Having said this, it should be mentioned that the COS record must also be accessed using the subscriber's directory number as the reference. When exchange staff update a COS record, they want to do so in the most convenient manner, and this implies using the directory number rather than an equipment number. Moreover, when an incoming call arrives, its addressing mechanism is the called subscriber's directory number. In this case, there is indeed no other means of accessing the COS record, and access to the semipermanent and transient data is essential, as already described. The need to access it demands that the COS record is stored, together with the SLTU equipment number and the subscriber's directory number. These constitute the subscriber's record.

14.3 The subscriber's record

In summary, a subscriber's record requires four main data fields, as in Fig. 14.11. It may need to be accessed either when the equipment number or the directory number is known. Subscriber's records can thus be arranged either in equipment-number order or directory-number order, according to the manufacturer's preferred design. If they are in directory-number order, a search using the directory number as the key is fast, because a key transformation can be employed; then, if the equipment number is the key,

| equipment number |
| directory number |
| semi-permanent data |
| transient data |

Fig. 14.11   *A subscriber's record consisting of 4 main data records*

scanning must be used and the search is slower. If the records are in equipment-number order, it is the other way round.

14.4 The call record ˙

As soon as it is confirmed that a call is to be processed (i.e. when the off-hook signal has been detected), it becomes necessary to co-ordinate the data concerned with that call. This is achieved by creating a call record,[5] which then becomes the point about which processing of the call hinges and in which all data pertaining to the call are stored. The records of calls in progress are stored as a linked list, each call record being a node (see Fig.14.12) and, typically, each record, or node, consists of about 40 words. A call record is essential to a call. It is a software equivalent of an item of common equipment. Without a free call-record node, a new call attempt must be rejected or made to queue; thus, the total available stack of nodes must be large enough to accommodate the maximum number of simultaneous calls in progress across the exchange.

Typical data stored in the call record are:

equipment number of the line-terminating equipment
time of origin of call
equipment numbers of all common equipment associated with the call
actions taken in processing the call
times of actions
identification of switch paths used in the call set-up
the current state of the call
the last signal to have been received
a count of the address digits so far received
the dialled digits
the type of call
a count of meter pulses during the conversation stage, if meter pulsing is used
time of answer signal
the time of clear-down of the call.

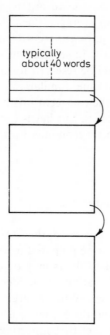

typically
about 40 words

Fig. 14.12   *Call records – kept for the duration of a call in a linked list*

In addition, the call record contains flag bits to indicate the status of processing at any time and pointers to the appropriate current position in the SDL diagram.

All the above data are not essential to call processing. Some are necessary for the efficient and effective running of the exchange and the network by the administration. Data used for network planning, network management, maintenance, billing and accounting are output from the call record and the SPC system after the call has been cleared down and prior to the return of the call record to the available stack for reuse with another call.

14.4.1 Access to the call record
One method of accessing a call record in a linked list is by following the links in the list and scanning a given word of each record until a match with the key is found. A record only needs to be accessed when a signal has arrived to initiate the transition of the call from its present state to some other state. Such a signal must arrive either from an item of equipment involved with the call or from a timer. The equipment number of any equipment concerned with the call will already be stored in an appropriate word of the call record. Which word is used in the scan will therefore be determined by which type of equipment has given rise to the input signal (e.g. SLTU or signalling receiver).

A second method of accessing the call record, which is used in many cases, is direct addressing. Even though the record is a part of a linked list, it possesses an absolute address in storage. Each item of equipment, e.g. a signalling receiver, has its own equipment record containing such information as whether or not the equipment is in service and whether it is busy or free. An extra field in this record could be reserved for a pointer to the call record with which it is associated at the time. Similarly, a software timer contains a field for the storage of a key to the call record with which it is associated (see Chapter12). When a time-out occurs, this key is used to access the call record directly. Indexing the record directly economises on the time taken to scan and process the list's links.

14.5 Connection of common equipment

When an off-hook signal has been received, the COS record checked and a call record created, the exchange must be prepared to receive the digits dialled or keyed by the subscriber. Receipt of digits may depend on the connection of a signal receiver, and this takes a variable time, depending on the availability and loading of the pool of receivers. When a receiver of the appropriate type has been connected to the line, and the exchange is ready to receive digits, the exchange must inform the subscriber by connecting the dial tone to the line.

14.5.1 Loop-disconnect signalling receiver

Loop-disconnect signalling is detected and analysed by a permanently connected receiver, time-shared among a number of lines. The receiver consists of a microprocessor which scans the subscriber's line-termination units in the same way as was described for on- and off-hook signals. In this case, however, scanning is a great deal faster, because dial pulses of 33 ms duration must be detected. A scan rate of once ever 10 ms would be adequate, but modern systems usually provide a much faster rate than this. A further

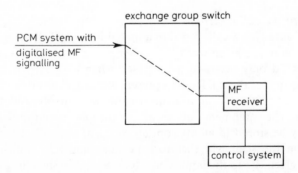

Fig. 14.13 *Connection of an MF receiver via the exchange group switch*

requirement of this equipment is that the duration of both the make and the break conditions on the line are timed. This ensures that the longer-duration interdigital pauses are recognised; it also enables the pulses to be summed correctly, in order to deduce the digits which they represent.

14.5.2 Multifrequency signalling receiver

Multifrequency (MF) signalling is not detected by a permanently connected receiver. When the COS record reveals that MF signalling is used on the line initiating the call (See Fig. 14.8), an MF receiver must be connected to the line via the SLTU. There are two types of digital MF receiver, one which is used for incoming trunks and a simpler device for connection to a local subscriber's line. The former is connected via the exchange group switch (see Fig. 14.13), and it must handle signals in both directions. The control of path set-up through the group switch is described in the following chapter. The subscriber-line MF receivers are also kept in a pool and are usually accessed via the subscriber-concentrator switch. In both cases, the receivers are represented in memory by equipment records, each containing fields for indicating if the equipment is in service and whether or not it is free. Selecting a receiver involves finding one which is both in service and free, busying it by setting its busy/free bit, and recording its number in the call record.

For local subscribers, the MF receiver may be connected via the group switch. However, in most modern systems it is connected locally at the concentrator stage (see Fig. 14.14). The receivers, each capable of handling a number of simultaneous calls (typically about eight), are multiplexed and

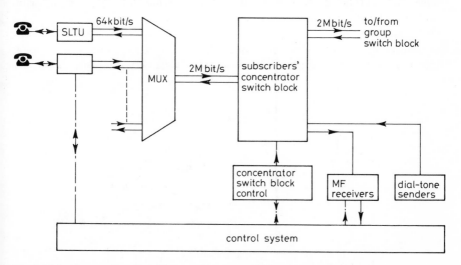

Fig. 14.14 *Connection of auxiliary equipment to SLTUs via the subscribers' concentrator switch block*

connected to the switch via one or more PCM highways. If dial tone is returned from the receiver, both 'transmit' and 'receive' paths are necessary. However, dial-tone sources are usually separate; then only a unidirectional path is required to handle the signals from the subscriber to the exchange.

In setting up a connection, the control processor will already have detected the off-hook signal at an SLTU and will have determined that the call requires an MF receiver; it will have interrogated the equipment records of the receivers and selected one which is in service and which has a free time slot. It will also have a record of which time slot on which highway the SLTU is connected into. The control system is thus able to direct the setting up of a switched path between the SLTU and the MF receiver.

When the receiver transmits the incoming digits to the control system, it attaches a label to identify the time slot in which they arrived. The control processor translates this label into the appropriate SLTU number, and thus associates the digits with the correct call. When dialling is complete and the dialled digits have been transmitted from the MF receiver to the control system, the latter resets the busy/free flag in the receiver's equipment record. This frees the receiver (or at least that particular one of its time slots) for another call.

14.5.3 Dial tone
The final actions to be taken before the call is left in the steady state of waiting for the first digit are sending dial tone to the caller and setting a timer. The latter is done in case the caller fails either to input digits or to clear-down within a given time. In such a case, the timer sends a time-out signal, the call is aborted, and the line parked, i.e. disconnected from all exchange equipment

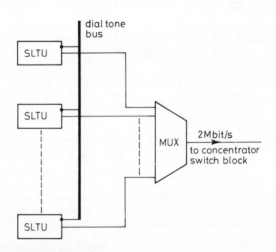

Fig. 14.15   *Dial tone provided on a bus*

(see Fig. 13.3). (The principles of setting a software timer are described in Chapter 12.)

There are two ways in which the connection of dial tone to the line may be achieved. The first is by connecting a digital source via the group switch or concentrator switch, as for multifrequency receivers. However, it must be remembered that, when analogue-to-digital conversion takes place at the SLTU, a four-wire system becomes necessary, with separate transmit and receive paths. Fig. 14.14 shows that, whereas an MF receiver needs to receive signals from the SLTU, a dial tone source needs to send a signal to it and must therefore be connected in the receive path.

The second method of connection is via a bus which is permanently terminated on every SLTU, as in Fig. 14.15. This, too, is a digital source, and the method of connection to line is simply by closing logic gates in the SLTU, under the control of the control system. When the first digit (or time-out) has been received, the dial tone is disconnected.

14.6 References

1 JACOB, J.B. and TRUBERT, R. (1984): *The Third Generation of Subscriber Connection Units for the E10 System*, ISS'84, Florence, May 1984

2 GOEBERTUS, H.J. (1984): *ESS-PRX Architecture*, Philips Telecommunication Review **42**(3)

3 WARD, R.C. (1985): *System X: Digital Subscriber Switching Subsystem*, British Telecommunications Engineering, **3**

4 HARR, J.A., HOOVER, E.S. and SMITH, R.B. (1964): *Organisation of the No.1 ESS Stored Program*, Bell System Technical Journal, **XLIII**

5 ANDREWS, R.J. et al. (1969): *Service Features and Call Processing Plan* (of No.2 ESS), Bell System Technical Journal, **48**

Chapter 15

Call processing

The last chapter described the data which a control system must hold in order to control line terminations and calls. It discussed digit receipt and call processing up to the point of digit analysis (see Fig. 15.1). This Chapter deals with processing the digits, setting up an appropriate path through the exchange, monitoring the call and, finally, clearing it down.

A call may come either from a subscriber terminated on the exchange (an 'originating call') or from a distant exchange (an 'incoming call') — see Fig. 15.2. Considering an originating call, it may be to another subscriber terminated on the exchange (a 'terminating call'), to a subscriber terminated on another exchange (an 'outgoing call'), or it may be setting up or cancelling a customer's facility (a 'facility call'). An incoming call may be a terminating call or a call to a subscriber on a third exchange (a 'transit call'). It is recognised that a satellite exchange may use the parent exchange for customers' facilities and, in this case, some incoming calls may appear to break the convention of Fig. 15.2. However, for the purpose of this description, such calls are considered to be originating calls.

The main actions to be carried out by the control system depend, primarily, on whether the call is outgoing or terminating. However, within each of these categories, there are certain minor variations, dependent on whether the call is originating or incoming. A number of functions: (1) digit analysis, (2) digit translation, (3) switch-path set-up, (4) charging and (5) some customer facilities, are described in detail under separate heading in this chapter. The remaining actions necessary for an outgoing call are, in the main, the same as those for a terminating call, so only the latter will be described in detail, under the heading of 'Setting up a call', in Section 15.3.

15.1 Digit analysis

Digit analysis is required for an incoming call in order only to determine whether the call is terminating or transit. The main analysis will have occurred

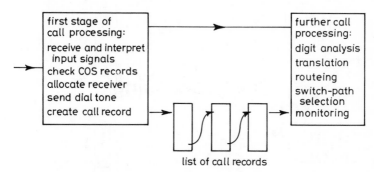

Fig. 15.1 *Constituent parts of call processing, as considered in Chapters 14 and 15*

at the exchange where the call originated. On the other hand, an originating call requires a considerable amount of digit analysis. In particular, the first digit is significant:[1] in nearly every case, it defines the type of call being initiated. Fig. 15.3 illustrates the possible interpretations of the first digit and will be used as the basis of a description of digit analysis for an originating call. The Figure also shows that 'A-on-hook' may occur while digits are expected. In the example, 1 and 9 are reserved for operator, emergency and other service calls; 0 is the prefix for nonlocal calls (i.e. trunk and international); * and # (represented in software as 11 and 12) are the prefixes for setting up, using and clearing down customer's facilities; digits 2 to 6 initiate local calls; digits 7 and 8 are inadmissible (which might be the case if the exchange were not fully provisioned). These incoming signals are listed, not in ascending numerical order, but in order of likely frequency of occurence. When the Specification and Description Language (SDL) diagram of Fig. 15.3 is represented in software as a linked list, which must be searched each time an input signal arrives (see Fig. 15.4), optimising the order can result in significant run-time efficiency.

The SDL diagram of Fig. 15.3 shows that, for any input signal, the first tasks to be performed are storing the signal, setting a timer and removing dial tone from the line. All digits are stored in the call record for routeing purposes and for later data analysis. Even illegal signals are recorded, because they

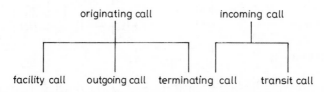

Fig. 15.2 *Types of call*

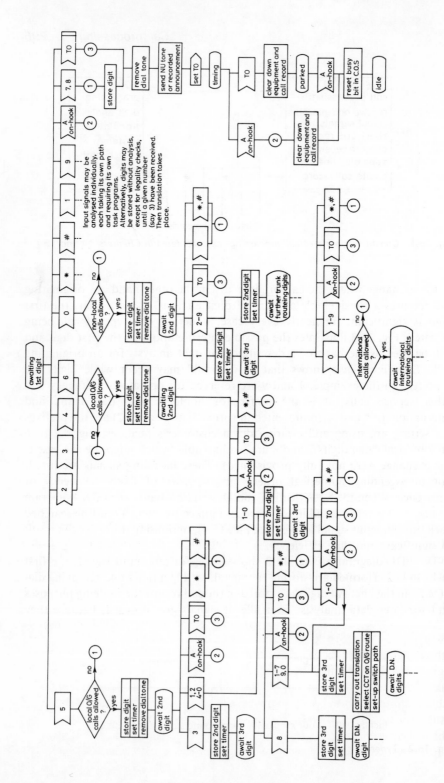

Fig. 15.3 *Analysing subscribers' input digits*

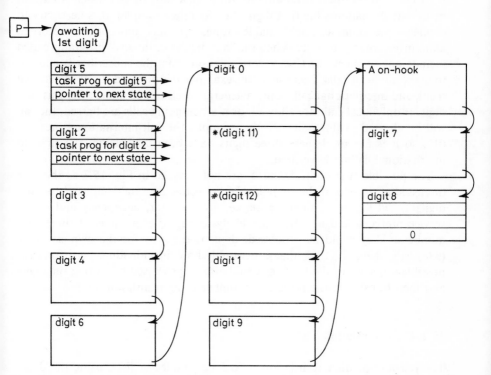

Fig. 15.4 *Input-signal list for first digit*

might indicate faults and because they can contribute to statistical analyses. Timers are set between steady states; thus, if an input signal does not arrive within a predetermined time, the call can be aborted. This frees common equipment (inclduing common memory) for other calls. (The methods of setting software timers are described in Chapter 12.) Removing dial tone from the line is carried out by reversing the process described in the previous chapter for connecting it to the line. The task program which carries out the above three functions would not be replicated in store for each of the paths of Fig. 15.3, but would be stored once, and called as required.

Although Fig. 15.3 offers a self-contained description of digit analysis, a few points are worth making. The exchange in the example is assumed to have a three-digit exchange code of 538. For this reason, the first digit, 5, the second digit, 3 and the third digit, 8, are placed first in their input-signal lists and are analysed along individual paths. Recognition of this sequence allows the subsequent digits to be treated immediately as the directory-number digits of a subscriber terminated on the exchange; this obviates the need for translation which, because it involves searching tables, is time-consuming.

For every call attempt, the calling subscriber's class-of-service (COS) record is checked. When there is an attempt to make a class of call which is barred, or to dial an inadmissible first digit (7 or 8 in the example), it is necessary to connect the number-unobtainable tone, or an appropriate recorded announcement, to the line. When the first digit is 0, the task programs called include one to check the COS record to ensure that the subscriber is allowed to make nonlocal calls. Because, in the example, 0 is prefixed for both national trunk and international calls, only a general check can be made after this first digit. If only local calls are allowed, or if all outgoing calls are barred, the call will be aborted at this point. In the example, the full international prefix is 010, so it is not until these three digits have been dialled that the right to international calling is verified.

The possible first digits * and # are not pursued in Fig. 15.3 as they are discussed at the end of the chapter. Neither is the diagram completed for first digits 1 and 9. These, in the example, are operator, emergency and other-service prefixes, and are therefore likely to be the beginnings of three-digit numbers. They could be processed, digit by digit, as for the own-exchange (538) call; then, the SDL diagram would show an individual path for each possible sequence of digits. They could also be processed by storing the digits and then translating the three-digit number using translation tables.

15.2 Digit translation

The purpose of digit translation is to access tables or lists, using the digits which have been input, so as to find:

(i)   A circuit on which the call can leave the exchange.

(ii)   The routeing digits which must be sent to the next exchange (if the call is not terminating) so that it can route the call correctly.

(iii)   The rate at which the call should be charged.

For an originating call, digit translation follows digit analysis. Indeed, it is a form of digit analysis and, in most cases, it is carried out in the manner already described, i.e. using each digit as a key to the next stage of translation, until the ultimate goal is achieved. An incoming call, however, enters the translation process directly, as digit analysis will have been carried out at the call's originating exchange.

An originating own-exchange call, if given a special path through the analysis phase (as in the 538 call in Fig. 15.3) requires no further translation. The SDL diagram's interpretation in software guides it through the data structures to the point where the outgoing circuit to the called subscriber can be determined.

However, a call to another exchange needs translation. It may be destined for a neighbouring exchange, or one which requires a connection involving several links in the network. Routeing digits must therefore be dispatched, so that the next exchange in the chain can determine its own routeing of the call. The requirement on link-by-link interregister signalling systems, including common-channel signalling, is for the dispatched routeing digits to consist of the input digits. Then, it is not necessary to carry out a translation to find routeing digits. However, the process is still required in order to determine on which route the call should leave the exchange and what rate of charging should be applied. But on older types of exchange and on routes using end-to-end signalling, the routeing digits dispatched for calls to the same destination differ in accordance with the route on which the call leaves the exchange.

The principles of routeing determine the contents of the routeing tables within the SPC memory. However, this chapter is concerned only with how the data are stored and accessed within the control system, and how the necessary translations are carried out in software. In general, translation depends on accessing the contents of translation tables, the principles of which are described in Chapter 12.

15.2.1 A terminating call

Translation, in the context of a terminating call, consists of two stages. The first deduces that the call is terminating on the exchange. The second translates the received directory number into the called subscriber's equipment number, which defines the outgoing circuit to the called subscriber.

An own-exchange call is recognised during digit analysis, as shown in Fig. 15.3 and discussed earlier. When the final digit of the exchange code has been dialled, the data structure of the SDL diagram leads the call into the state of 'waiting for directory number'. The directory number is then used as the key for translation into the called subscriber's equipment number.

Incoming calls may be either terminating or transit. Depending on the system, a terminating call may or may not contain the exchange code among its address digits. In older systems, the exchange code will already have served its purpose in routeing the call and will not have been transmitted to the terminating exchange. The fact that the call is terminating is deduced from the junction on which it arrives. The received digits are then known to be the called subscriber's directory number. In common-channel signalling, however, it is usual for all address digits to be forwarded. Digit analysis and translation are then more consistent between exchanges and do not differ according to the incoming route. Thus, transit calls are not recognised from the route on which they arrive, but after an analysis of their initial digits.

15.2.2 The principles of translation

The general principles of translation are illustrated in Fig. 15.5. A controlling program receives the input code, either by extracting it from the call record or by transfer from a task program during, or at the end of, digit analysis. The controlling program uses the code as a key to a translation table (or series of tables)[2,3] which translates the code into a pointer. This points to a route table

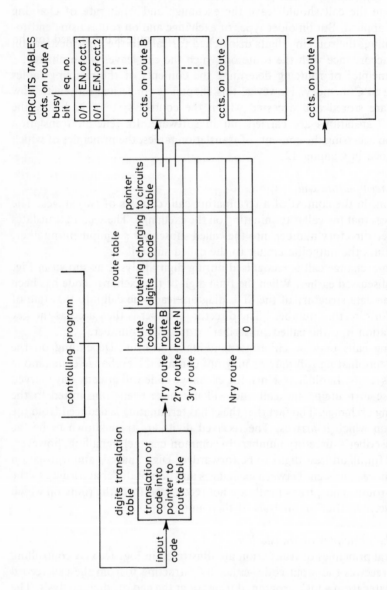

Fig. 15.5 *Digit translation – highlighting the route and circuits tables*

which contains the necessary information on all the outgoing routes which may be used to transmit the call to its destination. The route table in Fig. 15.5 shows typical stored data, though, depending on the methods of routeing and charging in particular networks, it may not be necessary to store all this information. The data comprise:

(i)   The name of the route, or a code which represents it.

(ii)   The digits which must be transmitted when the route is being used for the particular destination.

(iii)   The charging rate, or a code representing it, for that route to the given destination.

(iv)   A pointer to a further table which stores the equipment numbers of the circuits on the route.

The controlling program uses the pointer which has been 'translated' from the input code to address the route table, in which the primary route, secondary route, etc., are stored. The pointer to the table actually points to the first entry in the table, i.e. the primary route. This controlling program, having thus identified the destination's primary route, uses the pointer stored in the table to access the route's table of circuits. In this, each circuit possesses a record containing its equipment number and a flag to indicate if the circuit is in use.

The controlling program examines the flag of each circuit in turn until it finds a free circuit. If there is no free circuit on the route, it returns to the route table, using the pointer which it stored at translation; it then repeats the procedure for the secondary route, and so on. The controlling program is written to access the table in sequence, beginning at the entry addressed by the translated pointer, and ending with the first route on which there is an available circuit, or at the end of the table. Because the number of alternative routes offered by an exchange varies according to the destination, the size of this table cannot be standardised; thus, the table must have a terminating character (e.g. 0). If no outgoing circuit has been found when the terminating character in the route table is reached, a congestion tone or a recorded announcement must be returned to the calling subscriber.

When a free circuit is found, it is reserved by setting its busy/free flag to busy. Its equipment number is stored in the call record, where it will be accessed when the switch path is being set up. The necessary outgoing digits, the charging code, an the route name or code from the route table, are also stored in the call record.

The digits are transmitted on the outgoing circuit when a transmission path has been set up to the next exchange. The routeing digits are necessary in systems which use end-to-end signalling with control vested in the originating exchange. In link-by-link systems in general and, in particular, on modern common-channel signalling links between SPC exchanges, the input digits are dispatched as the routeing digits. Each exchange along the route then uses the same digits as the key to its translations, and the storage of routeing digits in the route table becomes unnecessary. Because exchanges have different outgoing routes, the contents and sizes of their translation tables are different, although the original input digits are the key in each case.

The route name, or its code, may be used for charging or for later statistical aggregations. The charging code is used much later, perhaps off-line, when the call charge is being calculated. In many networks, all calls within a local area (say, all originating calls not having the trunk prefix) may be charged at the same rate, which may vary only according to the time of day. In such cases, therefore, it is not necessary to store the charging code in local route tables. However, more and more administrations are using the flexibility of SPC to provide an increased range of charging methods. It is therefore unlikely for this information to be considered superfluous.

15.2.3 Translation with a consistent numbering scheme

In networks, translation is somewhat simplified by using a standard length for the individual field of a number. For example, the standard may require a three-digit area code followed by a three-digit exchange code and a four-digit subscriber's number. For either the area code or the exchange code, the digit translation module of Fig. 15.5 can consist of a single table. Run-time efficiency can be achieved if this consists of 1000 locations, so that it can be addressed directly using key transformations (see Fig. 15.6). Alternatively, for a network with few nodes, the table can be reduced to fit the data exactly, although scanning must then be used to access it (see Fig. 15.7). Time and storage efficiency tradeoffs are discussed in Chapter 12.

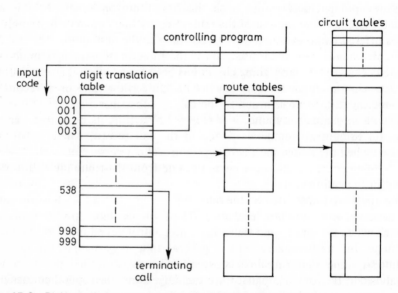

Fig. 15.6 *Digit translation – using a single digits translation table*

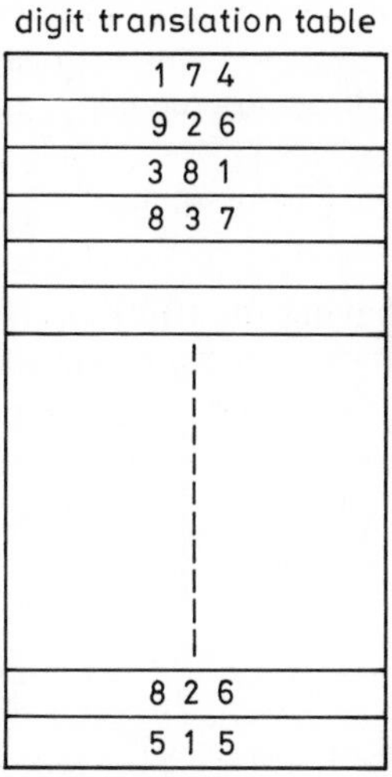

digit translation table

| 1 7 4 |
| 9 2 6 |
| 3 8 1 |
| 8 3 7 |
| |
| |
| |
| 8 2 6 |
| 5 1 5 |

Fig. 15.7 Size of table exactly matches number of entries. Access is therefore by scanning and order is irrelevant

The example address-code standard, given above, may be even more amenable to translation if it is preceded by a trunk prefix for nonlocal calls, and if the area code is omitted for all local calls. If the trunk prefix is input, it is deducible that the next three digits comprise the area code; if the trunk prefix is not input, the first three digits must comprise the exchange code. Although the use of a trunk prefix facilitates the choice of digit-translation table (local or trunk), a consistent numbering scheme, such as in the example, does not require a prefix except for international calls. However, without a prefix, it could not be deduced (except in some specific cases) whether the first three digits comprised an area code or an exchange code without the subsequent digits being counted. The need to count the digits would delay translation until it will clear whether more than seven digits had been input. However, since digits are counted anyway, to ensure that circuits are not occupied by incompletely defined calls, the processing involved is not increased.

A local digit-translation table will need to carry out translations of incoming terminating calls. In this case (for exchange code 538 in the example), instead of a pointer to a route table, the translation table provides an indicator to the controlling program that the call is to be connected to a subscriber's line rather than a route. To keep the controlling program general and applicable to all exchanges (of a given type), an appropriate unique table entry is standard for all exchanges.

15.2.4 Translation with a nonconsistent numbering scheme

The most usual way of implementing the digit translation is by multistaged tables, as illustrated in Fig. 15.8. This method is essential if the number of

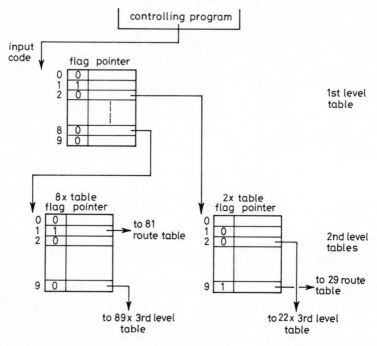

Fig. 15.8 *Digits translation by multistage tables*

digits in area or exchange codes is variable. The principles of multistaged tables in software are described in Chapter 12, but a number of points must be made about their use for digit translation in a network with a nonconsistent numbering scheme.

Because it is uncertain, in general, how many digits will be required in any particular instance to define routeing, any one table (say, a second- or third-level table) is likely to contain some pointers to route tables and some to the next-level digit-translation tables. Each table entry must therefore process an associated flag bit to distinguish between the two. The convention of Fig. 15.8 shows that, if the controlling program finds the flag set to 1, the associated pointer is to the appropriate route table. If the flag bit is zero, further routeing digits are required, and the pointer is to the appropriate next-level digit-translation table.

It is also important to note that, in most networks, fewer digits are required to define the routeing out of an exchange than to define the actual destination of the call. This is most obviously so when the national numbering scheme corresponds to geographical location. Yet, both the charging code and the routeing digits dispatched must be based on the destination, rather than on the outgoing route. Thus, if the route table is accessed before the destination has been deduced, the controlling program must return to the digit-translation tables and continue translation for the purpose of determining the charging code and routeing digits. This in turn would mean that these quantities could not be stored in the route table but would require further tables. This would create a more complex data structure than that of Fig. 15.5 and would also require more instructions at run-time for its execution. It is therefore more economical for routeing (i.e. accessing of the route table) not to take place until sufficient translation has been performed to accommodate all three functions: route choice, routeing-digit transmission and charging.

Single-digit tables based on key transformations (discussed in the preceding Section) are not as common as multistaged tables. However, two factors are likely to reverse this. The first is that memory has become miniature and cheap; thus, more and more frequently, economy of storage loses the battle with run-time efficiency. The second is that all-digital networks are bringing with them huge changes in network design. Fewer trunk exchanges and more consistent numbering schemes are two of these. In consequence, as networks become all-digital, routeing will become a simpler process. On the other hand, flexibility in charging is being seen by administrations as a prime requirement in the emerging more competitive environment. Thus, rather than being simplified, charging will become more complex. This is discussed later in the chapter.

15.3 Setting up a call

As stated at the beginning of this chapter, the main control actions required for setting up a call are best exemplified in the terminating call. The main difference between this and an outgoing call is in the method of selecting the termination to which the call must be connected. The method of choosing an outgoing circuit has been described under 'Digit translation (Section 15.2)'. This Section concentrates on the handling of a terminating call. An understanding of the actions which have already been explained (e.g. accessing COS records, connecting auxiliary equipment, setting up a switch path) is assumed.

A call terminating on the exchange may also have originated on the exchange, or it may have arrived on an incoming junction. Most of the actions (see the SDL diagram of Fig. 15.9) are independent of the call's origin, but, where dependence exists, the distinction will be made. The following description should be read with reference to Fig. 15.9.

When analysis or translation has shown the call to be terminating on the exchange, the directory number is used as the key for accessing the called subscriber's record for the purposes of looking up the equipment number of the called subscriber's line-terminating unit and interrogating the busy/free bit in the transient data to ensure that the line is not engaged.

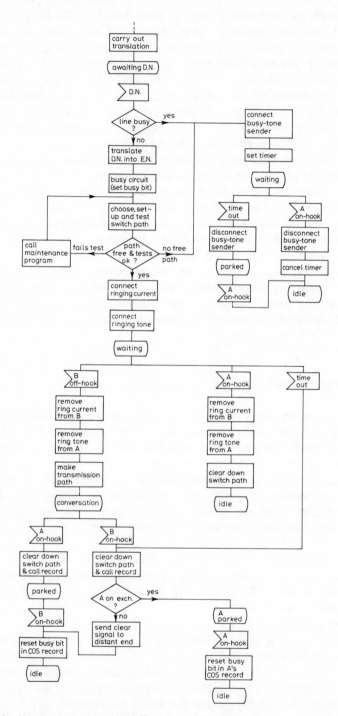

Fig. 15.9 *Handling a terminating call*

If the called subscriber's line is busy, it is necessary to send the busy tone to the calling subscriber. For an originating call, a tone sender may be connected to the calling subscriber's line via the concentrator switch (as described in the preceding chapter in the context of a dial-tone sender). However, for an incoming call, this method of connection is inapplicable, and two other options apply. The first is the connection of a tone sender (via the group-switch block) to the outgoing channel of the circuit, so that the busy tone originates at the terminating exchange and is transmitted over all the links of the circuit to the calling subscriber. The second option is to use the signalling system to send a 'called subscriber busy' indication back along the circuit to the originating exchange, which then connects a busy-tone sender to the calling subscriber's line. This allows the interexchange connections to be cleared down and not held ineffectively. This method is a feature of common-channel signalling systems.

If the called subscriber's line is free, its busy bit in the COS record is now set. This, in effect, seizes the line, so that it would give the busy indication to subsequent callers. The equipment number is copied from the subscriber's record into the call record being created for the call.

Using the called subscriber's equipment number, and the equipment number of the terminating equipment of the calling line (whether a local subscriber's line or a circuit on a trunk route), an attempt is made to set up a path through the group switch. This has been described in detail in Part II. If there is blocking in the exchange and, thus, no available switch path, the busy tone must be sent to the calling subscriber by one of the methods already described.

When a switch path has been reserved for the call, a ringing current must be sent to the called subscriber, and this is done from the subscriber's line-terminating unit. The ringing tone is connected to the calling subscriber's line via the concentrator switch for an originating call, or via the group switch for an incoming call.

At this point, the call is placed in the steady state of 'waiting', from which only three valid input signals can occur. If neither the called (B) subscriber's off-hook, nor the calling (A) subscriber's on-hook, signal is detected with a predetermined time, a time-out occurs. The controlling program then disconnects the ringing-current and ringing-tone senders from the called and calling subscribers' lines and resets their busy bits in memory to free them for use with other calls. It also disconnects the switch path. Then, if the call has originated on the exchange, the calling subscriber's line is parked. For an incoming call, the disconnect signal is sent back to the originating exchange, which parks the calling subscriber's line, thus freeing the circuit. If the calling subscriber clears down, the on-hook input signal is received. The ringing-current sender, ringing-tone sender and switch path are then disconnected and the circuit cleared. The third legitimate input signal is 'B off-hook'. Then, the ringing-current and ringing-tone senders are disconnected, as before, but now transmission is provided on the switch path so that conversation can take place.

When conversation is terminated by one or other of the subscribers inputting the on-hook signal, the call is processed as shown at the bottom of Fig. 15.9. The lines are left in either the idle or the parked states, depending on the actions of the subscribers. At this point, the software call record (i.e. the storage which it occupies) must be made available for use with another call. Before it is released, however, some of the information in it must be extracted for output from the exchange for use for maintenance, network management, billing, planning, marketing, etc., as discussed in Chapter 19.

15.4 Charging

Charging is carried out by a local exchange for originating calls. When a call is set up, the times of significant events in its progress are inserted into fields of the call record designed for the purpose. Thus, charging may be carried out by analysing the appropriate contents of the call record. This may be done off-line, by a computer, long after the call has been made. Traditionally, the process requires knowledge of the call's destination (or the charge group within which the destination falls), the duration of the call, the time of day and day of the week during which the call is in progress, and the tariff rate or rates applicable.

Traditionally, there have been a limited number of charge groups and a limited number of tariff periods during a day. In spite of this, the number of combinations of charge groups and tariff periods has been considerable. Implementation of these in software has required tables to crossreference destination to charge group (as shown in the route table of Fig. 15.5), and charge group to tariff period.

In the current environment, competition is increasing, new telecommunications carriers are coming into being and carriers are sharing each others' networks, so there is a need to base charging on a greater number of parameters. For example, the chosen route out of the exchange may become a basis for charging. It is also becoming increasingly attractive to charge according to the type of termination on which the call originates; for example, domestic subscribers and business subscribers may be charged at different rates. An administration (or carrier) thus gains flexibility in charging (and marketing its products) at the expense of the storage of large numbers of tables and the time it takes to access them. Although this storage and processing may be carried out off-line, the information needed to make the calculations possible must be collected at the time of setting up the call. The identity of the outgoing route, determined during digit translation, must be stored in the call record and output at the end of the call, as must line-type information obtained from the COS record. Charging is an example of an area in which SPC has allowed flexibility but, in doing so, has increased the demands on itself in terms of the increased data and processing requirements.

15.5 Customer facilities

When a subscriber's digits are analysed (see Fig. 15.4), it may be found that they signify an attempt to set up, use or clear down a facility. How these aims are achieved in SPC will now be described, using as examples the implementation in software of short-code dialling, alarm call and three call-transfer facilities: transfer all calls, transfer on no reply and transfer on busy.

For these descriptions, the following conventions will be assumed. When the first digit is '*', a facility is to be set up or used; the subsequent digits determine which is the case and which facility is required. If, in setting up a facility, there is the need for a terminating digit, '#' is used. When a facility is to be cleared down, the first digit must be '#'.

15.5.1 Short-code dialling

In this example, it is assumed that the use of the facility depends on the following digit sequences:

'* 21 ab DN #' sets up the short code. Thus, when 'ab' is subsequently dialled, it will be translated into the full directory number DN.
'# 21 ab' cancels the short code 'ab'.
'* 1 ab' uses the short code 'ab', translating it into DN.

The processes are shown in Fig. 15.10. It should be noted that, for the sake of clarity in this description, the Figure omits some details which would be necessary for a full specification.

15.5.1.1 Short-code set-up. When the digits '*21' have been received (see Fig. 15.10), the task program (called into play as a direct result of the digits dialled) must examine the calling subscriber's COS record to check if use of the short-code facility is allowed. If it is, the COS record contains a pointer to a table which is allocated to the subscriber. This contains a number of nodes in a sequential linear list (the number of short codes which the subscriber may have at any one time). Each node consists of two fields: one for the short code and the other for its equivalent directory number (DN) (see Fig. 15.11).

The program may be designed to check, at this point, if the table is full and, if it is, to connect the subscriber to an appropriate announcement. On the other hand, such a refinement may not be provided; the system may work on the basis that, if attempts are made to set up short codes after the table is full, previous entries will be overwritten.

When the short code itself, 'ab', has been received, the system awaits the directory number of which 'ab' is to be the short code. This number is of uncertain length because it may be a local, trunk or international number; it must therefore be followed by a recognisable terminator, '#'. When the terminating digit has been received, the directory number and its short code are loaded into the table (see Fig. 15.11). It should be noted that, although

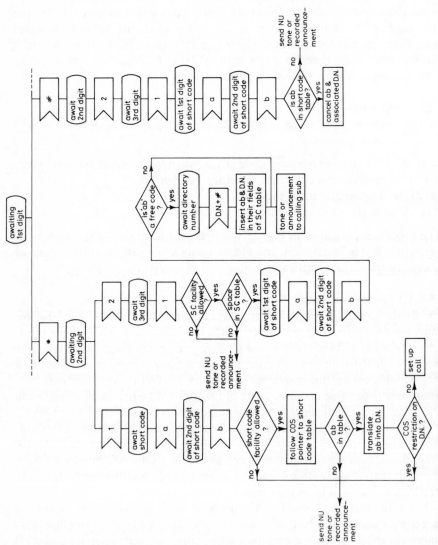

Fig. 15.10   The short-code dialling facility

short code 1	directory number 1
2	2
3	3
4	4
5	5
6	6
7	7
8	8
9	9
10	10

Fig. 15.11 *Short-code translation table*

there may be no restriction on the type of directory number for which a short code may be set up, when it is translated for set-up of the call itself, it is subject to the usual checks of the COS record for allowable outgoing calls.

15.5.1.2 Use of the short code. When the short code '*1ab' has been received (see Fig. 15.10), the subscriber's COS record is checked. If short-code dialling is allowed, the pointer to the subscriber's short-code translation table (see Fig. 15.11) is followed, and the table is scanned for the code 'ab'. If this has not been inserted, the call attempt must be aborted and a number-unobtainable tone (or an appropriate recorded announcement) returned to the subscriber. If the code is found, it is translated into its equivalent full directory number. An additional check is made of the COS record to ensure that the number is not of a type for which outgoing calls are barred. When it has passed this test, the call is set up. If it is found that the call type is barred, a number-unobtainable tone (or an appropriate recorded announcement) is connected to the subscriber's line.

15.5.1.3 Clear-down of short code. When the digits '#21ab' have been received (see Fig. 15.10), checks similar to those described above are made of the subscriber's COS record and short-code translation table. If the code 'ab' is found, it is removed by setting the code field to a null value, or one which is impossible to achieve with an allowable short code, or by resetting a flag.

15.5.2 Alarm call

The set-up and cancellation of an alarm call are illustrated in Fig. 15.12. It is assumed that the dialled code '*54abcd' achieves set-up of an alarm call, and that '#54abcd' cancels it. The digits 'abcd' define the time to the minute (on a 24h clock) at which the alarm call is to be made. It is also assumed that a subscriber can program the system in this way for more than one alarm call.

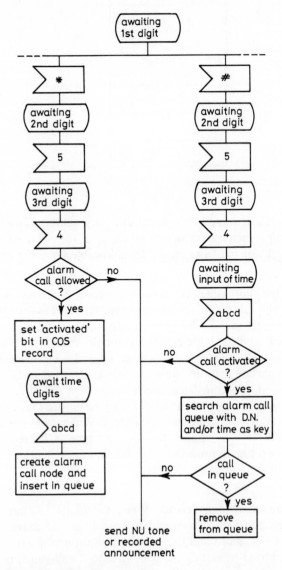

Fig. 15.12 *The alarm call facility*

If this were not the case, cancellation of a call could be achieved by the code '#54' only.

15.5.2.1 Alarm call set-up.

After the digits '*54' have been received (see Fig. 15.12), a task program may have to ensure from the COS record that the subscriber is allowed to set up alarm calls. Alternatively, because a subscriber is automatically billed for an alarm call when it is made, the facility may be made available to all subscribers; in this case, no check would be necessary.

When the time digits, abcd, have been received, a timer must be activated. This is done by use of a linked list whose nodes consist of three fields: one for the time input by the subscriber, one for the subscriber's directory number and one for a link field. Nodes in the list are in alarm-time chronological order, so the new node may have to be inserted into the body of the list (see Fig. 15.13). A flag should also be set in the transient data of the subscriber's COS record to indicate that an alarm call has been activated. At 1 min intervals the control system activates a program to make any alarm calls that are due. It examines the first node in the queue, using as its key the actual alarm time in hours and minutes. If there is a match, the program must also check the next node in the queue, and continue checking subsequent nodes until a mismatch is found. Because the nodes are in chronological order, no further searching is necessary. For those nodes with a match, calls are set up to the associated directory numbers and, for successful alarm calls, the owners of the directory numbers are billed. If flag bits in the COS transient data have been set, they must be reset when the calls have been made.

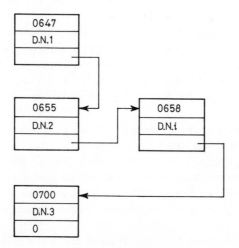

Fig. 15.13 *Insertion of a node into the alarm call queue*

15.5.2.2 Cancelling an alarm call. To cancel an alarm call (see Fig. 15.12), the linked list must be searched. To avoid an abortive search if no alarm call has been activated by a given subscriber, the transient data of the subscriber's COS record must first be checked.

If a subscriber is restricted to one alarm call at any one time, '#54' could initiate cancellation of the call. In this case, the search key must be the subscriber's directory number (acquired from the subscriber's record). If the subscriber needs to input '#54abcd' in order to cancel an alarm call, then the time (abcd) could be used as the key. When a match is found, directory numbers also have to be compared, as it is possible that more than one node in the queue would have the same time setting. In either case, cancellation of the call is achieved by removing the node from the queue and returning it to the available stack.

15.5.3 Call transfer
The following digits are used as examples for the setting up and cancelling of the call-transfer facilities:

transfer all calls (TAC) set up : *34DN#
transfer on no reply (TNR) set up : *35DN#
transfer on busy (TOB) set up : *36DN#
cancel TAC : #34
cancel TNR : #35
cancel TOB : #36

Because the directory number to which calls are to be transferred is not necessarily of standard length, the terminating character '#' is required. Although, in theory, any category of number, even international, could be specified, an administration may choose to allow transfers only to local, or perhaps, national, numbers. The question of charging for the call is of relevance here. Clearly, the calling subscriber, having dialled (say) a local number, will not be prepared to be charged for a trunk or international call. The called subscriber, who initiates the transfer, must therefore accept its cost. If nonlocal transfers are permitted, the software process which controls their use must contain tasks for billing the called number.

15.5.3.1 Call transfer set-up. Since the set-up processes of the three examples of call transfers differ only very slightly, a single diagram (Fig. 15.14) illustrates the principles of the algorithm for all three. As in previous diagrams, it shows the features essential to transfer facilities. It omits repetitive tasks such as storing digits and setting timers. Also, it ignores input signals (such as 'A on-hook' or erroneous digits) which are irrelevant to the examples.

When the input signals '*3' have been received, the SDL diagram leads the control to expect one of the three digits 4, 5 or 6. When one of these has been

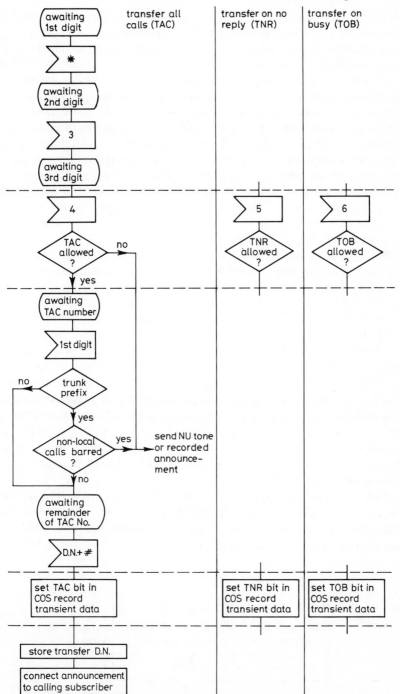

Fig. 15.14 *Call transfer set-up (diagrams the same except where shown)*

received, there is a check of COS-record semipermanent data to ensure that the facility requested is allowed. The COS record is found and the appropriate bit examined. If it is set to zero, then (by convention only), the number-unobtainable (NU) tone, or a recorded announcement, is connected to the line to inform the calling subscriber that the desired transfer facility is not permitted. Otherwise, the call is placed in the steady state of awaiting the first digit of the directory number to which calls should be transferred.

When the first directory-number digit is received and analysed, it is stored, pending the arrival of the whole of the transfer number, followed by the terminating character '#'. If transfers to nonlocal numbers are not permitted, the first digit must be checked to ensure that it is not the trunk-prefix digit. If it is, the calling subscriber is connected to the NU tone or an appropriate recorded announcement. However, such transfers could be allowed to certain customers; in this case, another check of the COS record, for a bit to indicate approval of the facility, is needed.

When the full transfer directory number, followed by the terminator '#', has been received, the COS record transient data must be adjusted, so that incoming calls are appropriately diverted. The appropriate bit for the transfer facility must be set, and the transfer directory number must be stored where it can be found when an incoming call requires it. The COS record may be designed to store transfer numbers. However, if only a portion of subscribers use the transfer facilities, this would incur a significant waste of storage. Moreover, subscribers may be allowed to set up more than one transfer facility at a time. Whereas 'transfer all calls' precludes the other two, 'transfer on busy' and 'transfer on no reply' can exist together. This requires storage of two transfer numbers, so two fields must be reserved for the purpose. Thus, although transfer numbers may be stored in the COS record, as in Fig. 15.15, it may be cheaper to store them separately and so use only as much storage as they require. In this case, they must be accessed via a pointer in the COS record, as in Fig. 15.16.

When the COS-record transient-data bit has been set, the transfer number stored and the pointer set, if necessary, a recorded announcement may be connected to inform the subscriber that the facility has been set up successfully.

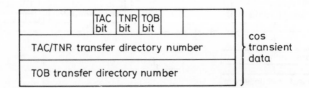

Fig. 15.15 *Transfer directory numbers stored in class-of-service record*

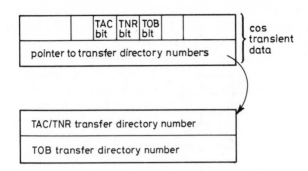

Fig. 15.16 *Transfer directory numbers stored remotely from COS record*

15.5.3.2 Use of call transfer. When the called directory number of an incoming terminating call is received, it is used as the key to access the called subscriber's record (see Fig. 15.17). This directory number is translated into the equipment number of the called subscriber's line-terminating unit. It is also necessary for accessing the COS record to check if the line is on transfer. The check involves examining the three transfer bits within the transient data. Depending on the number of subscribers on the exchange permitted to transfer their calls, it may be worth making a general check by considering the three bits together as one field. Then, only when a nonzero value is found is a further check made of the same bits to discover which facility is implemented. This can achieve run-time efficiency if transfer facilities are not heavily used.

If the 'transfer all calls' bit is set, the control program must find the transfer directory number, either from the COS record or by following the pointer stored there. It will then use this number to set up the call. This may be an own-exchange call or an outgoing call. If it is nonlocal, it must be monitored and charged to the subscriber on transfer. The calling subscriber's call, to the number actually addressed, is monitored and charged for at its originating exchange.

If the 'transfer on busy' bit is set in the called subscriber's COS record (see the TOB branch in Fig. 15.17), the above procedure is carried out if the line is found to be busy. For this, the busy/free bit in the COS transient data is checked.

If the 'transfer on no reply' bit is set, the call is set up to the called number in the usual way and, at the same time, a timer is set. If this times-out and issues an input signal before the called subscriber answers, the call is transferred, as described above.

15.5.3.3 Cancelling transfer facilities. Fig. 15.18 shows the cancellation of the three transfer facilities in the example. Cancellations are initiated (by convention) by the use of '#' as the first digit, and it is convenient and easy

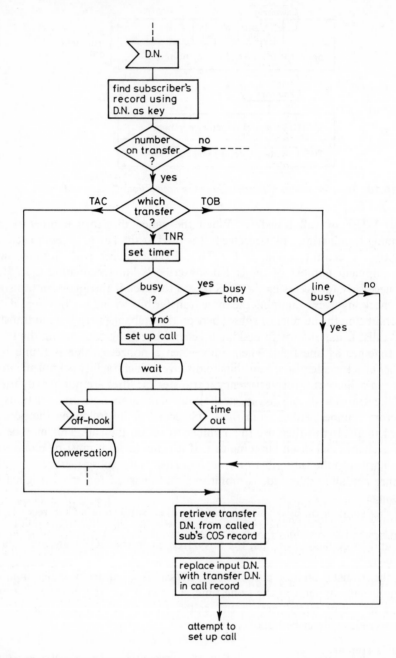

Fig. 15.17 *Use of call transfer facilities*

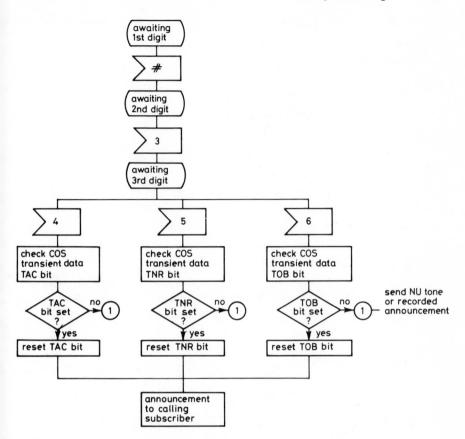

Fig. 15.18 *Cancelling call transfers*

for subscribers if the next digits correspond to those which were used for setting up the facilities themselves.

When the third digit has been received, the facility to be cancelled has been defined. The COS record is accessed, using the line-terminating equipment number as the key, and the appropriate bit checked. If it is found to be set, it is reset, and an announcement may be connected to the subscriber's line to confirm that the facility has been cancelled. If the bit is found not to be set, the subscriber may have made a mistake in specifying the transfer facility to be cancelled. An announcement would warn the subscriber of this error, and provide the opportunity for another attempt.

15.6 References

1 CARBAUGH, D.H., DREW, G.G., GHIRON, H. and HOOVER, E.S. (1964): *No.1 ESS Call Processing*, Bell System Technical Journal, **43**(5), Pt.2

2 ULRICH, W. and VELLENZER, H.M. (1964): *Translations in the No.1 Electronic Switching System*, Ibid.

3 HILLS, M.T. and KANO, S. (1976): *Programming Electronic Switching Systems*, (Peter Peregrinus Ltd)

Chapter 16

System Software

The preceding two chapters have described the applications software necessary for call processing in a telephone exchange. As in all computer systems, the processing of SPC applications software requires system software, and this will now be discussed. Since both applications and system software may be written specifically for exchange control, it is somewhat arbitrary how the division between the two is defined. The approach taken here is that the software which provides call-processing control and facilities and that which interacts with the telephonic equipment constitute the applications software; that which supports the applications software and provides more basic computing and communications facilities constitutes the system software.

The system software, in conformity with software-engineering principles, is designed as a collection of modules, with each module performing a given function (see Fig. 16.1). The totality of the system software may be referred to as the operating system.[1] However, in this chapter, the system-software functions necessary in an SPC system are described singly, without reference to their total structure. Their structure will be considered in Chapter 17.

16.1 Clock-pulse handler

The timing of an SPC system is controlled by a clock pulse generated at regular intervals within the hardware. (This is different from the much faster pulse used to drive the computer circuitry.) As the clock pulse is the very hearbeat of the SPC system, its source is secured by duplication or triplication, with hardware circuitry to check its function. The frequency of the clock pulse differs from system to system; once every 5 ms is adequate, but in practice its rate is faster than that. The pulse circuitry is such that the pulse activates an interrupt[2] of the type used in all computer systems. This causes the current activities of the processor to be suspended, and control to be transferred to a software module known as the clock-pulse handler.

This interruption of processing by the clock pulse has two main purposes. The first is ensuring the correctness of the software real-time clock by regular

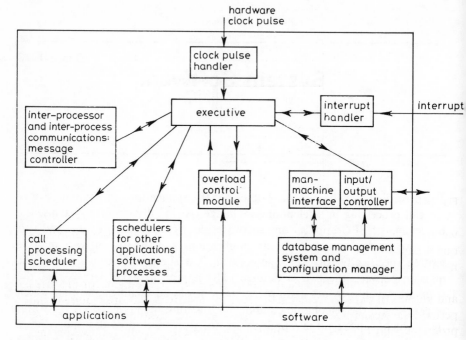

Fig. 16.1 *SPC system software*

updating. The second is creating the interrupt which allows a restructuring of the flow of processing to ensure that time-critical functions are carried out within an acceptable time.[3]

16.1.1 Real-time clock

A clock within the system is used for many purposes.[4] Each time the system provides an output of information, it needs to attach a header which contains the date and time. Events which are activated on the basis of actual time, such as alarm calls and some maintenance routine-test programs, depend on a real-time clock. The times of many events, such as a call's answer and clear-down signals, must be recorded for each call within the call record.

For the real-time clock to be used by software programs, its value must be interpretable by them and capable of being stored in software. Fig 16.2 shows that it occupies seven fields, the number of words depending on the word size in a given system. On receiving control, the clock-pulse handler's first task is to update the value of the real-time clock. It starts this process by adding one to the value of the pulse-count field and then testing the result. For a 5 ms pulse, a value of 200 indicates the passing of 1 s. Thus, every 200 clock pulses, this value is detected, the pulse count is restored to zero and 1 is added to the

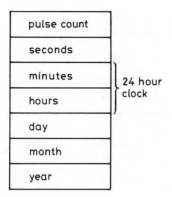

Fig. 16.2 *The software real-time clock*

value of the 'seconds' field. When the latter reaches 60, it is reset to zero, and one is added to the 'minutes' field, and so on. More elaborate algorithms are needed for correctly updating the 'month' and 'year' fields.

The clock itself may be used in various ways. On printouts, error messages and outputs of data on to magnetic tape, the last five fields are required, and perhaps the second if such detail is necessary. For activating daily routine-test programs, only the hours and minutes fields need to consulted; these together constitute a 24 h clock. Thus, the programs for using the clock must be written with a knowledge of the clock's structure and the purpose of interrogating it. To provide flexibility and increase the chance of correctness, the various types of interrogation of the clock are normally encoded once and retained as subroutines which are called, as required, by the applications or system software.

16.1.2 Initialising processing flow
The second purpose of the clock pulse is to ensure that the flow of processing is controlled, and that time-critical functions are carried out on time and not delayed by prolonged use of the processor by less important programs. Thus, the clock-pulse handler, having updated the real-time clock, passes control to the 'executive', which is the software module designed to control the flow of processing, as in Fig. 16.3.

16.2 The executive
The 'executive' is a term used for the program which provides control of the flow of processing in an SPC system. In systems with distributed processing, each remote processor, as well as the central processor, will have its own executive. Because the various processors perform different functions, the actual functions required of an executive in any processor depend on the

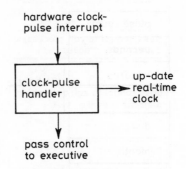

Fig. 16.3 *Operation of the clock-pulse handler*

system in question and on the particular functions of the processor within the system. The principles, however, will be illustrated by assuming a system within which a single processor carries out all the control functions. Examples of the functions activated by the executive are shown in Fig. 16.4.

The executive itself is a central controller, but it retains control within itself only for very short periods of time. It takes decisions, but only on if and when to activate other processes by passing control to them. A flowchart of those executive-program tasks shown in Fig. 16.4 is given in Fig. 16.5. The processes are seen to be activated in a fixed order designed into the system. The order shown, however, after the three call-processing functions, was chosen arbitrarily for this example. In an operational system, they may be allocated different priorities.

Whenever the clock pulse occurs, control is transferred to the clock-pulse handler, which immediately passes it to the executive. Then, the first processes

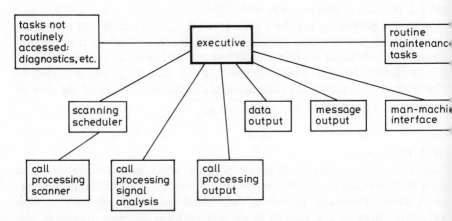

Fig. 16.4 *Example of processes activated by the Executive in a single-processor system*

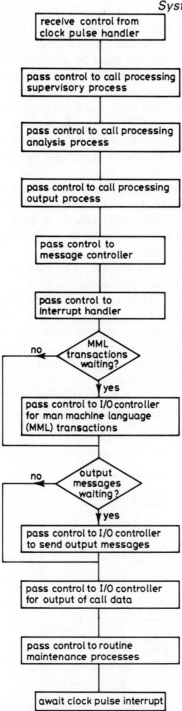

Fig. 16.5 *An example of executive program actions*

to be activated are those accorded highest priority: in particular, supervision processes, such as those which scan lines to detect input signals (which may persist for as little as 33 ms and must be detected within this time). Each process receives control, completes its tasks and returns control to the executive. Control is then passed to the next process, and so on. If time permits, all processes perform their functions, in which case the executive regains control and simply awaits the next clock pulse. However, if the clock pulse occurs before this, the process being executed at the time is suspended and control is returned to the executive, so that higher-priority tasks can be executed.

Under normal conditions, the interval between clock pulses is adequate for all processes to be completed. However, under overload conditions, this may not be so, as call processing then requires a greater proportion of the processor's time. Special actions may be necessary if overload persists, and these will be discussed in Section 16.9 below.

Calling the I/0 controller for the output of data is a function which may need to be activated only once per minute, or every 5 min or longer, rather than at every clock pulse. Time can be saved by buffering the data and activating the peripheral hardware less frequently. In this case, the executive needs only to increment a counter at each clock pulse and then test it for a given value. If the counter has reached that value (X, in Fig. 16.6), the I/0 controller is activated; otherwise, it is not. If data are to be oputput once per minute, and the clock pulse interval is 5 ms, then control would be passed to the I/0 controller when X = 12 000. Alternatively, the incrementation could be carried out by the clock-pulse handler, and the executive would only need

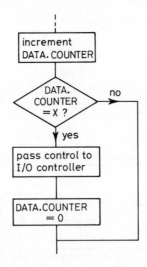

Fig. 16.6 *Executive calling I/O controller once in X clock-pulse intervals*

to check the counter. The advantage of this is that, if overload prevents the executive from activating the counter-check process, incrementation is not omitted. In case checking is delayed beyond the given clock-pulse interval, the process must be designed to test if the counter value is equal to or greater than X (in the example, if $X \geqslant 12\ 000$). Of course, when events such as I/0 are to be carried out at certain set times, activation may be accomplished by checking the real-time clock. This is convenient when the event should occur at widely spaced intervals, hourly or every 24 hours, say.

An executive is tailored to its system. The one described above is only an example. In most systems, the executive of the central processor would do far more than the example, and that of a remote processor would very likely do far less. It should also be noted that a process (defined in Chapter 13) may consist of a number of tasks. While a process is activated by the executive according to certain predetermined priorities, it may be the case that the tasks within a process are also prioritised. This can be achieved by extending the executive to allocate control to tasks, rather than processes. However, this would necessarily reduce flexibility of operation as well as the ability to make changes to the system involving the order of processing. If the tasks to be processed vary from clock-pulse interval to clock-pulse interval, it is usual for them to be activated by the use of a 'scheduler', after the process itself has been called by the executive. In this sense, a scheduler is a program which arranges that a given set of tasks is processed in the correct sequence.

16.3 Scheduling tasks within a process

In its most typical form, a scheduler has three components: a bit map which contains the information on which tasks are to be processed at each clock-pulse interval, a table containing the addresses of the task programs, and a

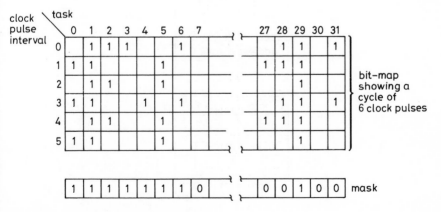

Fig. 16.7   A scheduler bit map and mask

program to interpret the bit map, find the address of the appropriate task from the table and transfer control to the task.

The bit map consists of a number of words in memory.[5] Each bit in a word represents a particular task, so the bit map can be used to schedule as many tasks as there are bits in a word. In the example of Fig. 16.7, there are 32 bits per word and there are 6 words comprising the bit map. Each word represents a clock-pulse interval; thus, the total cycle of time represented by the map is equal to 6 clock-pulse intervals. The map is in fact a matrix: all the bits in the same column represent the same task, and all the bits in the same row (a word) represent the same clock-pulse interval. A '1' anywhere in the map indicates that the task represented by that bit should be processed during the interval represented by the word which contains the bit. In the example of Fig. 16.7, therefore, task 0 is to be processed during intervals 1, 3 and 5, task 2 during intervals 0, 2 and 4, tasks 1 and 29 during every interval, etc. A table corresponding to this example is illustrated in Fig. 16.8. There are as many words in the table as there are bits in a word (i.e. 32); these are consecutive in memory, so that fast access can be gained by key transformations.

A flowchart of the basic operation of a scheduler is shown in Fig. 16.9. The program uses a counter (COUNTER.1) to indicate the current clock-pulse interval. When control is passed to the scheduler program, on each clock pulse, it increments the counter (resetting it to zero after the sixth) and then examines the bits in the bit-map word indicated by the value of COUNTER.1.

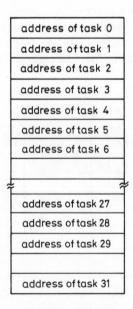

Fig. 16.8 *A scheduler table storing task addresses*

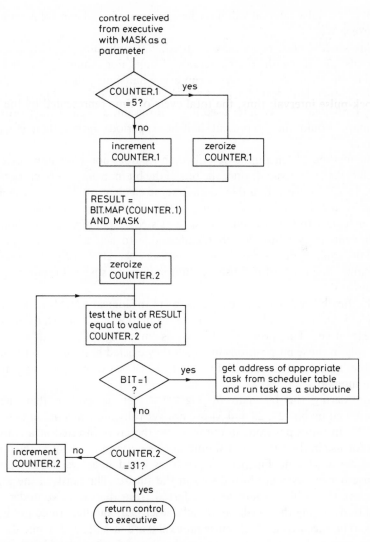

Fig. 16.9 *The basic logic of a scheduler program*

In this sense, BIT.MAP(0) equals word 'zero' of the bit map and BIT.MAP (COUNTER.1) equals the word indicated by the value of COUNTER.1.

To keep a check on which bit is being examined, another counter, COUNTER.2, is used. Wherever there is a '1', the program seeks the address of the corresponding task in the table and passes control to the task. When processing of the task is complete, control is returned to the scheduler, which then continues to interrogate the bits of the word in the bit map. When all 32 bits of the word have been tested, all tasks scheduled for processing during the given clock-pulse interval will have been processed, and control is returned to the executive.

This scheduling process may be used, not only directly with clock-pulse intervals, but also with larger intervals of time. For example, if a selection of routine-maintenance programs is run every half-hour, they can be scheduled in the manner described above, with the scheduler being called by the executive every half-hour. The counter used in the program, as well as the words of the bit map, would then represent half-hour periods rather than clock-pulse intervals.

The method of scheduling described above does not allow any flexibility in which tasks are activated, since the bits in the bit map are, in effect, permanent data. It is, however, desirable to have some element of choice in which tasks are activated, beyond simply having a '1' in the bit map. For instance, it may be convenient to load the bit map with '1's corresponding to the scanning of equipment not yet installed, but undesirable to activate the scan programs until the equipment has been brought into service. Also, during overload in the exchange, one form of defensive action may be to suspend the processing of certain tasks.

This flexibility is achieved by carrying out the logical AND function between the clock-pulse-interval word from the bit map and a word transferred from the executive.[6] This process is known as 'masking'. The result can contain '1's only in those bit positions in which they existed both in the bit-map word and the 'mask'; it is used to define the tasks to be executed during that clock-pulse interval.

Considering the bit map of Fig. 16.7, let us assume that the tasks represented by bits 27, 28 and 31 are not yet fully implemented in the system. The '1's in those positions in the map are therefore inserted at a convenient time for use in the future and should not be allowed to be interpreted as task activators at present. Further, bit positions 7 and 30 are vacant and available for use if new tasks are introduced in the future. The mask in the Figure is therefore that which would be transferred from the executive under normal conditions. Using this mask, as already described, and illustrated in Fig. 16.9, tasks to be executed at clock-pulse interval 0 are tasks 1, 2, 3, 6 and 29. Those to be executed at interval 1 are tasks 0, 1, 5, 29, etc. Under overload conditions, the executive may pass different masks to the scheduler, according to the stage of overload.

16.4 Interrupt handlers

In a commercial computing system, interrupts are used for a number of reasons, among them to ensure that programs which demand an inordinate amount of processor time, or of input/output time, do not preclude the handling of other tasks. In an SPC system, the primary task to be performed is call processing, and this does not rely on interrupts, other than the clock-pulse interrupt. It has already been shown that call processing is an event-driven, steady-state process and that a call remains in a steady state, not requiring either processor time or the use of input/output equipment until the occurrence of an event in the form of an input signal. There is no danger, therefore, of a call occupying the processor for an excessive time. Further, because call processing is itself the primary function of the system, there is no reason to protect it from itself by interrupts, other than the clock-pulse interrupt which is implicit to the system and which protects call processing against excessive processor occupancy by other tasks.

In the event of faults, having to detect messages arriving at the processor and take quick action may give rise to the need for interrupts. However, even the last statement must be qualified by a definition of 'interrupts'; in the context of SPC, the term may refer to two phenomena: an interrupt to the processing being carried out and an interrupt to the flow of processing.

An interrupt to processing involves the transfer of control to the 'interrupt handler'. The task being processed and its data are stored, so that processing can be continued at a later time from the point at which it was interrupted. The use of this type of interrupt during call processing impairs the system's efficiency in executing its prime function. It is essential only in situations of failure.

An interrupt to the flow of processing involves the executive, because this is the controller of processing flow. It can be arranged that a message arriving at the processor causes a flag to be set at the input port of its arrival, rather than initiating an interrupt to processing. Then, at predetermined points in the flow of processing, the executive calls the interrupt handler to check, or 'poll', the flags, as in Fig. 16.10. Any flag set to '1' requires attention, and the interrupt handler transfers control to whichever process is appropriate. For example, a flag denoting an incoming message requires the attention of the message controller or input/output controller; one denoting a fault may need

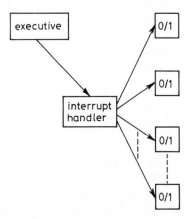

Fig. 16.10 *Interrupt flags polled by interrupt handler*

a particular maintenance program. The interrupt handler is programmed to call a particular process for each flag. When the interrupt has been dealt with, the flag is reset, and control is returned to the interrupt handler; this carries out any other necessary flag checks before passing control back to the executive.

The frequency with which the interrupt handler is called by the executive depends on the urgency of the interrupts. It can be arranged that a number of calls are made during each clock-pulse interval (as suggested in Fig. 16.11) and that the interrupt handler contains a scheduler to determine which flags are polled on the various calls. Certain flags, for example those involving messages from other processors, which may be awaiting acknowledgment, may be polled on each call. Others, for example those involving man-machine communication (in which a delay of a number of milliseconds is unimportant) may be polled only once per clock-pulse interval.

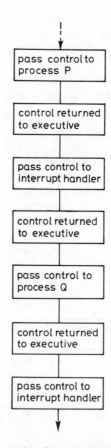

Fig. 16.11 *Executive calls interrupt handler several times during each clock-pulse interval*

16.5 Interprocess and interprocessor communication

In most modern systems, the transfer of information between processes, whether or not they are in the same processor, is achieved by message-based communication, the principles of which are described in Chapters 2 and 20. However, another method of communication, widely used, though less modern, is via the medium of common storage.

16.5.1 Processor interconnection

Where remote processors (RPs) need to communicate with each other, they are connected via a network which uses message-based communication, employing the principles of common-channel signalling. This network may take the form of a common bus over which the RPs are also able to communicate with the central processor (CP), as in Fig. 16.12. However, the majority of intercommunication takes place between RPs and the CP, and RPs may be directly connected to dedicated ports at the CP, as in Fig. 16.13. It is possible also to effect connection between RPs and the CP via processors which carry out the task of handling communications.[7] These may be known as message processors (MPs — see Fig. 16.14).

In interconnection configurations using message-based communication,[8] each message must be designed to be interpreted by the receiving processor; it must therefore contain fields which identify the originating and destination processors and processes and the type of message. Additional information which may be relevant is the message length.

Arrival of a message at a processor is signalled by the setting of a flag, as described in Section 16.4 above. When the message is accepted by the message controller in the destination processor, an acknowledgment signal (ACK) is returned to the originating processor. If this is not received within a given time, the message is retransmitted, and this is repeated a number of times before the attempt is abandoned. In some systems, acceptance of the message

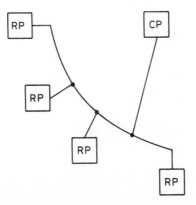

Fig. 16.12 *Interprocessor communication via common bus*

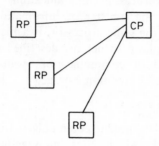

Fig. 16.13 *RPs communicate with CP over dedicated links*

depends not only on the message controller receiving it, but also on there being space in the input queue of the destination process within the destination processor (see Fig. 16.15). If there is no space in this queue, the message controller returns a message to the originating processor to signify this congestion. The originator then retransmits the message after a certain interval and repeats this until it receives an ACK.

Just as all interprocessor messages arriving at a processor are received and distributed by the message controller, so all interprocessor messages are sent by the originating processes to the message controller, which transmits them to their destination processor (see Fig. 16.16). The originating process addresses them only to the destination process, and the message controller (which contains a table relating each process in the system to its host

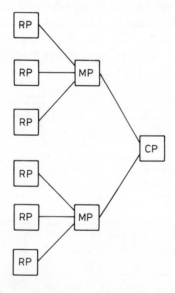

Fig. 16.14 *RPs and CP communicate via message processors*

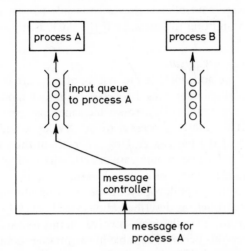

Fig. 16.15 *Message controller must find space for message in input queue of destination process*

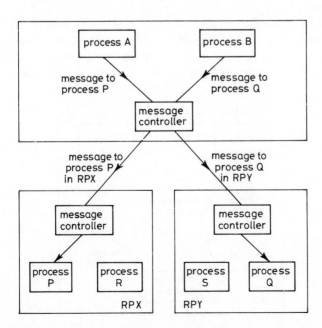

Fig. 16.16 *Message controller determines destination processor of each message*

processor) attaches the originating and destination fields and transmits the messages. The message controller is then responsible for detecting the ACK signal and for retransmitting the message if the ACK is not received.

16.5.2 Interprocess communication

Modern SPC design methods include the writing of software in self-contained modules with standardised interfaces. Processes form such modules, and these are considered to have boundaries across which message inputs and outputs are standardised.[9] This allows processes to be developed independently of other processes and of the hardware. Two processes in the same processor, therefore, may not be able to communicate directly with each other. They may not have been designed to do so, or even to be aware that they are in the same processor. Under this design philosophy, all interprocess communication must be via the messge controller, as described above (see Fig. 16.17). Recovery of an SPC system from a fault may involve hardware and/or software rearrangement. When this occurs, the table in the message controller must be updated. Updating is carried out by the software which controls the system's topology, and this is discussed in Section 16.6. It should be noted that the message controller need not be in an independent processor; it may be a process within any one processor or, indeed, distributed among many or all processors.

16.5.3 Communication via common memory

Not all interprocess communication is via a message controller. Even in very modern systems, there are processes which have access to common memory and which can be programmed to communicate using this common memory.[10] The sending process simply places the information, in a

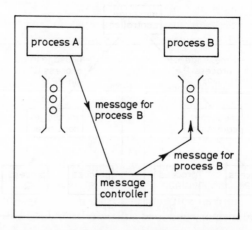

Fig. 16.17 *Interprocess communication via the message controller*

Fig. 16.18 *Interprocess communication via a queue in common memory*

predetermined format, into a queue, from which the receiving process extracts it (see Fig. 16.18). Message-based formats may be used, but this is not essential. The important thing is for the receiving process to be programmed to recognise the format encoded by the sending process.

16.6 Database-management system

Call processing depends on a great deal of data, including exchange-equipment data, routeing data, class-of-service (COS) data and transient data, arising from calls and monitoring. Much of these data are required during call processing and must therefore be stored so as to facilitate rapid access by direct addressing, key transformations or scanning. At the same time, they must be made accessible to operational staff, because data changes made by staff are frequent occurrences. Further, the vast majority of the data need to be stored, not only in fast random-access memory (RAM), but also, for security, on secondary (or 'back-up') storage. This may take the form of discs and/or bubble memory.[11] The program which controls the storage of and access to data in the system is known as the database-management system (DMS). The DMS must manage data so that application software and man-machine commands are independent of its physical structure, organisation and location.[12]

In distributed systems, storing data at the RPs which use them reduces interprocessor communications. Data such as COS data may therefore be stored in the RP which controls the termination to which the data refer. As well as saving message-transmission and queuing time, distributed storage of data allows parallel access, because a number of RPs can access data concurrently. It also allows a larger effective random-access data bank, the total storage being the sum of the memories of all the processors in the system. Indeed, with 32-bit microprocessors being used as RPs, the total effective storage is considerable. It permits all software, both programs and data, to be held in RAM. In the past, it was necessary for only frequently-used software to be held permanently in RAM, with the remainder being transferred from disc into RAM only when it was about to be processed. However, in spite of the convenience of RAM, security demands that back-up copies of both programs and data are held on secondary storage. Also, in those systems which do not yet store all software in RAM, it is a function of the DMS to control the virtual memory and load software from disc to RAM, as required during the operation of the system.

The database itself consists of data structures, such as linked lists, which store the data in a manner most appropriate to operational access. Of course, related items of data may be stored in different processors, depending on where the processes which access them are hosted. At the same time, managers and operating staff must have a simple, two-dimensional model of the data, and this (as described in Chapter 12) is known as a 'relation'. An operator, therefore, initiates changes to the database in terms of the externally-perceived relations and inputs the change instructions to the system in these terms. It is a function of the DMS to interpret the operators' inputs into the appropriate machine instructions to operate on the database.

To achieve this, the DMS contains tables which relate the various categories of data to their addresses in the system and to their references as understood by the operators. When data relating to subscribers' line-terminating units (i.e COS data) are to be updated, the usual key used by operators for accessing it is the directory number (DN). An example of an appropriate table, which may be stored in the centralised module of the DMS, is illustrated in Fig. 16.19. This centralised DMS module is directly accessible to the I/0 controller, which must determine to which RP the operator's instructions are to be directed. There must, however, be DMS modules in each RP to update the data. These access the data, using the DN as a key. The DMS updates the back-up copies of the data on disc as well as the RAM copies; for this, it must have additional tables which reference the back-up copies.

DIRECTORY NUMBER	HOST RP
0 0 0 0 0 0 0 1 0 0 0 2 ⋮ 0 4 7 9	RP C
0 4 8 0 ⋮ 0 9 5 9	RP F
≈	≈
⋮ 9 9 9 9	RP B

Fig. 16.19 *A DMS table relating directory numbers to the RPs on which their COS data are stored*

Reconfiguration of hardware and software in an SPC system is not uncommon. Whenever there is a processor or memory failure, the faulty equipment is taken out of service and replaced. Thus, software to replace that previously run on the faulty processor, or stored in the faulty memory, must be loaded from secondary storage on to its new host. It is the DMS's function not only to initiate these transfers, but also to update its own tables so that correct access can be made to the data in its new location.

There are times also when a software corruption occurs without there being an obvious hardware fault. The DMS must then have the ability to reload the memory already in use with the back-up copy of the data from disc. A further duty of the DMS is to ensure the consistency of software by performing regular checks on it. Some of these involve the assumption that the back-up copy on disc is correct. The copy in memory is compared with its back-up and, if a discrepancy is observed, the memory is reloaded with the back-up copy.

The DMS must also retain and manage copies of as much transient data as is practicable. In the event of a failure during which data are lost, an attempt is made to 'roll-back' by reloading the system with data which define the system as it was a short time previously. This requires a regular dumping to secondary storage of the tables which signify the status of the various parts of the system. These tables include the transient-data elements of COS records and busy/free and status maps for exchange equipment, switch connections and circuits. It should be noted, however, that all call data are not included in this; thus, roll-back usually implies that calls in the process of being set up are lost. The occurrence of further failure within a predefined short time suggests that the system was not adequately restored. Thus, further roll-backs, each going further back in time, or being more extensive in terms of data replaced, are attempted. Eventually, a full initialisation of the system is the inevitable last resort. Thus, the DMS must not only manage the dumps for roll-back, but must also maintain references to a full set of data which initialise the system.

Because the DMS controls the data in the system, it also maintains a record of the configuration of equipment in the exchange and assumes the role of configuration manager. In the event of the failure of a major exchange component (say, a processor), it must determine which item of spare equipment is to be brought into service and what the new configuration (of both hardware and software) should be. To achieve this, it maintains tables containing all the possible combinations of the items of equipment under consideration. One method of choosing a suitable configuration is to employ those in the table in a set order, excluding those which contain the faulty equipment, until a working system is achieved. The method is discussed in greater detail in Chapter 18.

As well as controlling the updating of existing data, the DMS is concerned with the introduction of new data as the system expands. As new equipment is introduced into the exchange, so its data must be inserted into the lists and tables of the database. The DMS must therefore be capable of extending the database in accordance with commands input via the man-machine interface. There are certain checks which accompany this function. For example, a command to introduce new equipment into service (i.e. to set a status flag to show that it is in service and thus available for use during call processing) might automatically initiate a test program to determine if the equipment has actually been wired into circuit. In some cases, tests may also be initiated to check the equipment's functionality. If a test is failed, the operator's command is not accepted and a message is output to advise of the failure. Such tests imply an ability of the DMS to interact with maintenance processes.

The DMS is an essential part of the SPC system software. It is not only a database manager; it is also indispensable to the security, recovery and effective operation of the exchange.

16.7 Man-machine interface

The main function of the man-machine interface (MMI) is to provide the interface between the SPC system as a whole and the outside world, which usually takes the form of operations and maintenance staff. It has been shown, above, that the DMS controls the system database, not only for operational purposes, but also when changes to it are required by the administration. These changes are input to the system as commands, using the MMI, whose main constituent is an interpreter. An interpreter is a program which translates the command input to the MMI, instruction by instruction, into the machine language of the computer, forming tasks which are sent to other processes in the system for execution.[13] It checks the commands for syntactical errors at run-time. Because the majority of the commands concern exchange data, the MMI and DMS interact closely. The boundary between them varies between systems, but the following discussion concentrates on the facilities normally expected in an MMI.

A priority of an MMI is that it should be easy to use. The dialogue, while conforming to certain syntactical rules, must be as close as possible to natural language. If it is too close to the machine level, it will be so unnatural that only a very experienced user can apply it without recourse to a manual. On the other hand, too high a level can necessitate dialogue which is too lengthy and frustrating to the experienced user.[14] The compromise is a high level with a composite-commands facility. The problem of having to cater for users of varying levels of experience also suggests the need for a 'help' facility which permits a user to input a request for help at any point in a dialogue. The system responds with a list or description of allowable responses, or other information which assists the user to continue the dialogue. Further, many MMIs employ screen menus, which are intended to minimise the need for the user to have either expertise or experience. An MMI which assists a user in these ways is said to be 'user-friendly'.

An important feature of an MMI is that it should protect the system, and also individual parts of the system, by restricting access to users of appropriate authority. Tables relating the various levels of the database to corresponding levels of authority are therefore necessary. Levels of authority of individual users may be indicated to the system by pass cards, passwords and user codes, and by the indentification of particular input terminals by their ports into the system.

An MMI must be able to deal with incorrect inputs. It must detect errors and inform the user that the command has not been accepted and (as far as possible) what the error is. The error may be in the syntax of the command, in which case the interpreter should detect and explain it. On the other hand, the error may be semantic or logical. It was shown in the preceding Section that a command to introduce equipment into service may be rejected if a test reveals that the equipment has not been wired into circuit. Such a test cannot be carried out at the interpretation stage, but only at the execution of the command. In order to provide useful feedback to users, the MMI must therefore interact with the DMS.

The above is not an exhaustive list of necessary MMI facilities but, rather, an indication of the types of functions which an MMI must perform. In general, the principles are those of many commonly-used query languages. However, the flexibility to expand the MMI language-command repertoire, by adding new commands to it, is a desirable feature. One MMI, known as the CCITT Man-Machine Language,[15] has been designed specifically for use with SPC systems.

16.8 Input/output controller

An MMI has been described as handling communication between human users and the system. However, the full spectrum of SPC input/output (I/O) requirements is wider than this and demands a more broadly-based I/O controller. Fig. 16.1 shows an I/O controller which controls all system inputs and outputs and an MMI which receives input commands and outputs responses via the I/O controller.

First, the I/O controller must handle all inputs to the system. These take the form of requests to run maintenance programs, requests to interrogate the database, commands to update the database, commands to initiate measurements or event recordings, and commands to output data. Communication between the I/O controller and the DMS and maintenance programs is thus a necessity.

A second task is to control all outputs from the system. These include responses in man-machine language, responses to requests for the results of tests or measurements with the output of data and the regular output of call and traffic data for administrative purposes. The latter output may be made to disc, magnetic tape or hard-copy printers, or directly to data-processing computers over data links. In some systems, a number of these options may be used.

A third duty is to form the interface between the processing system and the disc or bubble-memory backing store. As already explained in this chapter, there is a regular dumping of data to disc and a perpetual readiness to bring software (in the form of both data and programs) into the system from disc, for example, in the event of roll-back. A large part of the I/O controller's load is thus the handling of this communication between the system and its backing store, according to the commands of the DMS.

The I/O controller must ensure that it is directing output to the correct peripheral, e.g. disc, magnetic tape, printer or visual display unit. It must ensure that responses are returned to the units from which the relevant requests were received. In initiating communication, it must exchange signals with the intended peripheral to ensure that the latter is both in service and functioning correctly.

16.9 Overload control

A telephone exchange is dimensioned to provide a chosen grade of service while handling a given maximum traffic load. At the same time, it is recognised that traffic fluctuations are normal occurrences: at certain times, such as during the early hours of the morning, the traffic level may be negligibly small; at other times, during a radio 'phone-in' programme, for example, it may exceed the usual maximum. Each day, leading up to the busy hour, the load gradually increases towards a peak. On some occasions, such as when an emergency telephone number has been broadcast following a catastrophe, it may surge suddenly. A well designed exchange, therefore, should have means of dealing with a range of loads, between 'normal maximum' and true overload. Thus, 'load management' would be a more appropriate title than the better known 'overload control' for the process.

16.9.1 Causes of overload

Overload may have a number of causes:

(i)   Traffic surges in the network. Instantaneous increases, resulting from broadcasts, are now common.

(ii)   More gradual and prolonged load increases as a result of the ease and cheapness of national and international direct dialling. On religious festivals, such as Christmas, and commemorative occasions, such as Mothers' Day, traffic can reach high peaks.

(iii)   Exchange common equipment out of service for maintenance.

(iv)   Failure of a distant exchange or a transmission link, causing a build-up of traffic in other parts of the network.

(v)   Underdimensioning of exchange resources. Errors or poor decisions in design may lead to fundamental dimensioning problems in the system itself. Incorrect planning decisions may cause a purchaser of a system to underorder. (vi)   A mismatch between exchange subsystems, leading to unexpected traffic bottlenecks in the system. Modern modular systems are designed for easy replacement of subsystems, and this may occur because of improved design, cheaper components or new technology. If a replacement handles traffic at a different rate to its predecessor, bottlenecks may be created.

16.9.2 Detection of overload

It is the manufacturer's responsibility to validate designs[16] and detect problems (such as those mentioned in (v) and (vi) above) before a system is sold. However, an exchange system must also contain methods of detecting increasing in-service traffic loads.

Typically, all task queues in the system are monitored and their use compared against predetermined thresholds.[17] Other parameters are also monitored, and these may include the amount of spare processing power available. All software processes normally carry out monitoring of their own tasks and, when a threshold is exceeded, a message is sent to the overload-control module. In some cases, action to avert overload is taken immediately. In others, the situation is monitored, to discover whether the overload persists for a given time, before action is initiated. Moreover, even if the overload criterion is not exceeded, a number of offences within a defined time may result in a message being output. Staff may then investigate the problem; it could be, for instance, that certain equipment was out of service for maintenance.

16.9.3 Action against overload

The first action taken is in design. Because SPC systems handle traffic rapidly, and thus allow the quick development of overload, software queues are usually significantly overdimensioned, sometimes by a factor of ten. Then, thresholds may be set at 20–60% of queue capacity, and this adds security to a system.

Within the overload-control module, when an overload condition has been detected, action is taken in progressively more drastic stages; each successive stage is introduced only if the previous stage fails to abate the overload. The actual actions differ from system to system, but, in general, they conform to the following sequence.

First, certain tasks cease to be performed. These are normally 'background' tasks, or those which are at the bottom end of the executive's schedule of work (see Fig. 16.5). They include the running of routine-maintenance programs and the output of data for purposes other than billing and accounting: tasks which may be deferred until the overload has subsided. Cessation of tasks can be achieved easily by the overload-control module sending a message to the executive. This may cause the setting of a parameter or a flag which adjusts the flow of processing through the executive. The next message from the overload-control module can either reset the parameter or flag, and so return the system to normality, or initiate a more drastic reduction in the number of processes being activated by the executive.

The timing of overload-control actions is critical. Deferring action for too long compounds the problem and allows it to spread through the network. Acting too soon may anticipate a problem which would have cleared itself, and may cause unnecessary inconvenience to subscribers and loss of revenue to the administration. The 'tuning' of overload control may be system-dependent, or even location-dependent, and a familiarity with the network is required. Usually, action will not cease as soon as queues subside below their original thresholds. Recovery thresholds, lower than the originals, are set to avoid oscillation.

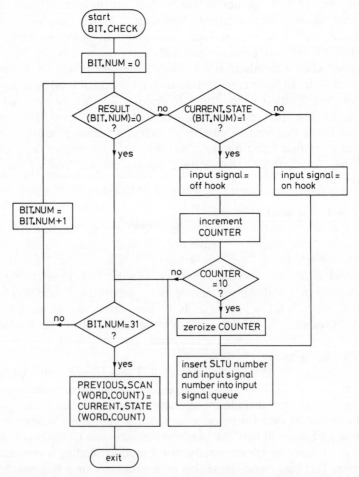

Fig. 16.20 *Using a counter to accept 1 in 10 new calls while scanning the state of lines*

Whereas many lower-priority activities may be suspended in order to alleviate overload, there are three which are not normally ceased. The first is the processing of existing calls. Without this, calls in progress could not be cleared down and overload would not be overcome. It would also result in the loss of signals and, if 'clear' signals were lost, subscribers would be overcharged. The second is the receipt of man-machine messages. These may be network-management actions designed to combat the problem. The third is billing. Moreover, certain facilities, such as calls to emergency services, may continue to be given priority.

The next stage of overload-control actions is the refusal to accept all, or a proportion of, new calls arriving at the exchange.[18] The principle involved in achieving this has been mentioned in Chapter 14 and illustrated in Fig. 14.6. When an off-hook signal is detected, the overload-control algorithm results in the signal being ignored, whereas the normal course of events is for the information to be queued for processing. This overload action thus abolishes the processing of all new calls, which reduces the load on the control system. All on-hook signals are still processed, as this may help to alleviate the overload to exchange equipment (as well as ensuring that subscribers are not overcharged). Alternatively, the system can process a portion of new calls. This can be achieved by inserting a counter into the program; Fig. 16.20 shows the principle applied to the acceptance of one in ten of new calls.

A method of avoiding the processing of new calls works by using the scheduler and its mask, as already described in Section 16.3 and illustrated in Fig. 16.7. The mask is designed to contain a '1' only in those bit locations corresponding to tasks which must be carried out under the particular overload conditions. When the mask is ANDed with a word from the scheduler bit map, the result contains '1' only for allowable tasks. Different masks may be stored and used according to the stage of action being taken.

If tasks 3 and 4 in Fig. 16.7 were for the detection of off-hook signals, a suitable mask to eliminate this function (consistent with the earlier example) would be that of Fig. 16.21. If the reduction in processing that results from not accepting new calls does not alleviate the overload, the next stage of action is to cease processing those calls already being set up. This too can be achieved by choosing a suitable mask for use in the scheduling process.

All the actions described are said to be initiated by an overload-control module. However, it is an arbitrary decision whether or not a system is designed with a separate overload-control module. The functions for which it

Task:	0	1	2	3	4	5	6	7		27	28	29	30	31
Mask:	1	1	1	0	0	1	1	0		0	0	1	0	0

Fig. 16.21 *An example of a mask to ignore new calls*

is responsible could be built into the logic of the executive. In this chapter it has been discussed as an individual entity in order to clarify its functions and because separating identifiable functions into distinct software modules is good software design practice.

16.9.4 Network management actions

The actions described above attempt to alleviate overload without considering the cause of the problem, except implicitly in their design. Network management, which is carried out externally to the exchange system, but relies on data output from it, is the discipline which attempts to discern the causes or sources of network problems and take actions directly related to these causes. Network management action include:

(i)   Automatic alternative routeing.

(ii)   Blocking traffic to certain routes, or altering the routeing options to given destinations, by making temporary changes to routeing tables.

(iii)   Sending recorded announcements to subscribers, and sending signals from a trouble point back into the network to block traffic to certain destinations.

The developing facilities of SPC exchanges and digital networks (such as common-channel signalling, more flexible routeing options and more reliable software) enable the above actions to be programmed into SPC systems. Soon, not only emergency overload-control actions, but also many of the more broadly-based network-management decisions, will be an integral part of SPC software.

16.10 Remarks on system software

The boundaries between processes or tasks or, more generally, software modules, are arbitrary and vary according to design objectives. The modules described in this chapter were chosen to facilitate the description of (a) essential functions and (b) how these may be achieved in software. No single system may be expected to exhibit all the above modules in the manner described, but all systems may be expected to possess all the functions mentioned. Thus, the relationships illustrated in Fig. 16.1 relate to general needs and to the descriptions of this chapter rather than to any particular system.

A collective title for the modules of this chapter, other than 'system software', has been avoided because, again, this introduces an arbitrary boundary. However, the term 'operating system' is common. Nevertheless, an operating system, as defined by any one manufacturer, may contain only some of the above functions, with others being included as applications software modules. The grouping of SPC software modules to form a given system is a matter of design choice rather than deference to arbitruary titles.

The fact that the system software is designed as modular processes with standard interfaces allows its necessary subsets to be used in different processors. Clearly, each processor, whether central or remote, requires system software, but each requires only those modules which support its application-software processes. Modern design, based on modularity and standard interfaces, permits such functional economy. This in turn reduces complexity, economises on memory and minimises the likelihood of software faults.

16.11 References

1 GOLDER, C.G., JOHNSON, D.W., KEMP, C.W., KRISHNAN, M.P. and OSTROVE, N. (1984): *The Multifunction Operations System*, Philips Telecommunication Review, **42** (3)

2 HARR, J.A., HOOVER, E.S. and SMITH, R.B. (1964):*Organisation of the No.1 ESS Stored Program*, Bell System Technical Journal, **43** (5), Pt 1

3 ANDREWS, R.J., DRISCOLL, J.J., HERNDON, J.A., RICHARDS, P.C. and ROBERTS, L.R. (1969): *Service Features and Call Processing Plan*, Bell System Technical Journal, **48**

4 TROUGHTON, D.J., LUMB, A.P, BELTON, R.C., GALLAGHER, I.D., BEXON, M.D., MOOR, S.R. and STEGMAN, S.C.J. (1985): *System X: The Processor Utility*, British Telecommunications Engineering, **3**

5 HARR, J.A., *et al.* (1985): op. cit.

6 HILLS, M.T. and KANO S. (1979): *Programming Electronic Switching Systems* (Peter Peregrinus Ltd), Chap. 2

7 GOEBERTUS, H.J. (1984): *5ESS-PRX Architecture*, Philips Telecommunication Review, **42** (3)

8 BOWDEN, F.G. and KELLY, M.B. (1984): *The Application of Local Area Networks to the Central Architecture of a Telephone Exchange*, ISS'84, Florence, May 1984

9 ARRANZ, R., CONROY, R. and KATZSCHNER, L. (1981): *Structure of the Software for a Switching System with Distributed Control*, ISS'81, Montreal, September 1981

10 JACKSON, D. and PATFIELD, K. (1981): *Impacts of Microprocessing on GTD-5 EAX Call Processing and Operating System Software*, Ibid.

11 TROUGHTON, D.J. *et al.* (1985): Op. cit.

12 SCHIEBER, W. and VANDER STRAETEN, C. (1981): *Database and Administration Facilities for a Switching System with Distributed Control*, ISS'81, Montreal, September 1981

13 BATY, R.M. and SANDUM, K.N. (1985): *System X: Maintenance Control Subsystem*, British Telecommunications Engineering, **3**

14 LAKSHMIPATHY, S. and HOLZBORN, O.C. (1981): *An Adaptive Man-Machine Interface for EPABX Administration and Maintenance*, ISS'81, Montreal, September 1981

15 CCITT (1981): Yellow Book, Volume VI, Fascicle VI.7, Recommendations Z.311 — Z.341, Geneva

16 DEN OTTER, J.J. (1984): *Overload Verification of Ultra Large Switches*, ISS'84, Florence, May 1984

17 TROUGHTON D.J. *et al.* (1985): Op. cit.

18 FICHE, G. and RUVOEN, M. (1984): *The E10 System: Functional Evolution and Quality of Service*, ISS'84, Florence, May 1984

Chapter 17

Software organisation

In the preceding four chapters, SPC software has been described from the point of view of its functions. This has provided something of an 'inside view' of the software. It is now time to take an overview of it, to step back and consider it from the outside.

17.1 Design objectives

SPC exchange software, consisting of several hundred thousand instructions, is among the largest existing software systems. It is also among the most complex because of its size, the fact that processing is carried out in real time and the vast extent of concurrent processing. In order to realise the flexibility which software offers, most exchange functions and facilities are now provided in software rather than hardware. This results in a large proportion of the development effort and cost of a new system being invested in software development. It is therefore important for the prime design objectives to be identified at the outset, and for these to be recognised as the main targets throughout the development of the system.

The traditional system-availability criterion is that an exchange should not be totally out of service for more than about 2 h in 20 years (the exact figure differs among administrations). This imposes the design constraint that any single fault should have minimal effect on service, be readily isolated (so as not to extend a malignant influence to other parts of the system) and be easily corrected. The need for these design objectives is amplified by the notorious error-proneness of software and the difficulty of finding errors unless the system is extremely well designed, documented and tested.

Further, the rate of change of hardware technology suggests that, within the lifetime of an exchange, it will almost certainly be necessary (or at least desirable) to replace one or more hardware subsystems with improved, more efficient or economic equivalents. If such upgrades of equipment necessitated wholesale modifications to the software, they would be costly and likely to induce errors. It is therefore a design objective that interdependence of hardware and software subsystems is minimised.

It is also recognised that facilities which are as yet unplanned, or even unimagined, will be future necessities, and they will need to be implemented in the software. Expandability of the software and minimal interdependence of software subsystems are therefore important design objectives.[1] Summarising these in more general terms, the system is expected to facilitate evolution and be 'future-proof'.

17.2 Design and development methods

The size and complexity of SPC systems have always decreed that their development must be strictly disciplined. Thus, modern systems have benefitted from the advances, during the last 15 years, in software-engineering techniques.

Essentially, software engineering is the discipline of applying engineering and quality-control methods to software development. Although it has been advocated since the late 1960s[2] and is universally accepted as pointing the way forward in software development, it is not yet as widely adhered to in practice as its 'acceptance in principle' might suggest. Exchange SPC software development teams, however, have not only been some of its strongest advocates,[3,4] but have also improved and extended its techniques.[5] The developments in techniques which most directly affect SPC software organisation are the use of software tools to create an off-line 'environment' for efficient software development, software modularity, and the extension of modularity to create 'virtual machines'.

17.2.1 Software environment

SPC software may be divided into three categories. The first consists of the operational programs, the second the exchange data (on which call processing depends) and the third is the off-site development software environment. The term 'software environment' refers to a composite of software tools which facilitate the development process.

Software tools may, for convenience, be regarded as being of four types. The first is 'conceptual tools', which includes methods and techniques. Defining the system-development life-cycle as a number of stages is thus a conceptual tool, as is top-down design. The second is 'symbolic tools'. These are used for representation and include languages. The Specification and Description Language (SDL) is therefore a symbolic tool, as is any high-level language or assembly language. The third is 'perceptible tools'. These are software entities (such as compilers, link-loaders and simulators) which are perceptible because they must exist on a real medium, such as a disc, a tape or hardware memory; but they are themselves intangible because they consist of software. The fourth is 'tangible tools', examples of which are documentation and software libraries.

An SPC development environment[6] (because of the size, complexity and cost of the project and the potentially high cost of failure) consists of a wide range of tools, comprising examples from each of the above four types. The development life-cycle is divided into stages so that control can more easily be applied. The stages are defined such that their conclusions form natural points for the verification[7] of work done. Rigorous standards are applied at each stage for specification, design, software coding, testing, documentation, etc. Such standards not only provide rules and guidelines of what is to be done, when and how, but also checklists against which quality assurance[8] audits may be carried out.

SDL is used by many manfacturers[9,10] for the specification and design of call-processing software. Top-down design is used for all software, resulting in modularity. The majority of task programs are written in a high-level language. However, when speed at run-time is critical, some are written in an assembly language appropriate to the computer in use. The software environment must therefore contain the appropriate compilers and assemblers, as well as the operating systems under which they function. The CCITT high-level language (CHILL)[11] has been designed specifically for programming telecommunications SPC systems and is used by a number of manufacturers.[12,13] Indeed, software-based tools are being produced for the automatic generation of CHILL programs.[14]

Part of the development environment is formed by a general-purpose computer and an operating system, which also may be general-purpose as its function is to support an appropriate compiler. When software modules are written, they are compiled on this system (see Fig. 17.1). Then, using simulated data, they may be tested[15] for functionality. Standardised message interfaces (see Chapter 16 for a discussion of these) make simulation tools quite easy to

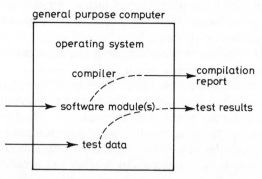

Fig. 17.1 *General-purpose computer and operating system in the development environment*

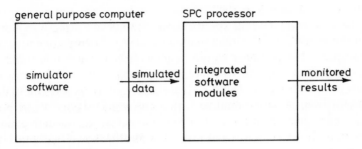

general purpose computer   SPC processor

simulator software — simulated data → integrated software modules — monitored results →

Fig. 17.2 *Software integration testing*

develop.[16] Integration testing,[17] as a number of software modules are combined, may be similarly performed.

When sufficient software is assembled to make time-critical testing necessary, it may be loaded into a processor of the type used in the exchange system, along with the necessary system software (see Fig. 17.2). Simulated call data are then input from a general-purpose computer, where they are generated within teletraffic constraints by simulator programs. These may be a wide-ranging set of data designed to test various operating conditions. More limited simulations may be used to input a large volume of simple calls, to test the throughput capabilities of a system, for example.

The above description is not an exhaustive description of an SPC software development environment. Rather, it is intended only to provide an idea of the range of development tools, techniques and disciplines that are necessary if the product is to be functionally correct and relatively error-free.

17.2.2 Modularity

While tools and techniques alone might lead to software which is functionally correct, a product (particularly one as large as an SPC system) will not be easily expandable and maintainable unless it is modular. In this sense, a module is a unit of software which performs a clearly defined function and has a clearly defined boundary. Modularity is achieved at the design stage by translating the functions of the specification into planned software modules. By this process, each module acquires a logical boundary. This renders the software more readily understandable to developers[18] and, later, maintainers.

By designing modules with logical boundaries and defined input and output interfaces, developers can work on different modules in isolation, combining them only when they are complete and ready to undergo integration testing. This parallel development reduces the total system development time; moreover, if the interfaces are standardised, the likelihood of integration errors (i.e. those due to interaction between modules) is reduced. Modularity also aids expansion: new facilities take the form of additional modules.

Maintenance also benefits from modularity. In diagnosis, a fault is most easily traced to a function; thus, if the modules of the design correspond directly to the functions of the specification, a fault can be traced to a module (software or hardware). It is then readily isolated, and corrective action may take the form of replacement of the module. In hardware, this could mean replacing a plug-in unit. In software, if the fault is due to a parity error, replacement of the module would be from secondary storage; if the fault is a logical error, replacement would be with a corrected version of the module.

As shown in Chapter 13, the CCITT SDL recommendations define modular software in terms of processes and tasks. It was also shown that a task is not necessarily the lowest-level software entity. If functions can be divided into subfunctions or sub-subfunctions, routines may be written to perform these, thus matching all levels of specified functions with software modules.

It should also be mentioned that the word 'process' is used with two slightly different meanings. The first suggests a well defined sequence of code such that the process performs a fixed set of functions when activated. The other meaning, which is more vague, is that a process is created out of the set of functions which are activated at a given time. In this context, call processing may be considered to be a process. However, the process which is created for a given call depends on the type of call and consists of the set of tasks which are activated for processing that particular call. A call-processing 'process' may thus assume a variety of forms.

17.2.3 Data abstraction and virtual machines

Owing to their size and complexity, SPC systems require well-structured software, and modularity is a feature of all systems. A further feature introduced into modern systems is data abstraction.

Data abstraction, which has been recognised as a tool in computing since the 1950s, is best explained by an example. A computer needs to access specifically defined memory locations in order to retrieve the correct instructions and data for the processing of a program. A machine-language programmer must understand these operations and write the program for their execution, defining the precise location in which each instruction or item of data is stored. A high-level language programmer, however, does not need to be acquainted with the intricacies of the machine: the details of the machine are abstracted by the use of the high-level language. As far as the programmer is concerned, the high-level language is responsible for processing: if the program is correct, the results will be correct. In this sense, therefore, the high-level language, with its well-defined protocols, forms a virtual machine.[19] In the same sense, the operating system of a general-purpose computer abstracts the details of the computer hardware from the high-level-language application software (see Fig. 17.3).

The adoption of message-based signalling as a means of communication, between processors and between software processes in general, has allowed

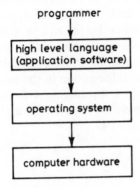

Fig. 17.3 *Levels of abstraction in a computer*

each process to present a constant interface (i.e. constant message input and output protocols) to all other aspects of the system. All other processes communicate with any given process via its interface and by using the defined protocols. They know it only by its required inputs or its produced outputs. No other process needs to be familiar with its internal arrangements, which are therefore abstracted from them. Each process in the system can therefore be designed as a 'virtual machine', from the point of view of all other processes.

The use of message-based communication and standardised interfaces has added flexibility to SPC systems, in the spheres of extendability, design and development, and the replacement of modules by equivalents in newer technology. An example of the use of the virtual-machine concept is depicted in Fig. 17.4, which is a simplistic view of the transmission of call data though a system. Because each process shown (P1 — P5) communicates with the next

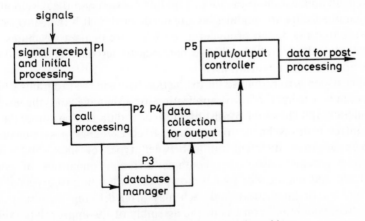

Fig. 17.4 *Software processes can be designed as virtual machines*

by messages which demand certain protocols and contain only certain types of data, the functional details of each process are hidden entirely from the others. Suppose then that a new signalling system is introduced. The exchange hardware requires some change and, no doubt, software changes need to be made to process P1 for the initial handling of the signals. However, if P1 is designed to pass the signalling information to the call-processing process, P2, according to existing interface protocols, P2 and other processes are unaffected by the exchange modification. The risk of introducing errors into the system is limited to process P1.

17.3 Software build

SPC software is not written for a single exchange, but to be replicated for use in large numbers of exchanges, perhaps in many parts of the world. Moreover, with processors having been reduced in price, the modern trend is for the same basic system to be marketed as suitable for a wide range of applications, from small local switches to large international exchanges.[20] Thus, while many aspects of a system (for example, basic call processing, data management and many maintenance functions) must consistently be present, others are only required in certain applications. For example, the handling of special customer facilities is not required in trunk and international exchanges. Consequently, while the software must be developed so as to form a single system in all cases, there must be a means of omitting those functions not required in a given application. This has the advantage of simplifying the application and reducing its storage requirement. It also allows a manufacturer to reduce the cost of a system by offering only what the adminstration requires, while permitting the introduction of additional functions at a later date.

The first step in achieving a system which can be configured according to its intended application is the use of top-down design and the development of self-contained software modules, as already described. It is then also necessary to ensure that the design produces a consistent set of data structures. Then, storage of and access to exchange data, customer data and call data are independent of the application.

The next step is to recognise the distinction between the programs which are written independently of any application or exchange and the exchange-dependent data. The latter consist of equipment data and customer data, and these differ from exchange to exchange. Each exchange possesses different numbers of lines, different ratios between types of lines, and different signalling systems; also, each contains different quantities of common equipment and numbers of switch crosspoints (according to expected calling rates and traffic intensities), and each may offer different facilities.

The 'software build' consists of the assembly of the appropriate exchange-independent software (the programs) and the exchange-dependent software

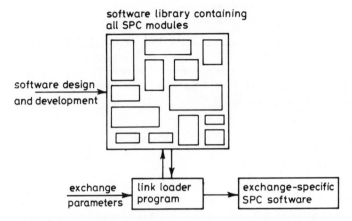

Fig. 17.5 *Building an exchange system using a link loader program*

(the exchange and customer data) to form a system suitable for a specific application.

The programs are selected from a software library, where the total set of exchange-dependent modules is held. Relevant parameters of the intended exchange (such as types of signalling system and the various types of line terminations) are input as data to a specially designed program.[21] These data provide it with a detailed description of the exchange. The program then selects the appropriate modules from the software library and links them together to form a system appropriate to the type of application, as in Fig. 17.5. This is then a standard system for that application. (It is referred to in the No.1 ESS as a 'generic'.[22])

Meanwhile, the exchange-dependent data are defined. These are the data which describe the exchange equipment and connections and the class of service of each line termination. Ultimately the software build is concluded with this data being combined with the selected software system for the exchange, as in Fig. 17.6. The resulting system is exchange-specific and ready to take its place in the network.

The software build also allocates common memory to the system in accordance with the expected calling rate and traffic intensity. Common storage, in the form of the numbers of nodes available for queues in call processing and message handling, is a software equivalent of common equipment in hardware.

The available stacks which form common storage must be dimensioned adequately if the exchange is to handle the varying patterns and intensities of traffic. In early systems, the exclusive use of centralised processing placed a restriction on the amount of available directly-addressable memory; thus, dimensioning of common storage was a critical issue. Now, the use of

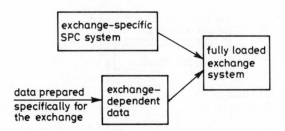

Fig. 17.6 *Creation of a fully-loaded SPC exchange system*

distributed 16- and 32-bit microprocessors, each with its own extensive consignment of random-access memory, makes it acceptable to overdimension. Indeed, it is advisable to do so, because traffic distortions can lead suddenly and rapidly to overload conditions. However, it is essential to calculate dimensioning requirements using sound teletraffic principles, so that all decisions are taken consciously and based on known criteria.

17.4 Configuration management

Any product which undergoes charge, either during its development or subsequently, requires careful configuration management. Because they are so easily changed, this is particularly true of software products.

Configuration management may be defined as a formal engineering discipline which provides developers and users with the methods and tools to identify the software developed, establish criteria, control documentation or software changes against these criteria, record status and audit the product. It is the means by which the integrity and continuity of the product are recorded, communicated and controlled.[23] Indeed, without configuration management, the reliability and quality of the software cannot be controlled.

17.4.1 Management during development
It was shown above that design modularisation allowed different modules to be developed concurrently by different people. Each module, therefore, has its own history. It goes through a number of versions. Then, when it is considered to be complete and correct, its 'final' version is integrated with the 'final' versions of other modules to form a subsystem.

The subsystem is then tested and, when faults are found, the modules containing the faults must be changed. Finally, correct subsystems are integrated with others in the construction of a total system. Integration testing at each stage is followed by full system testing. The integration plans for all stages of system building form an important part of configuration management.

Each system, subsystem and module must be uniquely labelled, not only with a title, but, because it is likely to change, with its version number. A given version of a system is, therefore, composed of specific versions of all its modules. A change to any module must result in a change of version of the system. Configuration management of a large software system under development includes the procedures, tools and quality assurance necessary for effective version control. It is, therefore, a hugely complex affair. It is further complicated by the fact that two teams of developers may be working on different versions of a system. While this is necessary to facilitate the regular production of new functions, it needs to be closely controlled. A method of managing the components of a large program with several releases being developed and maintained concurrently is given by Styma.[24]

17.4.2 Library management
The source of software modules, both for the software build of a particular exchange system and for further development work, is the software library. For evey entity in the library, there should be a file of associated information, including the title and version number of the entity and its author and date, all previous versions and the reasons for changes, and a description of each change and its author. In addition, software-build programs require information such as which version of a given module may be integrated with which version of another module. Versions which have not been tested together should not be brought together in a live system.

17.4.3 Change control
In order for versions to be numbered correctly and their histories to be recorded, there must be clear procedures to be followed when changes are necessary. Important among them are procedures for defining and documenting the change, obtaining the appropriate level of authority to make it, recording who makes the change and when, who tests the new version and who signs it off, ensuring that version control is carried out, updating the accompanying documentation, and making sure that it undergoes quality assurance.

17.4.4 Release control
Replacing a large software system with a new release is costly. It is, therefore, usual for a number of changes, either to correct errors or to add new features, to be made to an off-line system, and tested, before a new release is issued. Releases need to be planned, both in time and in terms of the changes from the previous release.

In SPC systems, a number of variations of a basic system may exist, depending on the requirements of the different exchanges in different parts of a country. A new version of a certain subsystem may, therefore, need to be released to some sites but not others. Release planning and control is, thus, not a trivial matter.

17.4.5 Documentation control

Every change in a module must be reflected in its documentation, which may consist of the specification, design, test plans, code listing and history record. Procedures need to exist to ensure that the necessary changes are made, quality assurance carried out and the new documents signed off. Numbering schemes must exist for document version control and to relate document versions to their corresponding design-module versions. It should be clear who is responsible for writing or updating a document, who is responsible for its quality assurance and who for signing it off.

Further, when documents have been updated, they need to be issued. Distribution lists must therefore exist for all documents, as well as procedures for distribution control.

17.4.6 Implementing configuration management

Configuration management provides an infrastructure for controlling software systems. Without it, control is impossible. With it, if its procedures are followed, control is automatic. The configuration-management database keeps track of all aspects of the system, so that 'the programmer now has only to supply the creativity and motivation'.[25]

Typically, a configuration-management database is easiest to use and, therefore, more frequently used correctly if it is held on the development computer. Then other configuration-management tools, provided by the computer manufacturer, can ensure that all new modules and changes to existing ones are recorded in the database according to defined procedures. Such tools are becoming more and more powerful and more and more able to ensure that all configuration activities are carried out as an integral part of development.

However, successful configuration management must be planned, and this demands the attention of project managers. The number and variety of tasks involved is always substantial, and, even though the details may differ, they are similar from company to company. A guideline for the production of configuration-management plans has been drafted by the IEEE.[23]

17.5 Software usage

Fig. 17.7 shows, as an example, the sizes and processor occupancies of the various elements of software in the HDX-10 system. This was published by Tokita, *et al.*[26] in 1978. It does not necessarily coincide perfectly with the software of different, or more modern, systems, but it does provide a reference guide.

In the example, software involved in call processing comprises about 27%

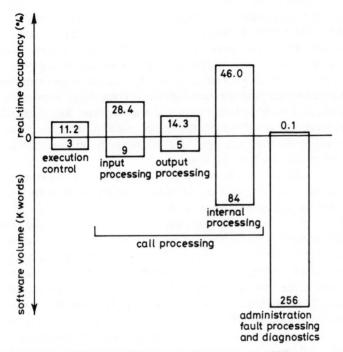

Fig. 17.7 *Quantity and usage of SPC software (Tokita, et al., 1979)*

of the total and employs about 88% of processing time. In fact, the average call exercises only about 20% of the software. At the other end of the scale is the maintenance and adminstration software. This is no less important to the system than the call-processing programs; without it, there would be no guarantee of system integrity or continuity of service. Owing to the numerous contingencies which it must cover, it constitutes by far the greatest proportion (71%) of the total software. However, being for the most part in reserve, it uses just 0.1% of processing time.

Execution control refers to such functions as have been defined for the executive, i.e. the scheduling of tasks, clock-pulse handling and the handling of interrupts and message transmission. As expected, it comprises a very small proportion of the software of the system (3%) and, being recurrently in use, requires 11% of real-time processor occupancy.

17.6 Structure of operational software

The software of the early SPC systems was fully centralised in a single processor. The consequent memory limitation demanded code that was designed for brevity and run-time efficiency, and this resulted in a lack of flexibility. Typically, the software was modularised, but there was a high degree of interdependence of the modules. In theory, modification and expansion could readily be achieved, because of the easily expandable data structures on which call processing was based; in practice, change was laborious, error-prone and expensive because of the numb≥r of modules affected. Early systems were also limited by the addressing capability of a single computer. Because the software comprised many hundred thousand instructions, and both exchange data and transient call data demanded huge volumes of storage, all software could not be hosted in random-access memory (RAM). Many of the programs had to be held on disc storage, being transferred to RAM when required.

In modern systems, the above problems have been overcome, or mitigated, by distributed processing and the software-engineering methods (such as the virtual-machine principle) already described in this chapter. Microelectronics has led to small powerful computers with relatively large addressing capabilities, and these are used to provide distributed processing in SPC systems. A processor 'unit' may have full control over its local environment and may even be responsible for its own security.[27] This is applied to varying degrees, but at least one manufacturer claims 'fully distributed control'.[28]

Software is being developed as processes (which, themselves, are composed of numbers of modules) whose internal functioning is abstracted from other processes by standardised interfaces. A process which interfaces directly with telephonic hardware may require modification if the hardware is altered, but it screens other processes from the hardware. This method of design results in unprecedented flexibility.

Earlier moves towards distributed processing, as shown in Fig. 11.1, had altered the balance of a system by transferring some control to regional processors. However, they had not fundamentally changed the design. Now, a fundamental improvement has been achieved by the use of message-based communication between processes in the system.[29] The goal is for processes to be entirely independent; thus, subject only to a table which describes its location, a process may occupy a dedicated processor or share a processor with other processes. Each process is designed and developed independently of other parts of the system; it exists within the system as a virtual machine and exchanges messages with other processes without being aware of their locations (see Chapter 16 and Fig. 16.16).

For such flexibility, each process must have, embedded within itself, the system software necessary for its scheduling, communication with other processes, and fault detection. This is an overhead in storage, but significant advantages are gained in the development and maintainability of the software. Indeed, even with the increased volume of software, the added addressing capability of the distributed microprocessors allows all software to be stored in RAM, with secondary storage needed only as a security medium.

The results are greater efficiency, both in development and operation. Development is better controlled and the product contains fewer errors and is more maintainable. Modification of the system is both easier and cheaper than previously. Good software-engineering principles in the development of SPC software pay dividends.

17.7 References

1 McDONALD, J.C. (Ed.) (1983): *Fundamentals of Digital Switching* (Plenum Press), Chap. 4

2 BUXTON, J.N. and RANDELL, B. (1969): *Software Engineering Techniques*, The NATO Software Engineering Conference, Rome

3 GOECKE, D. and HAAKE, G. (1984): *A Software Engineering Approach to the Complete Design and Production Process of Large Communication Systems Software*, ISS'84, Florence, May 1984

4 FRASSANI, L. and BORSOTTI A. (1984): *Production and Quality Assurance for Large Software Systems*, Ibid.

5 BOTSCH, D., FEICHT, E.J. and MIRUS, A.J. (1984): *EWSD Enhanced Software Engineering for the Deutsche Bundespost Network*, Ibid.

6 VANDERLEI, K.W. and WIRTH, R.D. (1981): *GTD-5 EAX: Development Language and Support Software System*, ISS'81, Montreal, Sept. 1981

7 European Workshop on Industrial Computer Systems, Technical Committee No.7 (EWICS TC7) (1988): A Guideline for the Verification and Validation of Critical Software; In REDMILL, F.J. (ed.), *Dependability of Critical Computer Systems — 1* (Elsevier Applied Science Publishers), Chap. 5

8 AT&T Journal: Special Issue on *Quality: Theory and Practice*, Vol. 65, Issue 2, March/April 1986

9 BAGNOLI, P., GIORCELLI, S., LONGO, F. and SARACCO, R. (1981): *Application of CCITT SDL for Software Development and its Maintainability*, ISS'81, Montreal, Sept. 1981

10 GAUTARD, G., PLOUZEAU, A. and DANAN, P. (1981): *Development Methodology of Call-Processing Software using CCITT's SDL*, Ibid.

11 CCITT (1984): Red Book, Vol. VI, Fascicle VI.12, Recommendation Z200, Geneva 1984

12 HIGASHIYAMA, F., INOUE, O., MATSUO, Y. and SATO, T. (1981): *New Technology in D-10 ESS Software System*, ISS'81, Montreal, Sept. 1981

13 BAGNOLI, P., CAMICI, A. and ROSCI, G. (1984): *Towards a Software Engineering Environment for Telecommunication Systems based on CCITT Standards*, ISS'84, Florence, May 1984

14 JACKSON, L.N., FIDGE, C.J. and PASCOE, R.S.V. (1984): *Computer-Aided CHILL Program Generation from System Specifications: Design Experience from the MELBA Project*, Ibid.

15 CLEMENT, G.F., MRAZ, W.L., WILSON, D.E. and YEH, K.M. (1981): *No. 1A/4 ESS — Program Test and Evaluation System*, ISS'81, Montreal, Sept. 1981

16 MAISONNEUVE, M., LEVY, J.P., PASCUAL, L. and STOKKING, M. (1981): *Aspects of E10.S-TSS.5 System Software: Architecture, Operating System and Software Testing and Debugging*, Ibid.

17 KATCHER, M.L. (1984): *Integration and System Testing: A Methodology*, ISS'84, Florence, May 1984

18 CHARRANSOL, P., HAURI, J., MERESSE, J. and TRENDEL, R. (1981): *Structure of a Time Division Switching System for Distributed Control and Linear Growth Capability*, ISS'81, Montreal, Sept. 1981

19 ARRANZ, R., CONROY, R. and KATZSCHER, L. (1981): *Structure of the Software for a Switching System with Distributed Control*, Ibid.

20 GITTEN, L.J., JANIK, J., PRELL, E.M. and JOHNSON, J.L. (1984): *5ESS System Evolution,* ISS'84, Florence, May 1984

21 FOX, M.J. and STOREY, M.H. (1985): *System X: Build Control,* British Telecommunications Engineering, **3**, Pt.4

22 HARR, J.A., HOOVER, E.S. and SMITH, R.B.: *Organisation of the No.1 ESS Stored Program,* Bell System Technical Journal, **43** (5), Pt.1

23 IEEE (1983): *Standard for Software Configuration Management Plans,* IEEE Std 828-1983

24 STYMA, R.E. (1984): *Configuration Management for the Concurrent Development of Multiversion Shared Resource Projects,* ISS'84, Florence, May 1984

25 SEAGRAVES, D.A. and SAGAN, J. (1981): *Configuration Management in Large Software Products,* ISS'81, Montreal, September 1981

26 TOKITA, Y., SUZUKI, T., SHODA, A. and HIYAMA, K. (1978): *ESS Software Architecture for Multi-Processor System (HDX-10),* Conference on Software Engineering and Telecommunications Switching, Helsinki, June 1978

27 MERESSE, J. MASSIOT, J.F., JOSSIF, J. and TRENDEL, R. (1984): *MT 35 Time Division Switching System: Distributed Architecture, Distributed Software,* ISS'84, Florence, May 1984

28 DENENBERG, J.N. and NIGGE, K. (1981): *Fully Distributed Control: An Approach Leading to Flexibility in Application,* ISS'81, Montreal, Sept. 1981

29 BOURGONJON, R.H. (1984): *5ESS-PRX Software,* Philips Telecommunication Review, **42**, (3)

Chapter 18

System integrity

The prime demand on a telephone exchange is to provide a continuous service to the customer. An exchange, therefore, needs to be 'available'. 'Availability' is a measure of delivery of the specified service as a fraction of a given time interval, and is, thus, the probability of delivery of the service. The stated requirement for the availability of electromechanical systems was that they should not be out of service for more than 2–4 h in 20–40 years (the exact figure depending on the administration and the system). Subscribers in most parts of the world are therefore accustomed to a high availability and are unimpressed by new technology if it decreases this. Further, the number of services being offered via the PSTN is growing. Many advisory, entertainment and data-information services, as well as the traditional operator and emergency services, now depend on exchange availability. Moreover, the trend in many countries is towards the abolition of monopolies hitherto enjoyed by telecommunications adminstrations; thus, the result of poor availability of service, or even of degraded service, is likely to be a loss of market share, and, hence, revenue. It is therefore essential for an exchange to provide continuous service. The keys to this are a reliable control system, maintenance software which contains automatic recovery procedures, and an easily maintainable exchange.

As well as high availability, subscribers expect a consistently high quality of service from a telephone administration. A high-quality telephone service, as perceived by a subscriber, implies quick and efficient call set-up, good transmission, no interruption to calls in progress, efficient clear-down at the correct time, and accurate billing. With modern stored-program-controlled exchanges, the control of all these aspects of telephony is carried out by an exchange's control system, which can monitor all activity within the exchange, as well as apply tests both on a routine basis and in response to predetermined events (Haugk et al.[1] offer a good overview).

Whereas the principles of call-processing software are described in detail, it is not intended to consider maintenance software to the same extent. However, the main functions will be introduced in the following Section, while control-system availability will be discussed in the remainder of the chapter.

18.1 Functions of maintenance software

In the context of maintenance software, maintenance implies:

- detecting faults,
- diagnosing the source of faults,
- verifying faults,
- taking steps, such as busying or rearranging equipment, to alleviate a fault's effect on service,
- recovering from a fault by restarting the system, for example,
- presenting an error message so that the necessary manual action (for example, changing a board) can be carried out,
- restoring service to subscribers (including, where appropriate, issuing advisory messages).

Because the provision of service depends on all aspects of the exchange, the maintenance software must perform the above functions on the switching and signalling equipment, the terminations on the exchange and on both the hardware and the software of the control system itself.

18.1.1 Exchange-equipment maintenance

Because control of the exchange equipment is achieved by means of software, it is possible and practicable to introduce programs to perform maintenance functions on exchange equipment. Tests are carried out, each exercising a defined module of equipment, according to a schedule which ensures that all equipment is exercised regularly. When a fault is found, it must be verified to ensure that it is not a transient condition. Then, the module of equipment is isolated by busying it.

In many cases, software-controlled reconfiguration of equipment (see Section 18.4 below) is possible. This ensures that service is unaffected, except, perhaps, for a short interruption. However, even when this is possible, manual intervention at the exchange is necessary to repair or replace the faulty equipment. Diagnostic programs are therefore activated to identify the type of fault and to locate it as exactly as possible within the module. Because the trend is towards a philosophy of 'return to service quickly' by replacement, rather than of carrying out *in situ* repair, the diagnostic programs are usually intended to locate the fault to the smallest replaceable unit of equipment. In the relatively few cases when software does not (or does not correctly) diagnose the fault, successive manual substitutions of replaceable units provides one means of fault location.

Hardware reconfiguration is only possible where equipment is replicated. Replication is provided in particularly vulnerable or important parts of the exchange, [for example, in the control system, switch blocks (see Chapter 6) and intraexchange trunks]. It is usually achieved by duplication or by providing a pool of equipment containing one or more items above the maximum number required at any time. Where replication is not provided, a fault may result in the loss of service to individual subscribers. Then, recovery software[2] reattempts to run the system, using the same data which were being processed at the time of the failure. If this is successful, the conclusion is that the fault was transient. If the attempt is not successful, diagnostic software initiates tests, traces the fault to a replacement unit, and outputs an error message. After manual replacement of the faulty unit, verification by software tests is carried out before it is brought into service. This is done by setting the unit's status bit in its equipment record. In many cases, the consequence of a fault in unreplicated equipment (e.g. trunk terminations) is minimised, or even eliminated, by the use of alternative routeing through the network. Similarly, common-channel signalling is protected by either replication of the signalling links or alternative routeing of signals, or both.

When all possible diagnostics have been performed to identify and trace a fault, an error message is output, using the system's man-machine language and, if the condition is serious, an audible alarm may also be activated to alert the operations staff. When the operations staff have replaced the faulty unit, the maintenance software retests it. Only if it is verified as being in working order will it be returned to service, either in the operational or the standby mode. The system's configuration plan, stored in the form of tables in memory, is then brought up-to-date with the inclusion of the unit.

18.1.2 Line-termination maintenance

The control system handles the connection, through the switch blocks, of circuits on the transmission links that terminate at the exchange. The circuits may be local, trunk or international, depending on the exchange. With access to the exchange terminations, the control system is able to:

- detect alarm signals on faulty circuits or transmission systems,
- carry out tests on transmission equipment,
- busy faulty circuits or transmission systems,
- reroute offered traffic so as to avoid faulty equipment,
- output information which would lead to repair.

Fig. 18.1   *The three subattributes of availability*

Not only are the tests, which are programmed into stored-program-controlled systems, activated selectively in response to faults; they may also be scheduled to occur on a routine basis or be initiated manually. Those on transmission links, i.e. trunks and subscribers' lines,[3] include circuit tests, transmission tests (using various tones with variable power), subscribers' apparatus tests, and exchange equipment (such as dial-tone applicator) tests. With more and more digital subscriber access to exchanges, subscriber digital-path tests and digital-terminal functional tests are also carried out by some systems.[4]

As well as these real-time actions by the control system, most SPC exchanges contain a network-management software module which collects data, as calls are set up, for performance monitoring[5] on a route and destination basis. Results are transmitted to a network-management centre at regular short intervals (5 to 15 min) for comparison against predetermined norms. When norms are exceeded, appropriate action is initiated at the discretion of experienced network managers. However, it is forseeable that, in the near future, many network-management actions will be programmed into the exchange-control system and, thus, taken automatically.

18.1.3 Control-system maintenance
Throughout call processing, the control system makes checks on its own integrity. The control system is crucial to the operation of the exchange and its availability must be guaranteed as far as possible. As shown in Fig. 18.1, availability depends on its three subattributes: reliability, recovery and maintainability, and these are discussed below.

18.2 Reliability

Reliability may be measured by the 'mean time between failures (MTBF)', and it depends on a number of features of the system as well the quality of management of the development project which produced the system, and the quality of manufacture. In practice, it is determined by the extent to which fault avoidance and fault tolerance have been designed into the system. 'Fault tolerance' is usually defined as the ability to continue operation, with no effect on service, in the presence of a single fault. This implies the ability to tolerate multiple faults in some circumstances, but it is not guaranteed. Fault tolerance is achieved by providing redundant hardware and recovery and diagnostic software.[6]

In the first place, the level of reliability of a system depends on the degree to which designers and implementers succeed in achieving fault avoidance. In hardware, it has always been recognised that total fault avoidance is impossible: component failures must occur sooner or later and, however carefully the system may be designed, design faults will almost certainly reveal themselves, even after the system has been in service for a long time. Further, the more complex the system, the more likely are these faults to arise. However, reliability is improved by selecting components of good quality, employing modular design and applying quality assurance procedures and checks to every stage of the manufacturing process and to all end products of each stage.

In software, the attempt to approach fault avoidance as closely as possible takes the form of adherence to software-engineering principles, good project management, thorough verification and validation and quality assurance.

When a fault is discovered, it may already have caused a system failure, in which case recovery procedures need to be initiated for service to be restored. On the other hand, the fault may be found during routine testing, or as the result of the failure of a component rather than the system. Then, if fault tolerance has been built into the system (e.g. the ability to reconfigure equipment so as to replace the faulty component), service can be maintained.

18.2.1 Control-hardware architecture

Recognising that hardware failures will inevitably occur eventually, fault tolerance is designed into systems in the form of redundancy. A number of hardware 'architectures' have been designed to provide redundancy and many have been employed in SPC. The simplest form of control system is a single control unit. This is adequate in applications which do not justify the added cost and complexity of redundancy. A public telephone exchange is not such a system. Even if the fault rate was acceptably low, the cost of an interruption to service and the risk of an interruption being of extended duration decree that redundancy is essential.

A dual-processing configuration involves the duplication of the central processing unit (CPU) and its storage. This can be employed in a number of different modes. Working synchronously, as in the No.1 ESS[7] and the AXE10 system[8,9] (see Fig. 18.2), the processors perform the same functions simultaneously and compare their results. Agreement allows further action to be taken by both CPUs; discrepancy implies that at least one is in error and leads to the running of test programs to discover which. The faulty processor is then taken out of service while the other proceeds with the task. The main

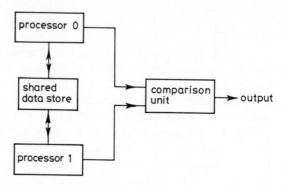

Fig. 18.2   *The principle of synchronous dual processing*

disadvantage of this mode is that since the two CPUs are simultaneously running the same programs, an error in the software will cause the failure of both, resulting in loss of service.

There is greater immunity to software faults in the dual-processing configuration if the load is shared between the two processors. In the load-sharing mode, one processor takes over the full load if the other fails. However, a failure of one processor results in a degradation of performance, and, if the processors are each more than 50% loaded, in lost calls. It is for this reason that, in SPC, the principle of load sharing is employed in the form of multiprocessing rather than dual-processing. In multiprocessing, a number of CPUs share the load. A configuration with at least one CPU more than the minimum number required allows a faulty processor to be withdrawn without degradation of service. Each processor repeatedly runs diagnostic programs to check peripherals, input-output channels, storage media, etc., and the CPUs check each other continuously by exchanging signals according to strict protocols. In the E10.S System[10] (see Fig. 18.3), a number of microprocessor control units (CU) are interconnected via an interprocessor data link (IDL) which is duplicated for security. Also attached to the IDL are a number of interface microprocessors (IPs) which communicate with the operations staff, using a man-machine language. The CUs are connected to the switching network and, by means of these connections, control the exchange.

A further configuration is the arrangement of processors in 'clusters'. A cluster is a tightly coupled group of processors, working in the multiprocessor mode, each having an allocation of dedicated random-access memory as well as access to common memory available to the whole cluster. There is a fixed limit to the number of processors in a cluster, but the power of the control system may be increased by the addition of clusters. Thus, the total control system is made up of a number of multiprocessor units. In System X, up to four processors (CPU) form a cluster[11] (see Fig. 18.4). In this system, the

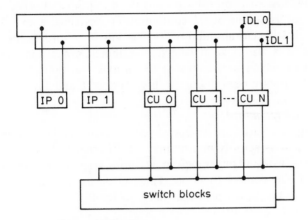

Key: CU: control unit
     IP: interface processor
     IDL: interprocessor data link

Fig. 18.3 *Multiprocessor architecture (as in the E10.S system)*

CPUs are interconnected by a synchronous bus, duplicated for security; an input/output processor interfaces the CPUs in the cluster to an asynchronous bus; this connects them to other periphals, such as backing store which is composed of bubble memory. A small exchange may comprise one cluster with a minimum of 2 CPUs. The maximum available power is 8 clusters of 4 CPUs each. Communication between clusters is via a controller which is connected to the asynchronous bus, but is minimised so that the clusters are almost independent of each other.

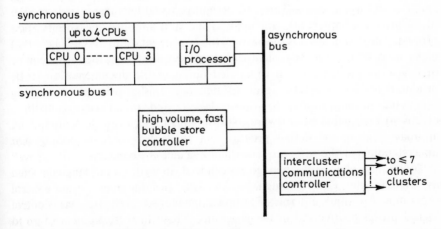

Fig. 18.4 *Cluster architecture (as in System X)*

In early SPC systems, control was centralised in powerful processors which were replicated, using one of the configurations described above. In modern systems there is still centralisation of some essential call-processing functions. However, there is also a great deal of distributed processing, provided by microprocessors which are dedicated to various functions. The principles involved have already been discussed in Chapters 16 and 17.

18.2.2 Software reliability

Whereas computer hardware is general-purpose, application software is not. By performing satisfactorily in a number of different tasks over a period of time, hardware, working under a given operating system, can show itself to be suitable for a new control function. Thus, given its known failure rate, an architecture can be chosen such that the availability for the job in hand can be calculated with confidence.

This is not so for application software. For each new task, unique programs must be written, and they may contain errors introduced at the specification stage, the design stage or the implementation stage of the project. Nor is it easy to test software under all possible conditions. Indeed, if the program, or the task, is complex, as in the case of a telephone exchange SPC system, it may be impossible to do so in a practicable or even a finite time. Thus, in spite of the fact that the number of errors revealed by successive tests decreases, perhaps to zero, there can be no certainty that there is not a set of circumstances, albeit an improbable set, which may give rise to a fault during operation.

Fault avoidance, therefore, cannot be assumed in software. However, it is strived for. Its aim is to offer the highest possible guarantee that software is error-free when introduced into service. The process begins at the specification stage, when it is necessary to state the requirements correctly and clearly: a system accurately designed to an incorrect specification may perform as specified but not as desired! In SPC design, pictorial languages, such as SDL (described in Chapter 13), are used to aid and simplify the specification process. The design stage is the basis of the implementation, and it is here that plans are drawn up for the structure of the proposed system, the future course of development, the tests to be carried out, and the documentation to be produced for later operation and maintenance. Design principles,[12] such as functional decomposition, top-down design, and modularity, contribute greatly to the quality of software. Not only do they constitute a disciplined approach during the design process, but they also facilitate coding and understanding of the software when maintenance is necessary.

Coding in a high-level language and, indeed, in a structured language such as CHILL (the CCITT language for SPC programming),[13] allows the programmer to think in terms of a spoken language as well as in terms of the logical processes involved in the programs. These factors, along with good documentation, allow the programs to be read and understood at a later date during testing and maintenance, almost certainly by different programmers. Whereas this may not seem a great advantage for a simple module, it should be remembered that a control system often consists of a large number of interworking modules comprising tens or hundreds of thousands of lines of code. The large number of modules and their interworking create a complex software system.

In early SPC systems, programming was carried out almost exclusively in assembly language. However, the recognition of the importance of maintainable software, the increase in the power of computers and the decrease in the price of computer memory have all led to high-level-language programming becoming normal in SPC development.

Testing takes a number of forms and is carried out at all stages of the development. Verification is the comparison between stages of the development life-cycle to determine that there is a faithful translation of one stage into the next. In accordance with its definition, verification is performed at all stages of the software development life-cycle, and a number of methods have been devised[14] for verifying the design, coding, module integrity, integration of modules, and finally the integration of the software and hardware subsystems.

Validation is the process of determining the level of conformity between an operational system and the system requirements, under operational conditions. The tests involved depend on the size and complexity of the system, but, as far as possible, they must be designed to check output correctness for all possible inputs. At all levels of testing, it is important to verify not only correctness but also completeness. The aim is to detect and eliminate errors and not to try to prove that there are no errors.

18.2.3 System reliability

Having referred to the technical system aspects of achieving reliability, the project-related aspects should be mentioned. In any project, particularly in a complex, software-based, development project, the quality of project management is crucial to the quality of the product.[15,16]

SPC design is aimed at producing an error-free system, with well structured design, high-quality components in the hardware, and fault avoidance achieved in the software. In addition, hardware redundancy provides a means of fault tolerance and, thus, of avoiding an interruption to service even when a failure occurs. While it is not guaranteed that a system is tolerant to more than one fault, it may be so, depending on the nature of the faults, their locations and interactions.

When certain types of fault occur, such as software faults or multiple hardware faults, service ceases. Then, rapid automatic resumption of service depends on the system being designed to recover.

18.3 Recovery

The effect of a fault and the method by which the system recovers from it depend on the nature of the fault and where it is located.

18.3.1 Recovery from a hardware fault
In most cases, a hardware fault does not cause a system crash. In the case of a processor, the fault is detected by the mechanisms built into the system architecture. For example, in dual-processor synchronous working, a discrepancy in compared results indicates a fault, and further tests carried out by the two processors on each other determine which is faulty. Programs run on the good processor then cause the faulty one to be taken out of service and appropriate error messages to be output. In this instance, service is not degraded. However, in the case of load-sharing processors, it would be.

Similarly, a fault in switching, signalling or transmission equipment does not cause an exchange crash. It can, however, result in the loss of service or a degradation of the grade of service to one or more subscribers. The means of minimising or overcoming this were discussed in Section 18.1.

When a fault occurs in one of a number of items of similar equipment which form a pool, there is the possibility of replacing the faulty item if standby equipment exists. Such an arrangement may be found in processors working in the multiprocessor or cluster configurations, and also in pools of telephonic equipment. In either case, the item of equipment found to be faulty is taken out of service and an error message output. Then, the standby item of equipment is brought into service and the table in memory which shows the status of each piece of equipment is updated (see Section 18.2). Unless the diagnostic software was wrong in deducing the faulty equipment, or a second fault exists, these operations should result in recovery.

18.3.2 Recovery from a software fault
When processing encounters a software of data error, the program being run cannot proceed, and control is passed to the executive, which initiates a recovery procedure. This can occur either in a central or a remote processor. Recovery takes the form of roll-back (see Section 16.6), in which the system is reloaded with data representing the state of the system as it existed at some instant in the immediate past.

While it is true that a given software fault (bug) exists until it is fixed, it only manifests itself and causes a failure if it lies in the processing path during operation. The bugs which lie on the most commmon processing paths are usually found and eradicated during testing. Thus, it is those which rely on infrequent combinations of circumstances to activate them which cause failures during operation. Rolling back the system thus almost certainly ensures that, when the system is restarted, a set of circumstances exists which is different from that at the time of the failure. Most are, therefore, overcome by a single roll-back. At the same time, steps are taken to output not only an error message but also sufficient data to allow the failure to be analysed and the fault found and corrected.

There are usually up to three stages of roll-back programmed into the system. The first is to a state of the system only a few milliseconds prior to the failure. Owing to loss of the latest data, all calls in the process of being set up are cleared down. If a software fault occurs within a predetermined short time of the restart, it is assumed that the original fault has not been overcome, and a second-stage roll-back is initiated. This takes the form of clearing down all calls on the exchange and resetting the transient-data storage associated with them. If this does not result in error-free running, the third stage is envoked. This consists of a full reload of all software, including permanent and semipermanent data, from backing store. This can take up to half-an-hour because it is a full initialisation of the exchange. All calls are cleared down for it to occur.

All levels of roll-back can be activitated not only automatically under fault conditions, but also manually. Manual roll-back and restart are necessary, for example, at the initial start-up of the exchange and when new versions of software are introduced.

18.3.3 Equipment reconfiguration

Recovery from a hardware fault, either in the control system or other exchange equipment, depends on the availability of spare equipment of the same type. The fault, or its effect, is normally detected by the operational or routine-testing software, which then runs appropriate diagnostic programs in an attempt to trace the exact location of the fault. In many cases, the faulty item of equipment is identified. In others, the fault is traced only to a group of items of equipment.

When redundancy exists in the form of duplication, recovery takes the form of removing the faulty equipment from service and introducing the equipment which until then had been on standby. This may be achieved by appropriately changing the status fields in their equipment records because, before the control system employs any equipment in call processing, it checks its status. A message is then output to maintenance staff to let them know that the equipment which should be on standby is now out of order and requires attention.

When redundancy takes the form of a pool with surplus equipment, and the faulty item of equipment is identified, this is replaced with the spare item in the same way as in the case of duplication.

However, there may be some instances when the software diagnostics do not locate a fault to an individual item of replaceable equipment, but only to a subsystem consisting of two or more items working together. The recovery is achieved by replacing one item, restarting the system and determining if the fault persists. If it does not, it is assumed that the replaced item is the faulty

Table 18.1 *Combinations of duplicated processor and memory units*

State	Combination of units
0	P0 + M0
1	P0 + M1
2	P1 + M1
3	P1 + M0

unit, and an appropriate error message is output. If the fault persists, another change is made and the system restarted. For each subsystem in which this procedure takes place there is a predetermined sequence of changes stored in a table in memory. The method is described by Takamura, *et al.*[17] and referred to as the 'rotation method'.

An example is of a central processing system consisting of duplicated processors (P0 and P1) and memory units (M0 and M1). There are, therefore, four ways of configuring a working system, as shown in Table 18.1. The sequence of changes, stored in memory, could be in the order shown in the table, i.e. State 0 → State 1 → State 2 → State 0. However, any other order could be adopted, as long as the sequence covers all possibilities. The state of the system at any given time is also stored in memory; thus, when a fault is detected, rotation to the next state in the sequence always takes place. It should be noted that, in the above example, recovery from a single fault requires one or two configuration rotations. The table is designed so that only one item of equipment is changed at each rotation.

Table 18.2 *Combinations of 1 from 2 processors with 3 from 4 memory units*

	P0	P1	M0	M1	M2	M3
State 0	+		+	+	+	
State 1		+	+	+	+	
State 2		+	+	+		+
State 3	+		+	+		+
State 4	+		+		+	+
State 5		+	+		+	+
State 6		+		+	+	+
State 7	+			+	+	+

In practice, rotation may have to take place in more complex situations than the one shown above. Suppose, for example, that a system requires one processor and three memory units to function correctly and that the processor is duplicated while the memory units are provided in a pool of four, of which one is always spare. The eight possible states, in order of rotation, are shown in Table 18.2, where active units are marked with ' + '.

18.4 Maintainability

Maintainability refers to the ease of returning the system to service and/or to its former state, after it has failed. It must therefore be designed into the system. Measures of maintainability are the average time taken to return the system to service and the mean time to repair (MTTR). Maintainability represents the third subattribute of availability and the third line of defence against loss of service.

Good maintainability demands that an error message is output for each fault, and this requires good diagnostic software. In cases in which recovery is achieved by the automatic substitution of faulty equipment, service is unlikely to suffer noticeably, and maintenance staff may depend entirely on error messages, even to know that faults exist. In most instances of hardware maintenance, the necessary action consists in replacement of a plug-in unit. This, however, requires clear, detailed and accurate error messages from the control system. Each message should be explicit in identifying the faulty equipment. Acquiring this information depends on extensive diagostic programs within the maintance software. Diagnosing a fault to a single plug-in unit requires that the system designers have located each function in its entirety on a unit. In many cases, diagnostic software locates a fault to a function; thus, if the function is distributed over a number of plug-in units, the error message may name all of these, perhaps with probabilities given for the fault being on each.

Likewise, isolation of faults in software depends on the program-module boundaries being defined on a functional basis. However, when a software failure has occurred, maintenance staff require more than an error message. A great deal of data concerning the state of the system and the program being run at the time of the failure must be made available. There are two principal reasons for this. The first is that, because the fault may only have come to light as a result of a rare combination of circumstances, the failure may not occur again for a long time. If the fault is to be found and corrected, all relevant data must be collected at the time of the failure. The second reason is that software faults are normally corrected centrally. The software-maintenance engineers are unlikely to be on site when the failure occurs to decide what data to collect. Thus, in order to maximise the probability of the essential data being available, a great deal of predefined system-parameter values and data are collected. A description of off-site software maintenance would be inappropriate here, but interested readers will find ample discussions in References 18 and 19.

It is now customary in modern digital telephone exchanges for staff to stay away from the equipment rooms unless repairing a fault. Indeed, there is a trend towards leaving exchanges unstaffed and monitoring the need for manual attendance from a remote operations-and-maintenance centre (see Chapter 19). However, rack alarms are still provided, and activation of these requires appropriately directed signals from the control system. Further, when a system failure has resisted all automatic recovery attempts, it is necessary to initiate an audible alarm in the maintenance centre to emphasise the emergency.

Finally, it is important to mention that maintainability has components which are external to the system. Examples of these are the level of training of staff, the availability and quality of procedural instructions (e.g. for restart of the system), the availability and quality of system documentation, the nature of the maintenance organisation and management, whether staff are on site or on call, and the availability of tools.

18.5 References

1 HAUGK, G., LAX, F.M., ROYER, R.D. and WILLIAMS, J.R. (1985): *The 5ESS Switching System: Maintenance Capabilities*, AT&T Technical Journal, **64** (6) Pt.2

2 BASSETT, P.W. and DONNELL, R.T. (1981): *GTD-5 EAX: Recovery and Diagnostic Aspects of a Multiprocessor System*, ISS'81, Montreal, September 1981

3 RADHAKRISHNAN, B.D. and YOUNG, J.S. (1981): *GTD-5: Line and Trunk Maintenance*, ISS'81, Montreal, September 1981

4 YOKOTA, S., HAMADA, Y. and KAWASHIMA, M. (1984): *Maintenance and Test for the Digital Switching System D70(D) and Subscriber Lines*, ISS'84, Florence, May 1984

5 REDMILL, F.J. (1985): *Performance Monitoring — A Telecommunications Maintenance Tool Using Live Traffic*, Reliability '85, Birmingham, UK, July 1985

6 MEYERS, M.N., ROUTT, W.A. and YODER, K.W. (1977): *No. 4 ESS Maintenance Software*, Bell System Technical Journal, **56** (7)

7 DOWNING, R.W., NOWAK, J.S. and TUOMENOKSA, L.S. (1964): *No.1 ESS Maintenance Plan*, Bell System Technical Journal

8 JONSSON, I. (1984): *Control System for AXE 10*, Ericsson Review, No.4

9 NILSSON, B. and SÖRME, K. (1980): *AXE 10 — A Review*, Ericsson Review, No.4

10 MAISONNEUVE, M., LEVY, J.P., PASCUAL, L. and STOKKING, M. (1981): *Aspects of E10.S — TSS.5 System Software: Architecture, Operating System and Software Testing and Debugging*, ISS'81, Montreal, September 1981

11 TROUGHTON, D.J., LUMB, A.P., BELTON, R.C., GALLAGHER, I.D., BEXTON, M.D., MOOR, S.R. and STEGMAN, S.C.J. (1985): *System X: The Processor Utility*, British Telecommunications Engineering, **3**, Pt.4

12 JACKSON, M.A. (1983): *System Development* (Prentice-Hall)

13 CCITT (1984): Red Book, Fascicle VI.12 — CCITT High Level Language (CHILL), Recommendation Z200

14 European Workshop on Industrial Computer Systems, Technical Committee No.7 (EWICS

TC7) (1988): 'Techniques for the Verification and Validation of Critical Software', in *Dependability of Critical Computer Systems-1* Ed. REDMILL, F.J. (Elsevier Applied Science Publishers) Chap.6

15 ROBERTSON, L.B. and SECOR, G.A. (1986): *Effective Management of Software Development*, AT&T Technical Journal, **65** (2)

16 REDMILL, F.J. (1988): *Considering Quality in the Management of Software-based Development Projects*, IFIP/IFAC Working Conference on Hardware and Software for Real Time Process Control, Warsaw, Poland, May/June 1988

17 TAKAMURA, S., KAWASHIMA, H. and NAKAJIMA, H. (1979): *Software Design for Electronic Switching Systems,* Ed. HILLS, M.T. (Peter Peregrinus Ltd), Chap.8

18 SWANSON, E.B. (1976): *The Dimensions of Software Maintenance*, IEEE Computer Society — Proceedings of 2nd International Conference on Software Engineering

19 MARTIN, R.J. and OSBORNE, W.M. (1983): *Guidance on Software Maintenance*, National Bureau of Standards, Special Publication 500-106. Washington, December 1983

Part IV
Network and Operational Aspects

Chapter 19

Network Control

Only a few years ago, this chapter might have been entitled 'Operations and maintenance' (O&M). Essentially, operations functions are those which are necessary for the installation, modification and reconfiguration of exchange components and data. They are carried out by controlling the exchange database containing the permanent and semipermanent data (see Chapter 14) on which exchange functions depend. Maintenance functions, are necessary to ensure service availability and they include supervision, fault detection, automatic recovery, initiating error messages and alarms, and fault clearance, as described in the preceding chapter.

During the past few years, network management has become an additional operational tool whose function is the control of the network. With the ability to take network-management actions (such as diverting traffic within the network) so as to overcome or minimise the effects of congestion or failure, it will not be long before maintenance itself is seen simply as one aspect of network control. An administration's goal is to make as many successful connections as possible and, if calls do not have to be lost as a result of a failure, maintenance can be planned, rather than always being regarded as urgent. With transmission based on 30-channel (or 24-channel) systems and the transport medium consisting of low-cost optical-fibre cables, it will be economic and strategically sound to use overprovision as a means of network control. This would not only provide the capacity for coping with traffic surges without the loss of calls or revenue, but also the flexibility for rerouteing traffic when necessary. Then, network control would be achieved, in the first instance, by the passive means of overprovision of equipment and, in the second, by the dynamic means of network-management actions. Maintenance would be a tool for control used in support of these.

In many administrations, the various functions which contribute to network control are automated on computer systems. This allows changes to be made rapidly and facilitates both exchange and network control. Although the functions are, in most networks, separate, centralisation of them is increasing. This allows the operation and maintenance of a number of exchanges to be

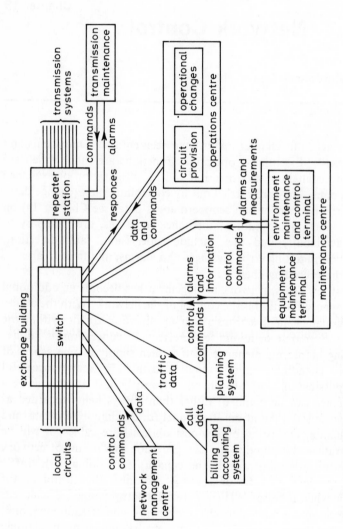

Fig. 19.1 *Functions for network control*

carried out from a single centre and so reduces the total number of staff needed. It is also possible to integrate the various functions, both within computer systems and by interconnecting computer systems. When this becomes prevalent, it will reduce staff further, as the facility will exist for full network control to be exerted from a single point. In this chapter, a discussion of the individual functions will be followed by an explanation of how they may be integrated.

19.1 Functions necessary for control

Each function discussed in this Section results in an input to the exchange or depends on an output from the exchange. The inputs and outputs are shown in Fig. 19.1 and, together, they form the means of controlling the operation and efficiency of both the exchange and the network. The desciptions below are brief only. They are not intended to be exhaustive definitions of the respective functions, but guides to their purpose in control and to how they relate to the exchange inputs or outputs.

19.1.1 Circuit provision

Circuit provision is the process of allocating the terminations on the exchange to destinations and adjusting their performance ('lining them up') against standard criteria. The destinations may be individual subscribers or distant exchanges in the same or a different country. The process begins with planning, when it is determined that one or more circuits to a defined destination are required. Interrogation of files (often in the form of a computer database) reveals whether or not spare terminations exist on a given exchange. When they do, planning is carried out for the allocation of exchange equipment to make up the circuit (for example, echo suppressors are necessary on some international circuits), testing (which may require manual co-operation at the other end of the circuit) and the choice of a date for bringing the circuit into service. A computer system to carry out these functions is described by Boot-Handford, *et al.*[1] In addition, manual intervention is sometimes necessary, for example, when wiring changes have to be made in an exchange or replaceable units have to be changed.

In an SPC exchange, circuit provision involves the input of data to the class-of-service records for the terminations involved, the input of data to the equipment records of all equipment involved (see Chapter 14), and the input of appropriate commands for the control system to store the necessary relationships between terminations and other equipment. It is still common practice to do this manually, from a terminal, but it is quite feasible for the exchange database to be updated directly from a computer system.

When a computer-system database, external to the exchange, has been developed for circuit and equipment allocation, it needs to maintain a record of all termination and equipment data stored in the exchange. It must, therefore, 'mirror' the exchange database. It can then also be used to store historical data for use in maintenance. It must be updated whenever a change is made to the exchange data during maintenance or as a result of network-management actions.

19.1.2 Operational changes

Operational changes are made to an exchange for purposes such as the introduction of new equipment, the allocation or reallocation of terminations to subscribers or trunks, changing the status of terminations (e.g. from 'in service' to 'unassigned'), and changing the facilities allowed to a given subscriber's circuit. In all cases, the exchange database must be updated (as in the case of circuit provision) and, in some cases (such as the introduction of new equipment), manual work in the exchange is also necessary. Typically, operational-change data are input from a terminal, perhaps the same one used for the input of circuit-provision data. If a computer system is used for circuit provision, it would be necessary to enter change data into that system. Then, all changes made to the exchange database would automatically be reflected in its external replica.

An example of the need for an operational change is the provision of a telephone line to a new domestic subscriber. Then, a number of changes must be made to the semipermanent data (see Chapter 14) in the exchange database. When a particular line has been selected, the data in its class-of-service record must be set appropriately. This occurs as a result of a 'dialogue' between the operator and the computer in use. First, the line type is recorded as being 'domestic subscriber's line'; then, for a normal subscriber, the barring level is set so that neither incoming nor outgoing calls are barred; finally, the customer facilities required by the subscriber are made available. As these data are entered, the subscriber's record is created within the exchange database and this sets up a crossreference between the subscriber's directory number and the subscriber's line-termination unit's equipment number. It is usual for the operator to use the directory number as a reference key while inputting the data.

19.1.3 Exchange maintenance

As explained in Chapter 18, exchange faults are detected by the control system. Diagnostic programs are then run to locate each fault to a specific software module or hardware plug-in unit or item of equipment and, finally, messages, giving details of the fault, are output. Usually, the station at which these messages are displayed is remote from the equipment areas and it is not uncommon for it to be remote from the exchange building. It is often referred to as the 'maintenance centre'. Staff are only sent to the equipment areas when manual maintenance is required. In most cases, recovery from loss of service is achieved by an automatic restart, with roll-back or a reload of the software.

However, in case automatic recovery is unsuccessful, the operator has the ability to initiate a manual restart and reload with commands from the remote terminal, and to initiate diagnostic routines and tests. In most systems, a whole-exchange failure would not only result in a print-out at the remote terminal, but also audible and visible alarms both within the exchange and at the maintenance centre.

19.1.4 Environment maintenance

In a telephone exchange, the power supplies and air-conditioning are monitored, as well as the symptoms of fire and building intrusion. Without power, availability of the exchange is lost. Thus, not only are power supplies duplicated, but they are also monitored so that failure causes an immediate alarm, in both audible and visual forms. The most usual cause of power failure is a blown fuse, and replacement cannot be carried out remotely. The alarm is thus given both in the equipment area, at rack and suite-of-racks levels, and in the maintenance centre. Loss of the external power supply is also alarmed remotely and locally, while its effect is minimised by the use of exchange batteries (which may be effective for periods ranging from a few minutes to several hours) and local generators. A typical mode of operation is for a set of batteries to be 'floating', i.e. providing electricity to the exchange while constantly being charged. In this way, they are already in circuit when a mains power failure occurs. Unfortunately, owing to the high current required by an exchange, the batteries can not substitute the mains supply indefinately.

Atmospheric monitoring is carried out for both temperature and humidity, because changes in either may affect the operation of digital equipment. Again, audible and visual alarms are provided in both the equipment area and the maintenance centre.

19.1.5 Transmission system maintenance

The cables leaving an exchange are monitored. When transmission is interrupted (for example, because of a severed cable) or is impaired beyond a given level, this is detected at the multiplexer, which initiates appropriate alarms. In analogue systems, there are alarms at the circuit level as well as at each level of multiplexing. In digital systems there are alarms at all multiplexed levels (e.g. 2, 8, 34 and 140 Mbit/s), but not normally at the channel level. Traditionally, transmission alarms were presented in the form of lamps on the equipment and as print-out. The tasks of analysing the alarms, so as to eliminate redundant lower-order alarms and to find the highest level of multiplexing at which the fault existed, was a manual one. Traditionally, too, the cables were terminated in a repeater station which was not necessarily collocated with the exchange. Transmission monitoring, rearrangements and maintenance were carried out in the repeater station, where the staff did not come into contact with the exchange switching staff.

Now, the distinctions between transmission and switching are not so clear.

Pulse-code modulation is fundamental to both transmission and switching, and multiplexed transmission systems (currently, only at the 2 Mbit level) terminate directly on to the exchange. Moreover, circuit and transmission-system rearrangements can be carried out simply by inputting the necessary commands to the exchange, via a terminal. This, in turn, allows rapid restoration of service when a fault is detected, even if maintenance of the equipment itself is lengthy. It is, however, important to have prompt detection, diagnosis and reporting of faults. This is achieved by extending the transmission-alarm monitoring points to ports on a computer system. The computer system scans the input ports and analyses any alarms which appear there. It deduces the origin of each alarm, suppresses lower-level (e.g. circuit) alarms already included in a higher-order alarm, and outputs appropriate messages to a display screen in view of transmission-maintenance staff.

19.1.6 Network management

A relatively new addition to network-control functions is network management. Indeed, it is still only used by a minority of administrations. Its purpose is to maximise the utilisation of the network at all times and under all circumstances. This is achieved by monitoring network performance and taking action to alleviate or circumvent the congested part of the network and prevent the congestion from spreading to other parts of the network.

Monitoring is usually carried out on a route or destination basis. Chosen parameters are measured and the results compared against norms or expected values. For example, 'the number of calls which are unsuccessful within a given period because all routes to their destinations are busy' is one initial indicator of network performance. The average numbers of ARBs (all routes busy) at the various times of day are known, and substantial deviations from these averages may indicate a problem in the network. In many cases, an even more telling test is the ABR (answer-to-bid ratio) on a given route or to a given destination. Examples of problems which lead to congestion are transmission failures, exchange equipment out of service for maintenance, and sudden surges in demand generated by momentous political events, national catastrophies, radio and television phone-in programmes, and religious or commemorative occasions such as Christmas and Mother's Day.

Monitoring periods need to be short enough for action to be taken quickly, but long enough to be statistically reliable. They are, therefore, usually from 5 to 15 minutes long. Problems resulting from faults require maintenance action and, if there is no resulting congestion in the network, even after the faulty circuits have been taken out of service, this may be all that is required. However, in many instances, some sort of network-management action is required to relieve or prevent congestion. When network-management action is taken, it may be either 'expansive' or 'protective' (often referred to as 'restrictive'). Both of these are intended to alleviate or pre-empt congestion or its effects. Expansive actions are those which provide alternative facilities for handling traffic and are intended to circumvent a problem and avoid inconvenience to subscribers. The provision of temporary alternative routes (TARs) to the affected destination, using a detour, is an example. Protective, or restrictive, actions are those intended to reduce traffic, and may be designed specifically to control the traffic which is most likely to compound the existing problem. Because this traffic is unlikely to be successful, the less equipment in the network that it uses, the better. Thus, it is blocked at its originating local exchange, if possible. Barring all traffic to a given destination is an example of protective action, and this may be accompanied by an appropriate recorded announcement to subscribers.

Those administrations which employ network management have one or more network-management centres (see Fig. 19.1) in which the relevant data from the exchanges are received and analysed by computers. Exception reports are then displayed for the attention of network-management staff who decide what action to take. Close communication is necessary between the network-management and maintenance centres, although, in most cases, both expansive and protective actions can be initiated from the network-management centre.

In some network-management centres, there is a wall-sized model of that part of the network which is important to that centre. The model may be computer-controlled, such that information on each route is automatically displayed. For example, congestion or traffic loading above a certain threshold may be shown by illuminations in different colours. Such a visual model assists network managers to anticipate problems and pre-empt them by taking early action; it also enables staff to visualise the distribution and extent of congestion, thus helping them to detect a pattern or predict the likely course and speed of its spread. This is often useful in choosing the most effective remedial action.

19.1.7 Billing and accounting

Although billing and accounting functions are not essential to the operation of an exchange, they are necessary for the running of an administration. In this description, 'billing' refers to the calculation of charges and invoicing of subscribers by an administration for services provided. Services range from the connection of simple telephone calls to the provision of special facilities and the connection of subscribers to information and entertainment services. 'Accounting' refers to the calculation of charges and the preparation of invoices for the settlement of accounts between adminstrations (and other providers of networks and services) for the use of each other's networks or facilities. Until recently, accounting was almost invariably carried out between nationalised administrations. Now, monopolies have been broken in some countries, with the result that the number of carriers and providers of services is increasing rapidly, both within countries and internationally.

The information required from an exchange for billing and accounting consists, in its simplest form, of the origin and destination of each call and its times of connection and completion. These times allow the call duration to be deduced and the appropriate tariff to be applied in the calculation of the charge. Hitherto, international accounting was based entirely on the sum of call durations ('paid minutes') over a given period between the two administrations concerned. Latterly, in a more competitive environment, adminstrations are using different tariffs in different periods for accounting as well as billing. Further, whereas both billing and accounting were previously based on the destinations of calls, they will in the future need to be based also on the routes taken by the calls, due to the number of network providers and the different accounting agreements between them.

The above paragraph is not an exhaustive description of the criteria on which modern billing and accounting are based. Rather, it is intended to demonstrate that the criteria are changing and that an administration requires flexibility in its billing and accounting system as well as in the ability of its exchanges to provide information on calls. (A more detailed coverage of telecommunications economics is given by Littlechild.[2])

When accounting was a transaction only between the administration of nationalised organisations, it was based on traffic measurements taken only at international exchanges which formed the gateways between national networks. With the possibility of there being more than one network provider in a country, traffic recording for accounting may be made at trunk or even local exchanges.

Similarly, in the past, billing has been based on metered or recorded information from local exchanges. More recently, administrations have become keen to provide better facilities, particularly to business subscribers. This has led to the possibility of subscribers being directly connected to trunk and international exchanges which must, in these circumstances, also possess billing facilities. Further, itemised billing is now a requirement on local exchanges.

The information necessary for billing and accounting is available from the call record (see Chapter 14) created within the exchange control system for each call. However, different exchange manufacturers have different principles on which they base the handling of the data. Some make it possible for all the relevant data to be output and, thus, processed externally to the exchange. This means that a minimum of control-system power is devoted to postprocessing of data and, thus, a maximum can be used for the main function of call processing. However, a large volume of data needs to be processed externally, and this requires a large and expensive computer system. On the other hand, the administration gains flexibility. Other manufacturers include software in the SPC system which carries out some preprocessing. This allows less processing to be done externally, but it reduces an administration's flexibility. If new criteria are to be introduced into billing and accounting, resulting in new data being required from the exchange, new exchange software must be developed.

Whether raw or preprocessed data are output from an exchange, billing and accounting require a great deal of processing power because every successful call connected through the exchange must be accounted for. Factors included in processing are subscriber type, type of service, whether supplementary features are used, destination, call duration, time of day, day of the week, and, perhaps, the route taken to the destination. Further, in itemised billing, the actual digits dialled must be stored and printed on the bill. The cost of the computer system is, necessarily, high. It is, therefore, an advantage to share this resource between a number of exchanges, rather than provide it on a per-exchange basis. Indeed, whereas subscriber billing could be carried out on a per-exchange basis, accounting needs to be a centralised function. Typically, the data are passed from an exchange to a 'billing centre' where the processing facility is located, as in Fig. 19.2. The future means of data transport will be data links, but at present it is usual for the data to be output from the exchange on to magnetic tape and transported physically.

The Figure shows the data from exchange 1 being input directly to the billing and accounting computer system, but that from exchanges 2 and 3 having to be preprocessed to arrange the data in a format acceptable to the computer system. This is often the case, because there are no international standards for the type and format of data output. For example, the preprocessing shown in the Figure would be necessary if the computer system was designed to accept the output from exchange 1 prior to the existence of exchanges 2 and 3, and if these were of different manufacture or model to exchange 1, or even later versions of the same system. The billing and accounting centre may be collocated with an operations and maintenance centre. However, it is more likely to be a part of the administration's centralised data-processing facility.

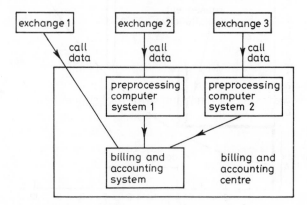

Fig. 19.2 *An example of a billing and accounting centre*

19.1.8 Planning

Whereas short-term congestion, due, for example, to equipment failures or sudden peaks of traffic, may be overcome by network management and maintenance, it may take a very long time to eliminate congestion which is due to a lack of equipment. This is because the process involves ordering, waiting for, installing and commissioning the necessary equipment. Planning is, therefore, an essential element of network control, although it is a process which involves a long lead time before its effect is experienced in the network.

To minimise congestion due to underprovisioning, of either exchange equipment or transmission plant, a continuing process of network planning is necessary. This, in turn, depends on regular measurements of relevant parameters, such as the frequency of all routes to any destination being busy and the traffic loading throughout the day on each route. Such measurements, taken regularly at each exchange, are used to observe traffic profiles with time, and thus determine when it is appropriate to increase the exchange equipment or transmission plant.

Because of the delay between planning an increase and implementing it and of the disruption at an exchange during the work, it is usual to overprovide

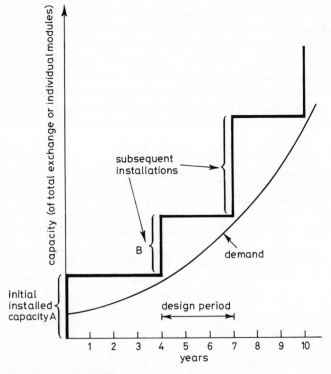

Fig. 19.3 *Exchange growth (from Reference 3)*

the exchange at each increase. Fig. 19.3[3] illustrates the step function of exchange growth to meet increasing demand. It represents an idealised situation in which the installed capacity is never quite overtaken by demand. Capacity A represents the initial installed capacity of a new exchange, planned to satisfy the forecast growth in demand over the first 'design period' of four years. This means that 4 years is allowed for the planning and installation of increase B in equipment capacity. The next design period is then shown to be three years. So long as the actual demand does not exceed the exchange capacity after each step increase, there is a period when the grade of service provided is better than that guaranteed, due to the surplus equipment.

In an SPC exchange there is usually a software module which performs the traffic-recording function when activated. The results are then stored and output at a suitable time. It is usual in current systems for the output to be on magnetic tape, although in the future data links will be used. Processing is then carried out on a computer system into which may be programmed the rules and criteria used in planning by the administration (see Chapter 1 for a discussion of this). Fig. 19.1 shows the planning data being processed by a dedicated computer system. However, in many cases the data share a magnetic tape with the billing or accounting data and are then processed by the same system.

19.2 Co-locating and centralising the functions

In the preceding Section, eight functions which contribute to network control and which constitute the exchange inputs and outputs were identified. While they are independent functions, two were shown to be located in a 'maintenance centre' and two in an 'operations centre'. These were natural

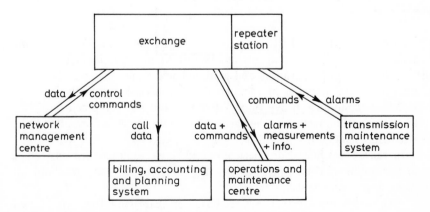

Fig. 19.4 *A model for grouping the functions*

combinations, as the two functions in each case were similar and lent themselves to collocation.

In practice, it is usual to centralise the four functions of these two 'centres' to form an 'operations and maintenance centre' (OMC). If the traffic-recording data for planning and the billing and accounting data are processed on a single system, the individual functions of Fig. 19.1 are seen to be grouped as in Fig. 19.4. The OMC is then the site where all switch-maintenance messages are output and where all operational data (for circuit provision, reconfiguration and modification to system components, trunks, subscribers' terminations and facilities) are input.

A feature of an OMC is that the different functions are performed at terminals by human operators who may have to consult operations or maintenance manuals in order to carry out their tasks. However, Sevcik[4] has described an 'expert operations system' for the EWSD switching system. It aims at a formal knowledge representation of operations and maintenance procedures, and uses the Specification and Description Language. (SDL, described in Chapter 13) 'enriched by a rule-based approach'. It is claimed that this system replaces the conventional operating manuals. This is a significant claim, because a great many of the faults that do occur on SPC systems are due to operators' errors. If manual interaction is minimised, reliability and system availability will be improved.

A feature of SPC is that all the functions may be performed remotely, because very few physical actions are required at the exchange. It is therefore possible to collocate the functions in whatever combinations are most convenient to an administration. However, the most important and advantageous effect of remote operation is that the same function of a number of exchanges may be centralised in the same place. With this ability, many administrations have collocated not one but many of the functions of a number of exchanges at single points (as in Fig. 19.5), which may themselves be collocated. The advantages of centralisation, as stated by Van Walle and Peeters[5] are:

(i)   More efficient use of the time and skill of human resources, which may increase both proficiency and job satisfaction of the staff involved.

(ii)   Collocation of operation and maintenance functions, and better co-ordination of the activities of various administrative departments.

(iii)   More economical provisioning of spares and test equipment.

(iv)   Easier attainment of the goal of round-the-clock monitoring of unattended exchanges.

(v)   Increased emphasis on network and area maintenance instead of exchange maintenance, which lowers the barrier between switching and transmission maintenance.

(vi)   Reports to management that are based on real-time information (due to automatic data analysis and processing), making possible immediate action to improve weak spots in the network.

To gain some of these advantages in the field of maintenance, the Deutsche Bundespost's principle (according to Maurer and Schuchardt[6]) is that 'all exchanges located at a maximum of 60 minutes travelling time around the operation and maintenance centre are operated and maintained by the staff employed at the centre unless their size exceeds a certain limit. More distant or very large exchanges may have a rgular maintenance staff of their own ... although directed from the operation and maintenance centre'. Leaving exchanges unstaffed is a benefit from the low incidence of hardware faults on SPC digital exchanges, the fact that these exchanges require a negligible amount of preventive maintenance, and the fact that almost all problems caused by software faults can be overcome by commands from a terminal at the OMC (although the faults themselves must later be corrected off-line).

19.3 Implementing centralisation

In handling exchange data, a number of activities must be carried out:

(i)   Collection of data within the exchange. This is done within the control software during call set-up, routine testing and supervision processes.

(ii)   Management of the data within the exchange. When the data have been collected, they are concentrated with other data of the same type and output at the appropriate time, on the correct port, in a standard format, and with the appropriate message.

(iii)   Transmission of the data to one or more points. Many systems use different output media for the various types of data. Billing and planning data are often output to magnetic tape, while maintenance and network-management data are usually dispatched on data links.

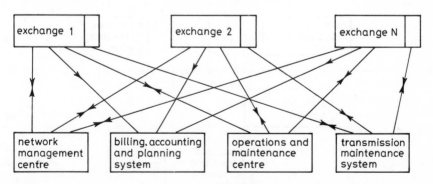

Fig. 19.5 *Centralisation of functions from a number of exchanges*

(iv) Storage and processing of the data at the points of receipt. Some of the data for network traffic management for example, must be analysed and the results displayed immediately. Other data, such as planning and accounting data, must be stored and combined with similar data from other exchanges prior to analysis.

For centralisation of functions, all data from a number of exchanges must be delivered to some common point (for example, to an OMC). In addition, operations data (for circuit provision, changes to customers' lines and facilities, etc.) must be input to the exchanges from the same point.

It is possible for all functions to be handled by a single computer. However, the tasks to be performed may conflict with each other. For example, the receipt of, say, a day's billing data may take 15 min or more and interfere with the receipt and immediate analysis of network-management data every 5 min. Although this may be overcome by installing more and more powerful computers, it is usual for different computers to be used for the different functions. At least one manufacturer offers 'integrated network operation and maintenance', wherein all data from a number of exchanges are output on to a computer-controlled network where they are directed to the appropriate

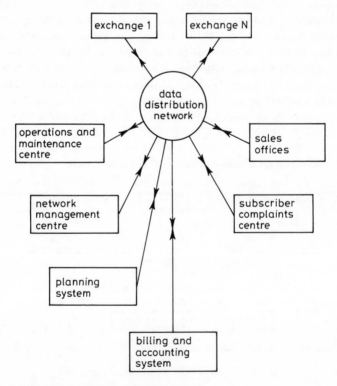

Fig. 19.6 *Intercommunication via a single network*

'centre' or individual computer system.[7] The principle is shown in Fig. 19.6 and the 'centres' may be collocated.

19.4 Interconnection of systems

So far in this chapter, remote and centralised computer systems have been shown only with their data flows to and from the exchanges. Yet, as the systems become more sophisticated and perform more integrated tasks, their interconnection becomes essential.

19.4.1 Network-management centre — operations system
In many OMCs, circuit-provision instructions and operational changes are input at a terminal connected directly to the appropriate exchange. However, it was shown in Section 19.1.1 that an operations computer-system (OS) database may be used to store a model of the circuit and equipment allocation within an exchange and for the preparation of modifications to the exchange. If it is, it then becomes important for this system to be 'in step' with the exchange at all times. Any changes made within the exchange must be reflected in the system. Thus, the necessity arises for a direct connection between the OS and the network-management centre. Modifications initiated as network-management actions are immediately communicated to the OS on the one hand; on the other, operational changes and transmission-system rearrangements are communicated to the network-management centre so that subsequent network-management actions are based on reality and not on an out-of-date model. The interconnections are shown in Fig. 19.7, and the fact that the OS may be an integrated-operations system is irrelevant in the present context.

19.4.2 Data for decision making
In these days of denationalisation, competition for the provision of telecommunications services and networks, the need to satisfy shareholders,

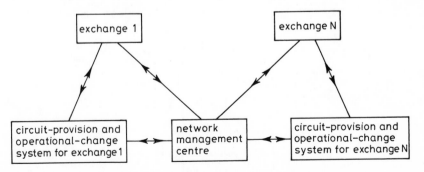

Fig. 19.7 *Interconnection of operations and network-management systems*

and an emphasis on quality, there is an urgent need for managers to have up-to-date information on the status of their networks. For example, directors have bilateral and multilateral meetings with their counterparts in other administrations and, for these, they require data on the loading of circuits, outstanding debts, maintenance problems, the effects of changes in tariffs, the use of new services, network-management actions, etc. Marketing and sales managers need to know the availability of circuits and the services being offered to given destinations, tariffs and tariff periods, maintenance facilities and histories, etc. It is not practicable to furnish each manager with all possible information each day. Ideally, managers require terminals via which they may access the desired information as and when it is needed.

For this to be possible, a number of steps must be taken:

(i) The raw data, output from the exchanges, must be stored in an easily accessible location and form. It is possible for this to be done on operation-and-maintenance systems, network-management systems, and planning, billing and accounting systems. However, if it were, the managers' terminals would need to be connected to a larger host computer, programmed to access the appropriate system in order to acquire the necessary data. A more likely solution is for those systems connected to the exchange to carry out some preprocessing and then to transfer only the essential data, in suitable formats, to one or more management-information systems, as in Fig. 19.8.

(ii) Processing of the raw data must be carried out, either prior to storage or when accessed. Some preprocessing would normally be done, as suggested in (i) above. The management-information system(s) would then need to do further processing in order to combine data from different sources and store

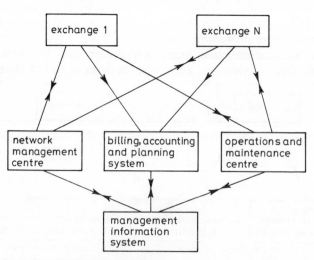

Fig. 19.8 *Providing data for decision making*

them appropriately. However, owing to the various requirements of the different managers, it would be almost impossible to store all data in forms which are ideal for everyone. Thus, further processing at the time of access is often necessary if managers are to be presented with the data they need in formats tailored to their requirements.

(iii) Software must be written to allow managers to access the sorted data from their terminals. With a large number of managers accessing the management-information system, it is advantageous to localise some processing in their terminals — which then need to be microcomputers. This allows the mangement-information system to deal with retrieving data in response to the requests from managers, and processing and storing newly arrived data.

Even with the above arrangements, the total amount of processing may be enormous. If managers' requests are to be met with rapid responses, the systems require careful design. Many administrations (and managers) do not plan their data requirements, and, in the absence of a clear understanding of what they need, decide to measure and store all possible data. This is not only hugely expensive, it also impedes rapid and effective access to the data which are needed. In developing management information systems, clear strategic goals and detailed planning are essential, as well as a flexible design to facilitate subsequent modifications.

19.4.3 A closed loop of automation

It has been shown in this chapter that traffic-recording data for planning is output from the exchanges in a defined area to a computer system (which may also receive billing and accounting data). This system may or may not cover all the exchanges which comprise the network for which a regional manager is responsible. If not, all planning data from the various systems must be combined in a single 'network planning system' (NPS). There, an analysis of the data is carried out, to deduce trends and thus arrive at forecasts of traffic intensities at chosen times in the future. In a national network, forecasting is usually based on data collected at local exchanges, where trends in the growth (or diminution) of traffic, on a destination basis, can be detected. Planning of how the forecast traffic should be routed in the network is then carried out.

The analysis by the NPS may result only in a print-out of forecast average traffic intensities. In this case, manual calculations are performed to deduce the changes necessary to the circuit quantities between the various nodes in the network during the ensuing period. On the other hand, the NPS may be programmed to estimate the circuit- quantity changes. Adequate planning requires a comparison of the requirements of individual routes against a model of the existing network. In a digital network, where circuits are installed in 32-channel (or 24-channel) systems and SPC allows easy alterations to routeing tables, increases in traffic between two exchanges may sometimes be

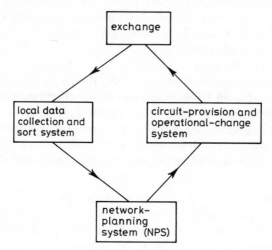

Fig. 19.9 *An automated closed loop for network planning and implementation*

accommodated by reallocating routes rather than by installing new plant. Thus, if the NPS contains an up-to-date model of the network, it may carry out this function automatically, given that the planning rules are programmed into it.

It was shown in Sections 19.1.1 and 19.1.2 that it is quite feasible for circuit provision and operational changes to be carried out directly from a computer (say, an operations-and-maintenance system). Where this is done, it is possible for the implementation of network plans to be achieved wholly automatically (as in Fig. 19.9, with documentation of the changes being output for information. The exception to this would be when new plant needs to be installed. Then, manual intervention is essential.

The above means of planning and implementation is fully automatic and forms a closed cycle of automated events. Fig. 19.9 has been made simple to illustrate this; in reality there would be a number of exchanges and computer systems forming a number of related closed loops. Such a closed-loop system would be relatively easy to implement for a very stable network. However, the planning rules for network modernisation and digitisation, restructuring the network, and the introduction and integration of new services are complex (and are considered in Chapter 21). In many cases they change as experience reveals the inadequacies of first proposals. Therefore, at present there is the need for a great deal of human input to planning and, although the loop of automation can be put in place, human intervention will continue to be essential.

19.4.4 *The connections between systems*

So far there has been no standardisation of the means of communicating data between the exchanges and the supplementary systems discussed in this chapter, or between the systems themselves. As seen in Section 19.3 above, it is possible to use a proprietary network, but this is not widespread. The result is that a great deal of development time and money are spent on designing and providing interfaces to allow intercommunications.

There has recently been a spate of work within the CCITT on a telecommunications management network (TMN) for the purpose. This is based on the 7-layer model used in open-systems interconnection, which defines standard interface protocols, by which computers of different types can communicate with each other. It is hoped that recommendations of the CCITT will encourage exchange manufacturers to standardise their interface protocols so as to facilitate communication with external computer systems. Meanwhile, Brandmaier and Axe[8] have described a model for using the public switched telephone network (PSTN) for the purpose, when CCITT signalling system No. 7 is fully operational. The work being done suggests that standardisation will be achieved and, when it is, network control will be greatly facilitated.

19.5 References

1 BOOT-HANDFORD, K.J., GRIFFITH, R.T. and KIMPTON, R.I. (1986): *Towards the Paperless International Telecommunications Services Centre — The Keybridge Engineering Records System*, British Telecommunications Engineering, **5**, Pt.1

2 LITTLECHILD, S.C. (1979): *Elements of Telecommunications Economics* (Peter Peregrinus Ltd)

3 REDMILL, F.J. and VALDAR, A.R. (1985): *Selecting a System* African Technical Review

4 SEVCIK, M. (1987): *An Operations System Based on Expert System Techniques*, ISS'87, Phoenix, Arizona, USA, March 1987

5 VAN WALLE, J. and PEETERS, F. (1984): *Centralized Operation, Maintenance and Call Charge Recording*, ISS'84. Florence, Italy, May 1984

6 MAURER, W. and SCHUCHARDT, I. (1984): *Operation and Maintenance Concept for Digital Telephone Exchanges*, Ibid.

7 STRANDBERG, K., SODERBERG, L. and VESTIN, C.G. (1984): *Experience with Integrated Network Operation and Maintenance*, Ibid.

8 BRANDMAIER, K. and AXE, M. (1987): *O&M Concepts for Public Networks Using ISDN, CCS7 and Personal Computers*, ISS'87. Phoenix, Arizona, USA, March 1987

Chapter 20

Common-channel signalling

20.1 Introduction to common-channel signalling

Common-channel signalling (CCS) was introduced in Chapter 2, where it was explained that several conversations may be controlled via one 'common-channel' signalling link. This is done by sending a stream of messages over the common physical signalling channel, each message containing the signals themselves as well as the information to identify the conversation or transaction to which they refer. The minimum form of a message was shown in Fig. 2.5, but this is inadequate for practical purposes. It must contain a field for the identification of the circuit to which the message refers, as well as a number of other fields for purposes which will be discussed later in this chapter. CCS is rapidly becoming the standard signalling technique in all modern national and international telecommunications networks, and its application to the subscriber-access network will increase with the demand for a larger set of customer services.

As shown in Chapter 2, CCS has several advantages over the earlier channel-associated signalling. Importantly, it obviates the need for most of the per-circuit signalling equipment which is necessary when channel-associated signalling is used. This in turn reduces its overall cost. A further key feature of CCS is its repertoire of signalling information. In the earlier analogue channel-associated systems, the number of different signals that could be sent was limited and the sending rate slow. For instance, in the multifrequency signalling system described in Chapter 2, there was a repertoire of only 16 different codes. A signal in this system is equivalent to 4 bits of information, and, as it would take about 100 ms to send it, the bit rate is 40 bits/s. The common-channel variant of subscriber-line signalling typically achieves 64 kbit/s.

This high speed (equivalent to a very high signalling bandwidth) is crucial to many new ISDN-based services, in which up to 1000 bits of information must be sent between two nodes in order for a call to proceed. If these had to be sent at the signalling rates of channel-associated systems, the call set-up delay would be excessive and unacceptable to subscribers.

CCS systems exploit the fact that the flow of signals required to control a telephony connection between two exchanges is intermittent in nature. Typically, a call requires a burst of signalling to establish the connection, little or no signalling during the conversation phase, and a further small burst of signalling for termination. The conversation phase of a telephone call is about 3min, although a sample of calls shows a wide distribution. Data calls may be much shorter or much longer, according to the user's application. Thus, provided that the various signalling bursts can be identified and associated with the appropriate calls, a single common-channel system can carry the signalling for a number of circuits by the controlled interleaving of their signalling bursts.

The number of traffic channels that can be supported by a CCS system is limited by the acceptable amount of message queueing for the signalling channel. Although the delays incurred follow a statistical distribution, a limit must be set if they are not to be intolerable. For a CCS channel of 64 kbit/s (the standard rate for most applications), the signalling for 10 000–20 000 traffic channels can be handled within acceptable delay limits, the exact figure depending on the signalling requirements of the services carried and the message structure of the CCS system.[1] (The figure is proportionally lower for lower-speed CCS systems.) However, in practice, administrations load the CCS links to lower levels in order to meet service and security targets. Because failure of a CCS link affects the calls of all its dependent traffic channels, performance monitoring is carried out to detect problems early and ensure a high level of reliability.

20.2 Applications of common-channel signalling

A CCS link between two communicating systems is a data link which has been chosen for the services it can offer, and optimised for its application against criteria such as throughput, performance, rerouteing capability and security of data. CCS is used in a number of applications, and these are outlined below and illustrated in Fig. 20.1.

(i) Inter-exchange signalling. The modern internationally agreed standard for this application, in both national and international networks, is the CCITT signalling system No. 7,[2] to which an introduction may be found in CCITT Recommendation Q.700, and which is now being deployed throughout the world. Inevitably, however, some national variants of the system were introduced in advance of a fully specified system becoming available.[3] The normal bit rate for CCITT No. 7 is 64 kbits/s, and it is carried in a dedicated time slot within a PCM system. Section 20.4 describes CCITT No. 7 in some detail.

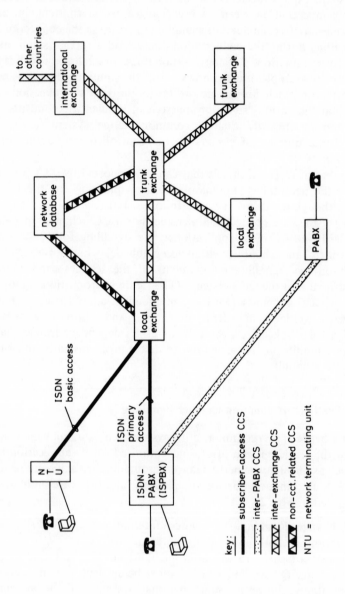

Fig. 20.1   *The applications of CCS to telecommunications networks*

An earlier inter-exchange CCS standard was CCITT signalling system No. 6.[1] This was developed in the 1960s, came into use in the early 1980s, and is still employed. It supports telephony and the transfer of some network-management information, but no ISDN-based services. CCITT No.6, which has a basic speed of 2400 bit/s, was deployed on many intercontinental routes and carried in a dedicated analogue speech circuit, using modems. This was necessary because when CCITT No. 6 was introduced there was very little digital transmission. Within the USA, a national variant, known as the common-channel inter-office system (CCIS), was deployed extensively and is still in widespread use.[4] This is transported over 24-channel PCM systems using a single bit per frame, as described in Chapter 5.

(ii) Subscriber-access signalling. Chapter 22 explains how the extension of common-channel signalling from local exchanges to subscribers' premises is an essential component of an ISDN. There are two forms of subscriber-access CCS systems: 'primary rate access' from PABXs and 'basic access' from single exchange lines, as shown in Fig. 20.1. Both forms are the subject of internationally agreed specifications by the CCITT. In addition, a number of national versions have been developed in advance of CCITT standards being finalised, and these have enabled the early deployment of ISDN-based services.[5,6] Section 20.5 introduces the principles of subscriber-access CCS systems.

(iii) Inter-PABX signalling. Modern PABXs offer a range of facilities to their users, e.g. the transfer of an incoming call from one extension to another. When several PABXs are linked by leased lines to form a network, the services can be extended by using a suitably powerful signalling system, such as the digital private network signalling system (DPNSS),[7] a common-channel signalling system described briefly in Section 20.6.

(iv) Non-circuit-related applications. The capacity of CCS makes it suitable for non-circuit-related applications, i.e., the transfer of information over a telecommunications network for purposes other than the direct control of a particular circuit. Examples of such applications are the control of intelligent networks (see Chapter 9), the transfer of location information in cellular telephone systems, and the transfer of billing or network-management data. As an example of this application of CCS, Fig. 20.1 shows a link between a local exchange and an intelligent network database. This concept is described in Section 20.4.3.

20.3 CCS networks

Unlike channel-associated signalling, CCS offers administrations the possibility of developing a separate signalling network. This results from the detachment of signalling from its traffic channels, as described in Chapter 2 and shown in Fig. 2.7. A signalling network[8] comprises signalling nodes, located in exchanges and normally forming a part of the exchange system, interconnected by signalling links which are provided by the transmission network. Chapter 21 describes the concept of a multilayered model of an integrated digital network (IDN) in which a number of auxillary networks, including a signalling network, support the switching and transmission networks.

The advantages to an administration of developing a separate CCS network are that it enables:

(i)   The capabilities of CCS to be fully exploited.

(ii)   The network of CCS links to be optimised for economy.

(iii)   A high degree of resilience to be achieved, using both the security aspects of the CCS system and the alternative-routeing possibilities of the line plant.

Signalling nodes (or, in CCITT No. 7 terms, 'signalling points' – see Section 20.4.4) may occur at the following locations:

(i)   Switching units (local, trunk and international).

(ii)   Operation, administration and maintenance centres.

(iii)   Intelligent-network database sites.

(iv)   Signal-transfer points (as described in Section 20.4.1.3).

In some cases, a node is common to two networks: for example, an international gateway exchange is connected to both a national network and the international network. As described in Section 20.4 below, all signalling points are allocated a code by which they are addressed in CCS messages. In the case described above, the international exchange will have two signalling point codes, one national and the other international.

There are three ways in which the CCS links are associated with their dependent traffic channels. Briefly, these are as follows:

(i) Associated. In the associated mode, the messages relating to the traffic circuits connecting two exchanges are conveyed over signalling links directly connecting the two exchanges (see Fig. 20.2*a*).

(ii) Non-associated. The non-associated mode is illustrated in Fig. 20.2*b*. The signalling messages between A and B are routed through several signalling links, according to the network conditions at the time, while the traffic circuits are routed directly between A and B. At other times, the routeing of the CCS messages may follow different paths. This method is not normally used because it is difficult to determine the exact routeing of signalling messages at any given time.

(iii) Quasi-associated. This is a limited case of the non-associated mode, and the signalling messages between nodes A and B (see Fig. 20.2*c*) follow a

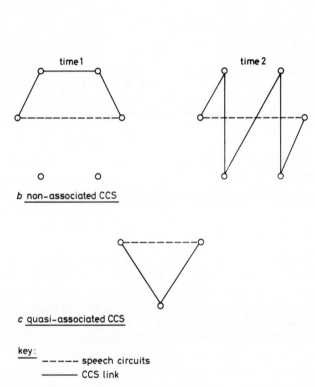

a <u>associated CCS</u>

time 1   time 2

b <u>non-associated CCS</u>

c <u>quasi-associated CCS</u>

<u>key:</u>
– – – – – speech circuits
———— CCS link

Fig. 20.2 *Signalling modes*

predetermined routeing path through several signalling links in tandem, while the traffic circuits are routed directly between A and B. Usually, different transmission bearers are used for the CCS and their associated traffic links.

Associated working is normally used where the traffic route between two exchanges is large. For example, in the case of 200 circuits between two exchanges, seven 2 Mbit/s digital transmission systems would be required, one of which carried the single CCS link in its time slot 16 (TS16). However, where a network has many small-capacity traffic routes, the overheads of providing a CCS port for each one may be reduced by quasi-associated working. The most important use of quasi-associated working, however, is as a security back-up, as described below.

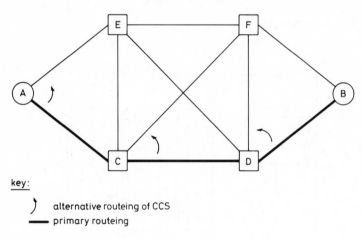

key:

$)$ alternative routeing of CCS
—— primary routeing

Fig. 20.3 *Example of signalling in a mesh network*

In the example above, failure of the PCM system carrying the CCS link would result in the loss not only of the 30 circuits of traffic but also of the signalling for the remaining 170 circuits, even though their transmission remained intact. In order to reduce to an acceptably low level the risk of this occurring, alternative means of carrying the CCS link are employed. The most common is to re-route the CCS link over quasi-associated routeings in the event of failure of the (associated) primary routeing.

In addition to re-routeing, which is programmed to operate automatically in the event of failure of the primary CCS link, network resilience is enhanced by permanently spreading the CCS messages over two (or more) parallel associated links. Thus, in the above example, the signalling for the 200 traffic circuits would be carried over two CCS links, in the TS16s of different 2 Mbit/s digital transmission systems. Ideally, they would be carried on physically diverse paths to ensure maximum security.

An example of a resilient CCS network is shown in Fig. 20.3, in which nodes A to F are interconnected by a mesh network. Then, if CCS messages from A to B are normally routed ACDB, and one of the links AC, CD or DB fails, the alternative paths AEFB, ACFB and ACDFB exist. The planning aspects of CCS networks are considered in Chapter 21.

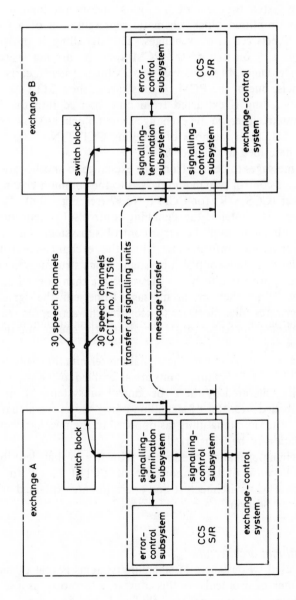

Fig. 20.4 *Block-schematic representation of CCITT No. 7 signalling system*

20.4 CCITT common-channel signalling system No. 7

CCITT signalling system No.7 is now the internationally accepted inter-exchange CCS system for use in both national and international networks. Fig. 20.4 shows a simplified block-schematic representation of a CCITT No.7 signalling link between two exchanges. The signalling is transported in a dedicated time slot on one of the PCM systems carrying speech channels between the exchanges. In the case of two 2Mbit/s systems, carrying up to 60 speech channels, one of the PCM systems carries the CCITT No.7 link in its TS16. The signalling is extracted from and inserted into the 2Mbit/s (or 1.5Mbit/s) system either via the exchange switch block, as shown in Fig. 20.4, or at the DLTU. Both methods of handling CCITT No.7 (and other CCS systems) are described in Chapter 8.

Signalling messages to be sent from one exchange to another are formulated by the exchange-control system and passed to the common-channel-signalling sender/receiver (CCS S/R) for CCITT No. 7 (see Fig. 20.4). The CCS S/R consists of three subsystems: the signalling-control subsystem, the signalling-termination subsystem and the error-control subsystem, all of which are microprocessor-based. Information from the exchange-control system is received by the signalling-control subsystem, which structures the messages to be sent in the appropriate formats. Messages are then queued until they can be transmitted. When there are no messages to be sent, the signalling-control subsystem generates filler messages to keep the link active. On the Figure, the dotted line labelled 'message transfer' indicates the source and destination of signalling messages.

Messages are then passed to the signalling-termination subsystem, where complete CCITT No. 7 signalling units are assembled, using sequence numbers and check bits generated by the error-control subsystem. In the Figure, the dotted line labelled 'transfer of signalling units' shows the actual source and destination of the signalling units (as opposed to the messages containing the basic information to be transferred).

At the receiving exchange, the reverse sequence to that described above is carried out. Because of the importance of CCITT No. 7 to digital SPC exchange networks, its architecture and operation are explained in some detail in the following Sections.

20.4.1 Traditional architecture

A technique used by designers of signalling systems is that of modelling the system as a stack of protocols. In this context, a 'protocol' is a set of rules by which the communicating entities abide. A protocol stack employs the principle of data abstraction described in Chapter 17: each protocol layer in

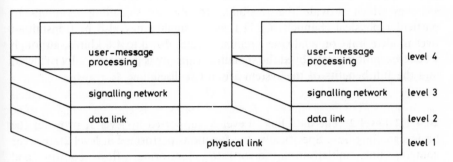

Fig. 20.5 *The 4-level architecture*

the stack is abstracted from those above and below it. A protocol stack thus allows the designer to divide the functions required for the signalling system into separate blocks, each of manageable proportions and with its information exchange with those above and below it defined, so that each block can be developed in isolation. A protocol stack splits the design into several layers, forming a conceptual tower. The upper layers are then designed on the assumption that each of the lower layers performs certain defined functions without error. For instance, if a lower layer is responsible for correcting errors in the transmission of messages, then the layer above it can assume that all messages it sends or receives via the lower layer are error-free. Thus, each layer depends on those below it, and, when layer N is referred to, the functions of layers 1 to N are assumed.

The traditional CCITT No. 7 protocol stack is represented by Fig. 20.5. It can be seen that the term 'level' is used to describe what was termed a 'layer' in the general description above. The traditional model has four levels. More recently, the open-system-interconnection (OSI) 7-layer model has been applied to CCITT No. 7, and, while this is discussed in Section 20.4.2, the 4-level model is explained here.

20.4.1.1 Level 1: The physical carrier. Level 1 is at the bottom of the protocol stack. In summary, it is the means of sending a stream of bits of information from one point to another, over a physical connection. At this level, there is no definition of, or requirement for, a structure to the information, other than that provided by the transmission equipment. Nor is there an error-detection mechanism. Such refinements are provided in higher levels.

CCITT No. 7 signalling information is normally carried in time slot 16 of a 2Mbit/s digital line system or time slot 24 of a 1.5 Mbit/s digital line system (see chapter 5). The link is bi-directional and is normally terminated on digital switches which provide the flexibility to allocate signalling terminals to particular physical channels. CCITT No. 7 signals may also be transmitted over modem links on analogue circuits, at rates down to 4.8 kbit/s, although when this option is employed the signalling capacity is proportionally reduced, and the full benefits of the system are not realised.

20.4.1.2 Level 2: The data link. Level 2, supported by level 1, provides the basic signalling link. Specifically, the functions performed at level 2 are error control, link initialisation, error-rate monitoring, flow control and delimitation of the signalling messages.

Error control is achieved by appending a defined cyclic redundancy check code to all messages and then checking for errors at the receiving end. If any error is detected, retransmission of the whole of the faulty message is requested by the sending end. If no error is detected, the received message is acknowledged explicitly.

Link initialisation prior to transmission is achieved by the use of defined link-status-control messages. It brings both ends of the signalling link into known states relative to each other.

Error-rate monitoring of the signalling link is achieved by the 'leaky bucket' principle. This means that a count is maintained of the number of messages in error, and the count is decremented at a fixed rate, according to the number of good messages. If the value of the counter reaches a predetermined level (indicating a critical error rate), the link is considered unfit for service. It is then removed from service and a report sent to level 3.

Flow control is required to prevent the overload of a receiving node which may be incapable of processing all of the traffic offered. Typically, this could occur between two exchanges made by different manufacturers. CCITT No. 7 offers limited flow control at level 2 by withholding message acknowledgments and explicitly informing the traffic source that the link is in congestion. It should be noted that this is only link flow control and does not guarantee that the processes at levels above the data link are able to support the traffic received over the CCS link.

Delimitation, that is, marking the boundary between one message and the next in a data stream, is achieved by using a flag consisting of a unique 8-bit code. The flag is inserted between the end of one message and the beginning of the next. The use of this flag has provided an example of how carefully specifications must be prepared if they are not to be misinterpreted. When the standard was first introduced, some designers' interpretations of the specification were that, in the absence of messages on the link, a continuous stream of flags should be sent. Another interpretation was that a continuous stream of flags could not occur under normal conditions, that such an event would indicate an error, and that the remedy was to take the link out of service. The two interpretations were incompatible, and this anomaly led to a more complete specification being produced by the CCITT.

Level 2 also ensures that a message cannot imitate a flag and cause an error in the message delimitation. This is done by checking to see if any stream of data has a bit sequence with more than five consecutive ones. If it does, an additional 0 is inserted before transmission, so that the flag sequence (01111110) can never be imitated. At the receiving end, the reverse process is employed to reconstruct the original data.

The formats for the CCITT No. 7 messages, known as signal units (SU), are shown in Fig. 20.6. There are three basic formats:

(i)   The message signal unit (MSU; see Fig. 20.6*a*), used for the transport of the higher-level information.

(ii)   The link-status signal unit (LSSU), used for link initialisation and flow control (see Fig. 20.6*b*).

(iii)   The fill-in signal unit (FISU), which is transmitted whenever it would be inappropriate to send any other unit (see Fig. 20.6*c*).

In summary, level 2 adds structure to the bit stream transmitted in level 1.

20.4.1.3 Level 3: The signalling network. The concept and advantages of a CCS signalling network have already been described. Within a national telecommunications network there may be several thousand public exchanges, and signalling messages must be routed flexibly and securely between them. The level 3 within the CCITT No. 7 system provides the functions necessary for the management of a signalling network.

Within a signalling network, each switching node is allocated a signalling point code, which is a 14-bit address. Every CCITT No. 7 message then contains the point code of the originating node (the originating point code or OPC) and that of the destination node (the destination point code or DPC). Here, it is important to note that the DPC indicates the destination of the message, not of the call, which may be in another network. Any node receiving a message compares the message's DPC with its own and, if they are not identical, passes on the message in a direction defined by routeing tables held within its storage. If a node can operate in this manner, it is referred to as a signal transfer point (STP). An STP acts as a tandem signal-unit switch and obviates the need for full interconnectivity between all nodes which may wish to exchange messages.

In order to support this flexibility, a set of complex procedures is defined

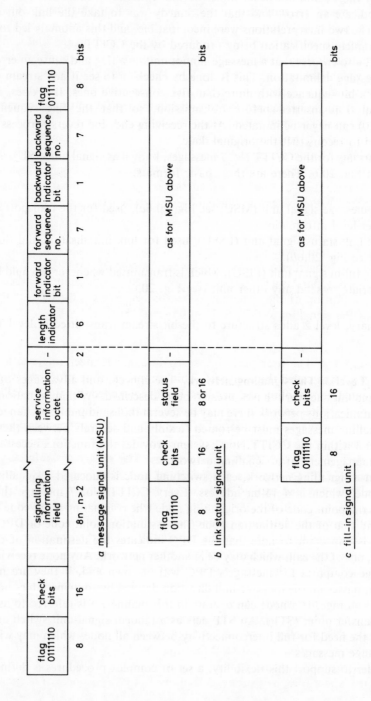

Fig. 20.6 CCITT No. 7 signal-unit formats

for the management of the signalling network. These procedures form level 3 and are designed to minimise message loss or retransmission under all conditions, particularly those of failure. The management of the signalling network depends on the architecture of the network and on the philosophy that the administration wishes to adopt. However, certain essential functions are those which concern the actions to be taken when link failures occur. If all of the signalling traffic is not to be lost in the event of a signalling link failure, a technique is required for re-routeing the traffic on to one or more alternative links. The method by which this is achieved in CCITT No. 7 is referred to as the 'change-over' procedure.

A problem with most systems which require the re-transmission of incorrectly received messages is that, when a signalling-link failure occurs, a node which has recently transmitted a number of messages which have not yet been acknowledged cannot be certain which messages were correctly received but not acknowledged and which were interrupted by the link failure. Most protocols cannot prevent message loss in this situation. CCITT No.7, however, has a procedure built into it which allows for the full recovery of all messages.

Levels 1 to 3 together form the 'message transfer part' (MTP)[8] of CCITT No.7. The word 'part' is a CCITT convention which, in this context, indicates that only a portion of the signalling system is being referred to. The other portion of the signalling system is the 'user part' (UP).

20.4.1.4 Level 4: The user part. In the traditional architecture, level 4 of CCITT No.7 was called the user part. This comprises those processes which are dedicated to handling the services being controlled by the signalling system. Three processes within the user part have been defined: the telephone-user part (TUP), the data-user part (DUP) and the integrated-services digital-network-user part (ISDN-UP), designed, respectively, to support telephony services, data services and ISDN-based services. The ability of the message-transfer part (MTP) to support a multiplicity of user parts is therefore crucial. With the flexibility and expandability which the user parts provide, CCITT No.7 will be able to handle new services as they are made available. This capability is essential if networks are to evolve into full ISDNs. Indeed, with new services controlled by the software of CCITT No.7, there will need only to be minimal change to the hardware of the exchanges, which will reduce the cost and complexity of the introduction of services. This was not possible with the earlier channel-associated signalling systems, which were tailored to match the transmission and switching systems with which they worked, and whose message repertoires were extremely limited.

The user parts define the call-control signals to be used between the communicating switching nodes. While the DUP has not yet reached an appreciable leve of use, and, indeed, is unlikely to, the TUP has. It is in use in its international form between several countries for the support of telephone

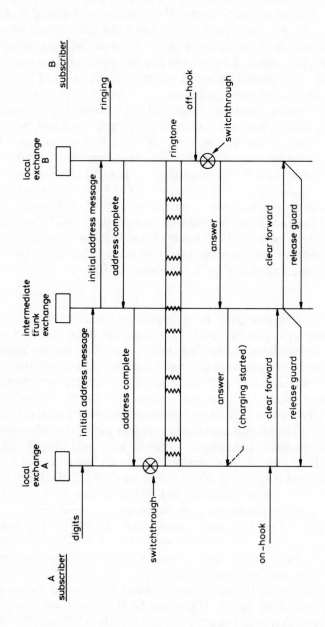

Fig. 20.7 A typical TUP call sequence

calls. Additionally, variants are in extensive use in many national networks,[9] and there have been some attempts to enhance it to support limited ISDN-based services within Europe, the enhanced version being called the 'telephone-user part plus' (TUP +).

A typical TUP message sequence between two local and one trunk exchange is shown in Fig. 20.7. In this example, the 'initial address message' contains the information necessary for the set-up of the call. This information includes the dialled digits, the calling-party category (for example, domestic subscriber) and the transmission-control parameters, such as echo-control requirements. When the terminating node is satisfied that all the required address information has been received, and that the call is proceeding successfully, it returns an 'address complete' message to the originating exchange. This causes the originating exchange to activate the speech path, which must be switched through at this point so that the calling subscriber is able to hear any tones or announcements indicating the progress, success or failure of the call. If the called subscriber answers, an answer message is returned to the originating exchange, the terminating-exchange speech path is connected through, and charging is initiated. Eventually, when the calling subscriber clears down, 'clear forward' and 'release guard' messages are exchanged, and the speech circuit is released.

In telephony, the convention has been that only the calling subscriber can clear the call. Thus, if the called subscriber hangs up first, a message is sent to the originating node, but the call is not cleared down until a time-out occurs or the calling subscriber also hangs up. However, there is a growing trend towards a first-party-release protocol, i.e. for the call to be cleared down as soon as one subscriber clears down, regardless of whether it is the calling or called subscriber. The ISDN-UP is a first-party-release protocol.

One cause of added complexity to the developing CCITT No.7 system has been the need to interwork with older signalling systems. For example, one requirement of older systems was to transmit individual address digits as soon as the calling subscriber had dialled them. This, in turn, meant that common-channel signalling systems had to be designed to transport digits either *en bloc* if they were all available, or individually (overlap) as they become available. An example of an initial-address-message structure is given in Fig. 20.8, which shows that, to allow for different numbers of digits in the message, there is a field for defining the number of digits being sent.

The ISDN-UP specification is now stabilised in the CCITT Blue Book, though many national variants exist. These are already being used to handle limited sets of ISDN-based services in some countries[10] and will be operated internationally in the early 1990s. The format of the ISDN-UP is far more structured than that of other user parts, and its specification ensures that the format of its messages follows the more modern practice of identifying every item of information by stating explicitly what it is and how long it is, as well as giving its value.

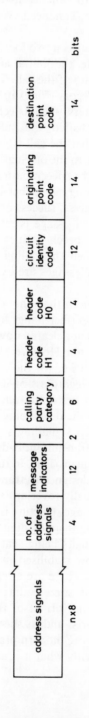

Fig. 20.8 *Format of an initial address message*

20.4.2 The 7-layer architecture

The 4-level architecture of the CCITT No.7 signalling system was originally developed in the 1960s. More recently, the International Standards Organisation (ISO) has defined a protocol model that enables communication between applications run on different computer systems. It is known as the open systems interconnection (OSI) model or the 'seven-layer model'. Although originally designed to provide protocols for data communication, the model can be applied to other forms of telecommunications with appropriate amendments. A brief description of the OSI model is given below. Readers seeking further information should consult the CCITT X200 series of recommendations.

Fig. 20.9 illustrates the principle. Communication between an application running on two different computer systems can be precisely defined at seven protocol layers. Each layer performs a fixed set of tasks which depend on the layers below and support the layers above. The seven layers are supported by the physical telecommunication medium between the two computers. This may be a simple pair of copper wires or a digital transmission system (e.g., over optical fibre), and is assumed to be transparent to the communication between the terminals. The protocol layers are briefly defined as follows:

(i) Layer 1: The physical-link layer. This layer of protocol definition gives the physical description of the interface between the terminal and the physical medium. Layer 1 definitions include the plug size and its pin connections, the logical meaning of the various pins on the plug, and the electrical characteristics (voltages, etc.) of the signals on the pins.

(ii) Layer 2: The link-control layer. This layer defines the link-control protocol (or procedure). Examples of layer 2 definitions include the envelope

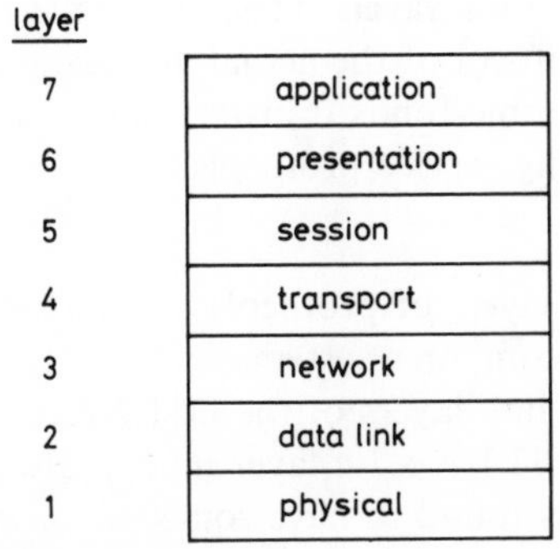

layer	
7	application
6	presentation
5	session
4	transport
3	network
2	data link
1	physical

Fig. 20.9 *The OSI 7-layer model*

structuring of information for purposes of error correction and detection, flow control, link initiation and recovery.

(iii) Layer 3: The network-control layer. The third layer of protocol definition covers the call-control procedures. These route the call (perhaps over several links in tandem) and initiate the flow of information. Signals such as 'request for service', 'ready for data', 'proceed to select' and 'call routeing address digits', as well as call-progress signals, are all examples of layer 3 activity.

(iv) Layer 4: The transport-control layer. This layer describes the process of the conversion of the user's data into a format suitable for being transported through the network (for example, the grouping of data into packets for conveyance over a packet-switched network).

(v) Layer 5: The session-control layer. The session-control layer relates to the management of the resources responsible for the data transfer. Examples of layer 5 activity include the synchronisation of the transmitting and receiving processes in the operating-system software of the two computer systems, and the scheduling of activities.

(vi) Layer 6: The presentation-control layer. This layer covers the structuring of the user's information into the form that ensures communication between two specific software files or terminals in the two communicating computer systems.

(vii) Layer 7: The application layer. This, the highest layer in the protocol structure, defines the protocol of the actual application. Examples of layer 7 activity include electronic-funds-transfer protocols, remote-job-entry protocols and time-sharing service protocols.

Finally, above the seven layers of protocols lies the actual information to be transferred between the computer systems.

The physical and data-link layers of the OSI 7-layer model align well with Levels 1 and 2 of the CCITT No.7 4-layer model. However, CCITT No. 7's message-transfer part was found to have some shortcomings when compared with the third layer of the 7-layer model. The result of this was that the message-transfer part of CCITT No. 7 could not be used as a general-purpose

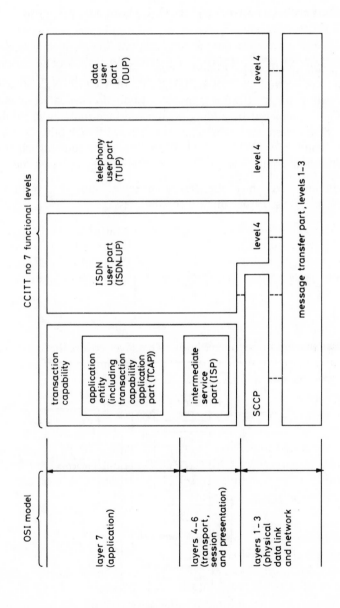

Fig. 20.10 *The relationship between CCITT No. 7 functional levels and the OSI 7-layer model*

tool for transferring messages across a signalling network, because it lacked certain key facilities. This shortcoming was overcome by specifying the signalling-connection-control part, which provides the equivalent of the OSI Network Service (layer 3).

The match between CCITT No.7 and the OSI 7-layer model is shown in Fig. 20.10.[11]

20.4.3 The signalling connection control part

The signalling connection control part (SCCP) of CCITT No.7 provides services which allow for the flexible transport of non-circuit-related signalling messages.[12] This is achieved by adding greater message-routeing capabilities and the facility for several messages to be associated with each other by setting up a signalling connection. While the MTP provides some routeing capability based on destination point codes, the SCCP offers routeing based on:

(i)   Subsystem number, which specifies a particular subsystem within the destination exchange as the destination for the message.

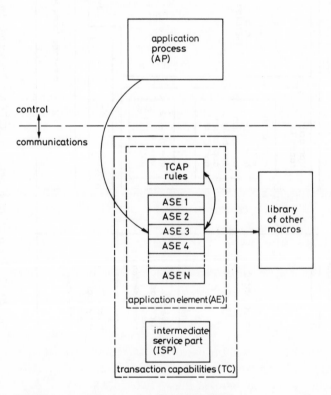

Fig. 20.11 *Capabilities supported by the SCCP functions*

(ii) Destination point code, as described in Section 20.4.1.3.

(iii) Global title translation (GTT), which is the most flexible of the three routeing methods. In this method, it is recognised that routeing in a CCITT No.7 network is based on point codes, and that numbers on which routeing has to be based may not be point codes. The method therefore provides translation of the original number into a point code. The most obvious need for translation is from dialled address digits. However, translation tables (see Chapter 12) may be built into the system to allow the translation of numbering schemes for data, telex, mobile communications, or even credit cards.

In practice, the functions of the SCCP are provided in a modular fashion, as shown in Fig. 20.11. In this Figure, the application process (AP) is a control module, independent of the communications functions, and the transaction capabilities[13] (TC) provide the communications functions in the form of a number of software modules. The AP determines the type of signalling to be carried out, and then selects the appropriate application service element (ASE) to initiate and control the communication of the signals.

The TC provides a basis for establishing communication between two points in the signalling network, without there being an associated circuit-switched connection. It may, for example, be used for the interrogation of databases, and it is defined in CCITT recommendations Q.771–775. It consists of two elements, the application element (AE) and the intermediate service part (ISP), which is still under study and not yet functional. Within the AE are the transaction capabilities application part (TCAP) and a number of ASEs. The TCAP comprises the general rules which must apply, regardless of the signalling application, while each ASE is a control module for a particular application.

When the AP deduces the type of signalling required in a given case, it selects the appropriate ASE to control the communications aspects of the signalling. The ASE itself consists of a number of macros, but, in order to be functional, they must be subjected to the TCAP rules (which govern all ASEs but are stored only once). The first task of the ASE is, therefore, to call the TCAP module. To carry out a given task, an ASE may need to call further functions, and these are stored as macros in a software library. Thus, to perform a function designated by the AP, an ASE needs to create an appropriate entity composed of its own macros, the TCAP module, and other macros chosen from a library. The assembly of the appropriate entity is a process similar to 'software build', which was described in Chapter 17.

A simplified example of the use of TCAP messages is shown in Fig. 20.12. This illustrates a call which requires access to a remote database in order to obtain digit translation, as used for services such as '0800' or 'Freefone'. In this case, messages are exchanged only between the originating exchange and

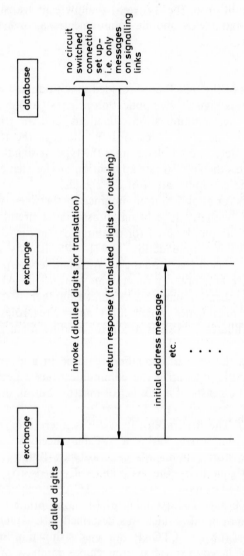

Fig. 20.12 An example of TCAP messages used to access a database

the database (both of which implement TCAP), and the messages are carried by the SCCP through the intermediate exchange.

In addition, TCAP is used to support the mobile application part (MAP) as specified in Europe to support the Pan-European Cellular Radio (sometimes known as the GSM system). In this case, the TCAP messages support information exchanges, such as requesting and receiving confirmation of the validity of a mobile subscriber making calls from a visited country. This mobile application part is probably the most complicated application of TCAP so far and contains many complex functions (see CCITT recommendation Q.1051).

20.4.4 *Management of signal point codes*

The concept of a signalling network was introduced earlier in this chapter. The MTP of CCITT No.7 allows messages to be diverted to other back-up signalling links on either the same physical route or a different one, using fully-associated or quasi-associated routeing. An important aspect of CCITT No.7 signalling networks is the location and control of signalling point codes (see Section 20.4.1.3 above). In general, each administration manages its own set of signalling point codes. Thus, DPCs are likely to be repeated in various countries. This raises the problem of how to overcome the duplication of point codes in countries between which interconnection is required. To solve the problem, an international signalling network has been defined. This has its own internationally agreed and administered set of codes. Each country is allocated a set of point codes and it allocates them, as required, to the international switching centres. These nodes are in both the national signalling network and the international signalling network, and can act as the bridge between countries.

20.5 The digital subscriber signalling system

The internationally defined subscriber common-channel signalling system, for use in an ISDN is called digital subscriber signalling No.1 (DSS 1). This is a common-channel signalling system in that the signalling for two traffic channels is carried over a separate signalling link and the signalling is message-based. However, because DSS 1 only connects a single subscriber's premises to a local exchange, it does not require the same networking capabilities as inter-exchange signalling systems such as CCITT No.7. Therefore it has neither the flexible addressing capability of CCITT No.7 nor the capability of quasi-associated routeing or non-circuit-related working. Nevertheless, DSS 1 does provide the capability to support services to the customer, e.g. closed-user-group operation and user-to-user signalling, as described later in this Section. It has been designed using a layered approach and has a physical level, a signalling or data-link level, and a network level.

The physical level[14,15] has two versions, corresponding to the basic and primary forms of ISDN access (which are described in Chapter 22). Signalling for the basic access is conveyed at 16 kbit/s within the D channel (known as 'D(16)') of the 144 kbit/s frame structure, usually carried by a digital transmission system over the local copper-pair subscriber network. This signalling channel is for subscriber-to-network and user-to-user signalling for the two B channels of basic access. The 16 kbit/s signalling channel may also be used to convey low-rate packet access between the subscriber and the local exchange. This is the so-called D-channel access technique.

Signalling for the ISDN primary access is carried at 64 kbit/s in a dedicated time slot on a 2 Mbit/s or 1.5 Mbit/s digital line system between the subscriber's premises and the local exchange. The digital line system terminates on a network termination (NT) at the subscriber's premises. This NT is usually associated with an ISDN PABX, known as an ISPBX (see Chapters 7 and 22). The 64 kbit/s signalling channel within the 2 Mbit/s or 1.5 Mbit/s digital system is designated the D channel (known as 'D(64)'). For 2 Mbit/s systems, the D channel is carried in TS16 and it conveys the subscriber signalling for 30 B channels. This is known as the '30B + D' format. For 1.5 Mbit/s systems, the D channel is carried in TS24 and it conveys signalling for 23 B channels. This is known as the '23B + D' format.

The data-link level[16,17] of DSS 1, also known as 'level 2', is based on the concept of a link-access protocol for the D channel, known as 'LAPD'. This layer provides for the secure transmission of messages (level 2) between user and exchange, using the medium of level 1 (physical). The LAPD functions include the appropriate labelling, frame structuring and error detecting of all signalling messages relating to its B channels. The information transfer between a user and the exchange, a 'data link,' may be point-to-point or broadcast. In the latter case, multiple addresses are used to identify the various destinations (known as 'signalling endpoints') to the exchange.

The frame format of the LAPD is based on the high-level data-link-control (HDLC) technique. A frame is divided into octets (8 bits) grouped into fields, with the beginning and end of the frame deliniated by flags of bit sequence 01111110. Fig. 20.13 illustrates the frame format. A two-octet field is used to indicate the destination address. This may be either a terminal end-point identifier (TEI) or a service-access-point identifier (SAPI), according to whether the message is to or from the subscriber. The control field, consisting of one or two octets, identifies the type of frame (which may be a command or a response) and contains sequence numbers for use in the error-correcting process. The information field contains the level-3 (signalling message)

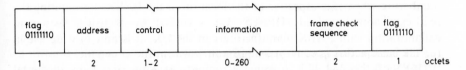

flag 01111110	address	control	information	frame check sequence	flag 01111110	
1	2	1–2	0–260	2	1	octets

Fig. 20.13 *Frame format for LAPD*

information. It consists of a maximum of 128 octets for normal applications, but this may be extended to 260 octets when the D channel is used to convey data packets. A 16-bit frame-checking sequence, operating on the address, control and information fields, is used to identify transmission errors. Compelled signalling is applied, and frames are re-transmitted continuously until acknowledgement of error-free reception is received from the far end. This, together with the frame-check sequence, gives a level-2 system highly tolerant to the higher error rates which can occur on digital transmission over the subscriber-access network.

The network level of level-3 element[18,19] of DSS 1 is conveyed in the information field provided by level 2. The first octet of the information field indicates the type of message. Message groups have been defined for call control and maintenance information. There is a wide range of message formats defined to support the many services and user facilities provided by an ISDN.

A number of digital subscriber-access signalling systems have been developed in advance of the full specification of the CCITT DSS 1. These non-standard systems are often similar to DSS 1 at layers 1 and 2 but differ at level 3 where the actual information content matches the particular functional design of the exchange systems used. One example of such a system is the digital-access signalling system No.2 (DASS 2) defined by British Telecom and used to support their basic- and primary-rate ISDN services.[6]

20.6 Digital private network signalling system

The digital private network signalling system (DPNSS[7]) is a common-channel-signalling system designed for private networks composed of PABXs linked by digital leased lines. It is comparable to CCITT No.7 except that, instead of linking public exchanges, DPNSS links PABXs located on subscribers' premises. Although the system does not have the networking capabilities of CCITT No.7, it does have a comprehensive message set which enables the advanced features offered by PABXs to be extended. For example, call diversion can be applied throughout a private network, using a unified numbering scheme, irrespective of the location of the PABXs terminating the two extensions involved. DPNSS was specified by British Telecom in collaboration with PABX manufacturers in the UK. It is now in wide use in the UK and is also used in other parts of the world.

DPNSS is defined as a three-layer system with a physical layer (layer 1), data-link layer (layer 2) and call-control layer (layer 3). Layers 1 and 2 are identical to those defined for DASS 2 and are thus very similar to the CCITT DSS 1 specification. The call-control layer defines messages for basic call set-up and a range of supplementary services for use between end (i.e., originating and terminating) and transit types of PABXs.

20.7 Pre-standard CCS systems

Signalling systems have always required a high level of standardisation, as they form the interface between swtiching systems which may be of different manufacture. This is particularly true in the international telecommunications environment. The international standardisation body, the CCITT, produces recommendations once every four years as a series of 'coloured' books. In 1980, there was 'the Yellow Book', in 1984 'the Red Book' and in 1988 'the Blue Book'. Within each colour series, the recommendations are subdivided in alphabetical order, signalling being covered in the 'Q' series.

The CCITT has defined many signalling systems, including the channel-associated systems in service all over the world. The predominant use of CCITT channel-associated signalling systems has been in the international telephone network, where the specifications need to be strictly adhered to. In national networks, the CCITT specifications often form the basis of signalling systems which then vary according to the application and the switching and transmission systems in use. However, with the advent of SPC, it became clear that a more integrated approach was required, and the first internationally agreed common-channel signalling system, CCITT No.6, was defined. It is interesting to note that even a signalling system as mature as CCITT No.6 (which has been in service since the early 1980s) is still undergoing minor development to eliminate errors as they are detected. Compared to CCITT No.6, CCITT No.7 is far more complex and therefore it is expected that its improvement will continue for many years. The CCITT is also defining the subscriber-access common-channel signalling system in the 'Q' (and 'I') series of recommendations.

The CCITT is a large worldwide organisation possessing considerable inertia which prolongs the development time of its recommendations. In the modern environment of fast-changing technology and rapidly developing markets for new services, the swift application of new developments is of the essence. Thus,initiatives are being taken in various countries to increase the rate of development of the specifications needed to support ISDNs and advanced services. A focal point of many of these initatives is the field of signalling, and, in particular, CCS. In this chapter, mention has already been made of the early implementation of prestandard CCS systems, e.g., DASS 2 and DPNSS. This is not only necessary to meet urgent customer-service needs, but it also provides useful experience with new forms of signalling. Administrations and manufacturers are then able to provide valuable input into the standardisation process. Thus, it is likely that early implementation of pre-standard CCS systems will be a continuing feature in future telecommunications.

20.8 References

1 WELSH, S (1981): *Signalling in Telecommunications Networks* (Peter Peregrinus), Chap. II

2 CCITT Blue Book: Recommendations in the Q.700 series. IXth Plenary Assembly, Melbourne, Australia, November 1988

3 FRETTON, K.G. and DAVIES, C.G. (1987): 'CCITT signalling system No.7 in British Telecom's network' *British Telecommunications Engineering,* **6**, Part 3, pp.160–162

4 *Bell System Technical Journal (1978):* Special edition on Common-channel inter-office signalling (10 papers), **57** (2)

5 BROWNE, B. and NORMAN, M (1988): 'PABX signalling in the ISDN', *Telecommunications,* pp.37–39, 70

6 BIMPSON, A.D., RUMSEY, D.C. and HIETT, A.E. (1986): 'Customer signalling in the ISDN' *British Telecommunications Engineer,* **5**, pp.2–10

7 HIETT, A.E. and DANGERFIELD, W. (1988): 'Private network signalling, *Computer Communications,* **11**, pp.191–196

8 LAW, B. and WADWORTH, C.A. (1988): 'CCITT signalling system No.7: Message transfer part, *British Telecommunications Engineering,* **7**, Part 1, pp.7–18

9 MITCHELL, D.C. and COLLAR, B.E.: 'CCITT signalling system No.7, National user part, *British Telecommunications Engineering,* Ibid., pp.19–31

10 DAVIES, C.G.: 'CCITT signalling system No.7: Integrated servies digital network user part, *British Telecommunications Engineering,* Ibid, pp.46–57

11 FRETTON, K.G. and DAVIES, C.G.: 'CCITT signalling system No.7: Overview, *British Telecommunications Engineering,* Ibid., pp.4–6

12 CLARK, P.G. and WADSWORTH, C.A. 'CCITT Signalling System No.7: Signalling connection control part', *British Telecommunications Engineering,* Ibid.pp.32–45

13 JOHNSON T.W., LAW, B. and ANIUS, P: 'CCITT signalling system No.7: Transaction capabilities, *British Telecommunications Engineering,* Ibid., pp.58–65

14 CCITT Rec. I430, Ibid.

15 CCITT Rec. I431, Ibid.

16 CCITT Rec. I440, Ibid (also Rec. Q920)
17 CCITT Rec. I441, Ibid (also Rec. Q921)
18 CCITT Rec. I450, Ibid (also Rec. Q930)
19 CCITT Rec. I451, Ibid (also Rec. Q931)

Chapter 21

Planning considerations

Earlier chapters in this book have described the characteristics of SPC digital exchanges and the principles of their operation. The introduction of such exchanges into a network requires consideration of the full range of network plans:[1] transmission, numbering, routeing, charging and signalling (see Chapter 1). It also requires the introduction of a new plan: network synchronisation. Many established telecommunications networks are in the process of modernisation, with the gradual replacement of analogue transmission links by digital systems, and of the existing analogue electromechanical and semi-electronic exchanges by digital SPC exchanges. Although within a national network these two processes are co-ordinated, practical and economic constraints on the network-conversion programme may cause some degree of mismatch between the technologies of the exchanges and the transmission links connected to them. Thus, this chapter reviews the planning considerations involved in the introduction of digital SPC exchanges in both analogue and digital transmission environments.

21.1 The multilayered model of a digital network

It is useful to describe the planning aspects of digital networks by reference to a multilayered model, as shown in Fig. 21.1. The model presents a digital SPC telecommunications network as a composite of switching and transmission basic networks supported by a number of auxilliary networks. The foundation of the model is the transmission-bearer network, which provides the transmission paths between all the nodes of the other layers. This network comprises the subscriber-line network between the telephone (or data terminal in the case of an ISDN) and the MDF at the local exchange, and all the transmission links between the local, junction tandem, trunk and international exchanges. (These terms are described in Section 1.1 and Figs. 1.3 to 1.6 of Chapter 1.) The transmission-bearer network is constructed from a pattern of links and flexibility points, the latter being the nodes of the

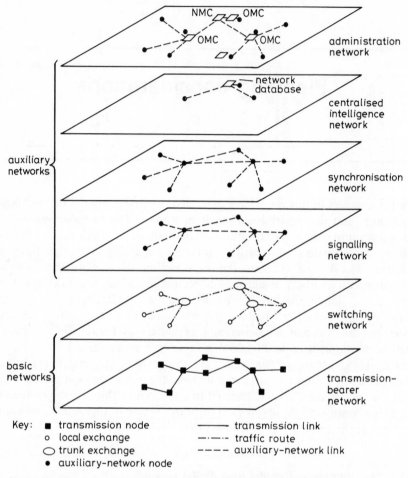

Key: ■ transmission node   ———— transmission link
       ○ local exchange   —·—·— traffic route
       ○ trunk exchange   ———— auxiliary-network link
       ● auxiliary-network node

Fig. 21.1   *The multi-layered model of a digital SPC telephone network*

network where semi-permanent connections are made between the channels in the transmission links. In the local network, the nodes are formed by local and area distribution points where connections are made between cable pairs using jumper wire or metallic straps (see Fig. 1.3).

In the junction, trunk and international parts of the transmission-bearer network, the flexibility points (nodes) are located at the ends of high-capacity transmission systems. Such nodes provide interconnectivity at the 2 Mbit/s (or 1.5 Mbit/s) level between two higher-order digital line systems (see Chapter 5). The interconnections are achieved using coaxial jumpers across a digital distribution frame (DDF) located between two sets of digital higher-order multiplexers in back-to-back formation (see. Fig. 21.2). Interconnection may

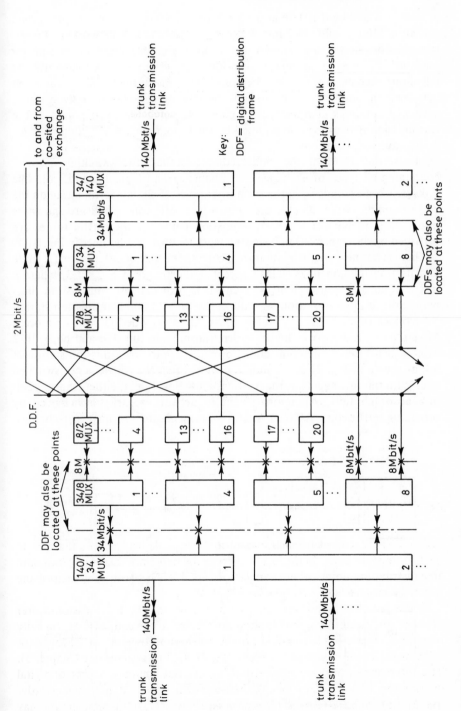

Fig. 21.2 *Trunk transmission station*

also be provided by the DDF at the 8 Mbit/s or 34 Mbit/s level, if appropriate (or at 6 Mbit/s and 45 Mbit/s for 1.5 Mbit/s-based networks). Trunk-transmission nodes are known as 'transmission stations' or 'telephone repeater' stations. Transmission stations are usually collocated with the telephone exchanges for which they provide access to the trunk transmission network. In addition, transmission stations may be located away from exchange buildings at nodal points chosen to serve the focal points of the transmission network, as determined by the topography of the country (e.g., microwave radio stations).

The second layer of the model comprises the basic switching network containing all types of exchange (see Chapter 1 and Figs. 1.5 and 1.6). The links between exchanges, in the form of bundles of circuits terminating on the corresponding switch blocks, are known as traffic routes and are shown as dotted lines in Fig. 21.1. The transmission paths for the traffic routes are provided by the transmission bearer network, shown in solid lines in Fig. 21.1.

There is not necessarily a one-to-one correspondence between a traffic route and its transmission links. This concept is illustrated in Fig. 21.3, where a traffic route between exchanges A and B is provided by transmission links between transmission station 'a' (collocated with A) and transmission station 'b' (collocated with B). The traffic route AB is provided via one or more 2 Mbit/s paths routed over higher-order transmission links ac and cb, with 2 Mbit/s or 8 Mbit/s interconnections provided via the transmission station at 'c', as shown in Fig. 21.3a. It should be noted that the traffic is not switched at c. With this arrangement, the traffic route AB will fail if either transmission link ac or cb fails. An improvement in network security is achieved by providing traffic route AB over two (or more) separate transmission routeings,

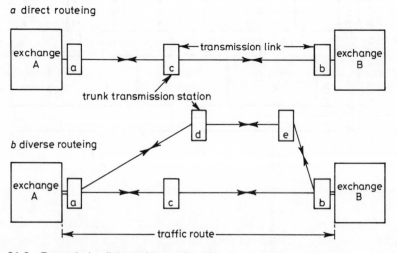

Fig. 21.3 *Transmission links and route diversity*

e.g. one 2 Mbit/s path routed acb, the other routed adeb (Fig. 21.3b). This splitting of traffic routes over parallel and separate transmission-link routeings is known as 'route diversity'.

Considering again the multilayered model shown in Fig. 21.1, the functions of the four auxiliary networks are summarised below (note that the transmission paths for all the links of the auxilliary networks are provided by the transmission bearer network):

(i) Signalling network: This is the network of common-channel signalling links between exchanges and signal transfer points (STP), as described in Chapter 20. There will generally be fewer signalling links than traffic routes within a digital network.

(ii) Synchronisation network: This network disseminates the timing from one or more reference sources to all the digital exchanges in the switching network, as described in Chapter 10. The network comprises synchronisation nodes formed by the timing units in digital exchanges and the synchronisation links between them.

(iii) Centralised-intelligence network: This network provides the access between the digital SPC exchanges and their service-control points (SCP). The SCPs contain the databases which provide centralised network intelligence, as described in Chapter 9.

(iv) Administration network: The adminstration network comprises the various operations and maintenance centres (OMC), network management centres (NMC), etc., together with their telemetry, control and communications links to the dependent exchanges, as described in Chapter 19.

This multilayered model illustrates that a digital SPC telephone exchange is a network element with a complex interrelationship with other elements. Thus, the planning of such exchanges must encompass the full set of network layers. However, some of the auxilliary networks, particularly (iii) and (iv) above, are still in the formative stages in many countries and therefore they may not exist as identifiable networks for many years. Although the model has been described in terms of a digital SPC telephone network, it is also applicable to an analogue SPC network, with the exception of the synchronisation auxilliary layer. Before considering the planning of digital SPC exchanges in a multilayered digital network, the effects of an analogue transmission-bearer network are discussed.

21.2 SPC digital exchanges in an analogue-transmission environment

An SPC digital exchange introduced to an environment of analogue trunk and junction transmission systems will incur problems of interworking and possible degradation to the quality of call connections.

21.2.1 Interworking

Interworking equipment is required at a digital exchange for the conversion of analogue signals into a suitable format for the digital switch block (Chapter 6). The interworking processes comprise:

(i)   Analogue-to-digital conversion (via codec).
(ii) Time-division multiplexing.
(iii) Signalling conversion.

The first two processes are invariably performed together, either within the analogue trunk-termination unit (ATTU, see Chapter 7) of an exchange, or within PCM primary multiplexers associated with the transmission links. Similarly, DC and 1VF in-band signalling is converted to PCM channel-

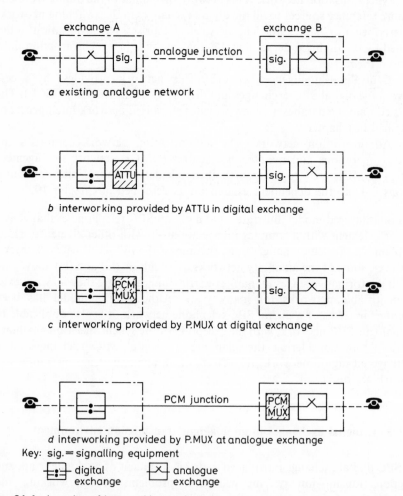

Fig. 21.4 *Location of interworking equipment*

associated form within either a PCM multiplexer or the ATTU (see Chapter 8). The location of these functions has important practical implications.

The importance of the location of the interworking processes can be explained by reference to Fig. 21.4. Consider a simple existing network arrangement consisting of two analogue local exchanges linked by a copper-pair junction route. The in-band signalling (DC or 1VF) is provided by complementary equipment associated with each exchange (Fig. 21.4a). When exchange A is replaced by a digital SPC unit the interworking processes may be provided by the ATTU (Fig. 21.4b), or by PCM multiplexers at exchanges A or B (Figs. 21.4c, d. respectively). The merits of the three options are described briefly below.

21.2.1.1 Interworking provided by an ATTU. This method has the advantage of the interworking equipment being integral with the digital exchange. Thus, it can be constructed in the same equipment practice as the new switching system, and be covered by the digital-exchange alarm and maintenance facilities. For example, this will enable primary fault diagnosis to be made from the exchange's centralised remote operations-and-maintenance centre (see Chapter 19).

However, a practical disadvantage of the method is that the accommodation space required for ATTU equipment is significant, perhaps more than doubling the exchange's floor area compared to a digital exchange without interworking. One of the consequences of such additional accommodation requirements is that 'turn-around' space restrictions may constrain the timing of the digital-exchange introduction. This concept is illustrated in Fig. 21.5a, which shows a plan view of an exchange building with an existing analogue switching unit. The vacant exchange floor is large enough for an SPC digital replacement unit without ATTU. However, with the ATTU, the total floor requirements for the new unit (although less than the existing analogue unit) cannot be accommodated simultaneously with the old unit. This problem can be alleviated by the staged replacement of parts of the analogue unit by parts of the new digital unit, provided that there is sufficient floor space for the 'turn-around' (Fig. 21.5b).

A further practical problem results from the difficulty of designing the appropriate signalling conversion processes into the ATTU. As Figs. 21.4a and b show, the ATTU must mimic the replaced signalling equipment sufficiently for the link to the existing signalling equipment to function normally. Although this may not be difficult for any one type of signalling system, it can prove onerous when there is a wide range of systems, some of which may be obsolescent or non-standard. The task is usually kept manageable by restricting the range of signalling types that the ATTU needs to handle. Signalling systems outside the ATTU category must then be replaced by an acceptable system, or handled by a PCM multiplexer (Fig. 21.4c or d).

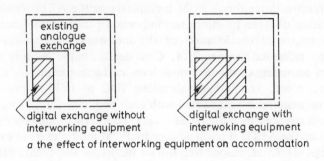

a the effect of interworking equipment on accommodation

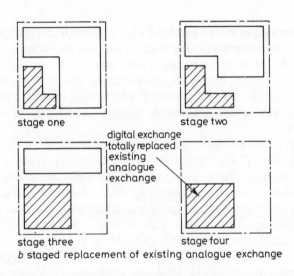

b staged replacement of existing analogue exchange

Fig. 21.5   *Interworking equipment and turn-around space*

21.2.1.2 Interworking provided by PCM multiplexers at the digital exchange. The provision of interworking by the use of PCM multiplexers at the digital exchange (Fig. 21.4c) offers an administration the advantage of flexibility. This results from the ability to shift multiplexers between digital exchanges, as the interworking requirements change with the gradual introduction of digital transmission into the network. With careful planning, such re-use of multiplexers may enable less money in total to be spent on interworking equipment for the whole network than with the ATTU method.

21.2.1.3 Interworking provided by PCM multiplexers at the analogue exchange. In this method, PCM transmission is introduced on the trunk or junction route between the analogue and digital exchanges. One advantage of this method is that the signalling system used previously over the analogue link may be replaced by the PCM channel-associated signalling (PCM/CAS) system. This usually results in accommodation savings at both exchanges, since many of the older analogue signalling systems are physically bulky compared to PCM/CAS equivalents. This method also has transmission-performance advantages, as described in the following Section.

21.2.2 Degradation of transmission quality

Although digital switching coupled with digital transmission provides significant advantages over an all-analogue network, digital local and junction-tandem exchanges can introduce additional loss if introduced in a 2-wire analogue local network. The impairment results from the 4-wire transmission characteristic of digital exchanges (see Chapter 6).

Fig. 21.6 illustrates how the 4-wire transmission of a digital local or junction tandem exchange introduces additional transmission loss. In the UK an own-exchange call connection on an analogue 2-wire local exchange experiences a

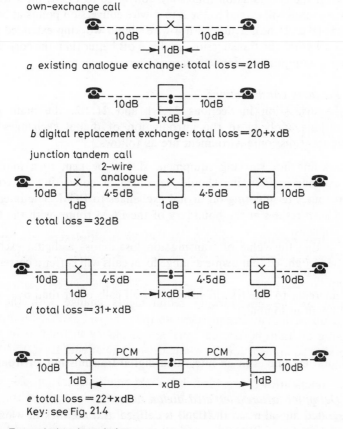

Fig. 21.6 *Transmission degradation*

maximum transmission loss of about 21 dB, comprising 10 dB on each subscriber's line (loop) and about 1 dB through the switch (Fig. 21.6a). When the local exchange is replaced by a digital system, the loss through the switch increases from about 1 dB to x dB. For stability, the minimum value of x, the loss between the 2-wire points on the subscriber line-termination units (SLTUs), must be at least 3 dB (see Chapter 7). In the UK, x is set to 6 dB.[2] The introduction of the digital local switching unit into a 2-wire local network thus degrades own-exchange calls by at least 2 dB and maybe as much as 5 dB. This drop in loudness may not be objectionable because local calls can be too loud anyway, and the introduction of digital switching creates connections which are far less noisy than those of analogue exchanges, particularly step-by-step exchanges.

A similar situation applies where an analogue 2-wire junction-tandem unit is replaced by a digital switching system. Then, overall transmission loss increases from about 32 dB (Fig. 21.6c) to $31 + x$ dB (Fig. 21.6d). In the UK this corresponds to an additional 5 dB end-to-end loss. (i.e. x = 6dB). However, if the two junction routes are converted to PCM, the x dB loss of the digital switch will extend to the 2-to-4-wire conversion points at both local exchanges (Fig. 21.6d). Thus, with 4-wire digital working extended over the junction network, the transmission level is 4 dB better than the corresponding analogue 2-wire junction tandem case.

21.2.3 Planning considerations

From the discussion in Sections 21.2.1 and 21.2.2, the main planning considerations applying to the introduction of digital exchanges into an analogue transmission environment are as follows:

(i)   Trunk-line interworking equipment should be kept to a minimum at digital exchanges. Planning should be aimed at converting all trunk and junction routes terminating on digital exchanges to PCM. The interworking function then resides at the boundary of the PCM routes with the analogue local exchanges.

(ii)   In setting the value of transmission loss across a digital exchange, it should be recognised that own-exchange local calls will suffer a corresponding reduction in loudness.

(iii)   Adherence to rule (i) will ensure that all calls other than own-exchange type will gain in loudness.

21.3 Digital exchanges in an integrated digital-transmission environment

21.3.1 Definition of an integrated digital network

An integrated digital network (IDN) is defined as a network in which all the exchanges are digital SPC units and all the trunk and junction traffic routes

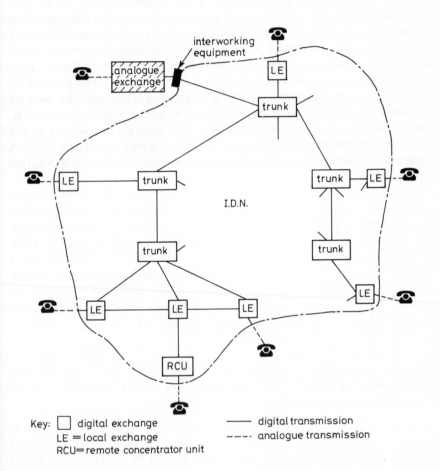

Fig. 21.7 *Definition of an IDN*

are carried on digital transmission systems. In addition, the signalling between exchanges in an IDN is assumed to be of the common-channel type. Within the IDN, all traffic channels are in digital (PCM) format; analogue-to-digital conversion is thus required only at the boundary of the IDN (Fig. 21.7). This boundary passes through the SLTU for the common type of analogue subscriber line. Any boundary within the trunk or junction portion of the IDN is provided by interworking equipment at the digital exchange or, preferably, at the distant analogue exchange, as described in Section 21.2.

The provision of digital transmission on subscriber lines extends the IDN boundary to the subscribers' premises. This is an essential element of an integrated-services digital network (ISDN), as described in Chapter 22.

21.3.2 Characteristics of an integrated digital network (IDN)

The characteristics and advantages of an IDN, introduced in Chapter 3, are important from a planning perspective. Undoubtedly, the major characteristic is that trunk and junction routes enter switching units at the multiplexed (30-channel or 24-channel) digital-group level (2 Mbit/s or 1.5 Mbit/s) rather than at the single-channel level. This results in the elimination of per-channel signalling and multiplexing equipment. Fig. 21.8 illustrates the channel equipment required at analogue exchanges with either analogue FDM or digital transmission (Figs. 21.8a and b), which may be compared with the group termination of digital links on a digital exchange (Fig. 21.8c). Channel equipment (signalling and line termination) represents a significant proportion

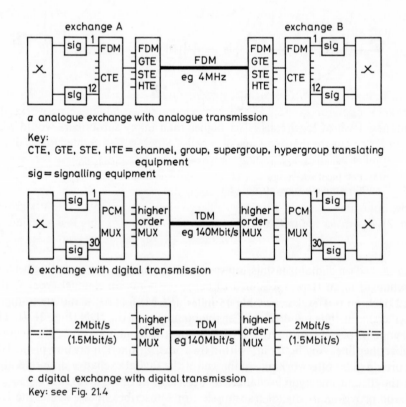

a analogue exchange with analogue transmission

Key:
CTE, GTE, STE, HTE = channel, group, supergroup, hypergroup translating
            equipment
sig = signalling equipment

b exchange with digital transmission

c digital exchange with digital transmission
Key: see Fig. 21.4

Fig. 21.8 The elimination of the channel equipment at the exchange-to-trunk boundary

of a telecommunications network cost, typically 60% of a link between two exchanges. Thus, its elimination ensures lower capital costs in an IDN than in an equivalent analogue network.[3,4] In addition, the elimination of the channel equipment contributes significant savings in floor area. The lack of channel equipment also facilitates quicker installation times, lower power consumption and heat dissipation, and a reduction in the manpower requirements for operations and maintenance.

All of the above consequences of the group-port entry to a digital exchange result in savings in both capital and running costs. Studies have shown that the overall annual charges of an IDN are 50% of an equivalent analogue network.[5,6]

The transmission quality of an IDN has three main attributes, as described in Chapter 3, namely:

(i)   Transmission loss is independent of the number of exchanges and links in a connection.

(ii)   Connections have lower levels of noise than over their analogue counterparts.

(iii)   Connections are more stable than analogue 2-wire networks.

The first attribute offers significant advantages to the telecommunications-network operator as well as the subscribers. With analogue networks, the loudness level of local calls is set higher than many subscribers would wish, in order to ensure that long-distance calls (which suffer greater losses due to the number of links) are audible. However, the transmission loss between the boundaries of the IDN is kept at a constant level for all types of connection (Fig. 21.9*a*). This loss may be set within the range of loudness preferred by telephone users (i.e., 4−18 dB (RE)), provided that echo and stability criteria are also met. In the UK, the value of 6 dB chosen for this loss provides an improved stability margin of 12 dB and an overall reference equivalent within the range of 6−20 dB (RE).[2,7,8]

For the telecommunications administration, the constant transmission-level feature of an IDN provides great flexibility in call routeing. For example, a call between two exchanges only 15 miles apart would be of the same quality if it were routed via distant exchanges as if it were routed directly (Fig. 21.9*b*). This feature enables greater use to be made of automatic alternative routeing (see Chapter 1), and offers full flexibility in designing the IDN structure and routeing plan. However, there are limits to the extent of circuitous routeing in an IDN; these limits are set by considerations of transmission delay and planning and control complexity.

Fig. 21.9*a* also represents a typical national transmission plan for an IDN. The plans for individual countries differ according to the losses allocated to

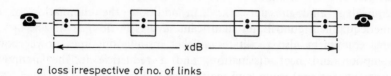

a loss irrespective of no. of links

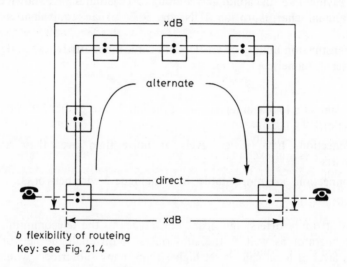

b flexibility of routeing
Key: see Fig. 21.4

Fig. 21.9 *Transmission loss and routeing flexibility*

the analogue subscriber lines ('local loop') and the value of x. Unlike the analogue transmission plan shown in Fig. 1.7 of Chapter 1, the allocation of loss in the national transmission plan for an IDN is independent of the number of levels in the traffic routeing hierarchy.

21.3.3 Network structure

21.3.3.1 Contributary factors. The optimum structure of an IDN is significantly different from an analogue network of equivalent size. The factors contributing to this difference are:

(i)   Size of digital transmission group.
(ii) Relationship between digital transmission and switching costs.
(iii) SPC flexibility.
(iv) Use of common-channel signalling.

Each of these factors is briefly considered in the following subsections:

(i) Size of digital transmission group: An IDN is based on the PCM standard group of 30 or 24 traffic circuits rather than the 12-circuit group of an analogue FDM transmission network. This increase of 150% or 100% capacity on the minimum size of a transmission link raises the level of traffic needed to justify an optional route between two exchanges. Thus, IDNs have fewer optional traffic routes than analogue networks of equivalent capacity. Since the minimum route size is 30 (or 24) channels, mandatory routes, which are required irrespective of traffic levels (e.g., between a local exchange and its parent trunk exchange) may have relatively low traffic loading per channel. Augmentation of both optional and mandatory routes requires traffic justification and is provided in units of PCM modules.

(ii) Relationship between digital transmission and switching costs: Both analogue and digital transmission costs have declined in real terms over the past 30 years, due mainly to improved multiplexing technology, whereas until the advent of digital exchanges, switching costs had been increasing. Digital transmission is cheaper than analogue transmission in many circumstances, particularly if signalling costs are considered,[4] and digital switching, particularly in large exchange units, is significantly cheaper than analogue switching. However, it is the ratio of switching to transmission costs, rather than the absolute costs, that influences the structure of a telephone network, as described in the following Section.

(iii) SPC flexibility: The catchment areas of SPC exchanges, digital and analogue, are constrained very little by the numbering and charging arrangements for the network. Thus, there is much flexibility in the way that local exchanges may be parented on to trunk exchanges, unlike the case of non-SPC exchanges, in which limits are imposed by the geographical topology of the charging groups and the numbering ranges allocated to local exchanges. These factors, together with the transmission advantages of an IDN, enable the coverage of trunk-exchange catchment areas to be optimised according to network economics.

(iv) Use of common-channel signalling: The use of common-channel signalling enables a single traffic route between local and trunk exchanges to be used for all classes of traffic. By comparison, exchanges without common-channel signalling rely on the identification of the point of entry to the switch block in order to discriminate between circuits requiring different call-processing or charging activities. In many cases, particularly with analogue step-by-step exchanges, this has resulted in several separate (and often small) routes being

provided. Therefore, common-channel signalling enables efficient use to be made of single traffic routes between local and trunk exchanges.

21.3.3.2 Trunk-network structure. The costs of a trunk network have two main components: those associated with the distribution network and those associated with the core network. The former include the transmission links between local exchanges and their parent trunk exchanges. (They also include links to non-parent trunk exchanges provided for reliability reasons.) The core network comprises the trunk exchanges, both parent (i.e., supporting local exchanges) and tandem, and the transmission links between them. Clearly, as the number of parent trunk exchanges increases, the cost of the distribution network decreases. However, the cost of the core network increases as the number of trunk exchanges increases. Thus, there is a minimum cost structure for the trunk network dependent on the relative costs of digital switching and transmission. As explained in Sections 21.3.3.1 (i) to (iv), the use of digital SPC exchanges in an IDN imposes few constraints on the network structure; therefore, the trunk network can be structured in a cost-optimised way.

It is informative to consider, as an example, the British Telecom network restructuring resulting from their introduction of an IDN.[9] The optimum size of the new trunk network is about 60 trunk exchanges. This should be compared to the previous analogue network of over 400 trunk exchanges established by the cost and operational constraints of predominantly step-by-step (Strowger) 2-wire analogue exchanges. The number of traffic routes in the trunk core network decreased from around 17 000 on the analogue network to about 1800 on the IDN. The traffic generated by the digital trunk exchanges is sufficient to warrant their full interconnection. This mesh structure simplifies network planning, but also establishes an infrastructure than can be exploited by automatic alternative routeings, as well as providing a high level of network resilience (see Section 21.3.4).

21.3.3.3 Local-network structure. There are also significant differences in the optimal local-network structure of an IDN compared to its analogue equivalent. Two major trends influence the structure. The first is the establishment of main digital SPC exchanges at the central sites with dependent remote concentrator units installed at surrouding sites (see Chapter 9). The second is the increasing trend to provide digital transmission on to subscriber lines.

The digitalisation of the subscriber lines is driven by two forces. The first, which is influenced primarily by market pressures, is the progressive introduction of ISDN which requires digital extension of the IDN to the subscribers' premises (see Chapter 2). This involves basic access at 144 kbit/s over copper-pair cable with jumpering at the exchange MDF on to the digital subscriber line-terminating unit (D/SLTU) (Chapter 7). In the case of primary access from a digital PABX, for example, the 2 Mbit/s (or 1.5 Mbit/s) link is routed via a DDF directly on to the local exchange switch block. The second force towards digitalisation of the local network is that of administrations seeking lower costs. The cost trends of technology are increasingly favouring the use of digital optical-fibre cable rather than copper-pair cable. This provides an economic bearer for either ISDN primary access or bundles of subscriber lines multiplexed into 2 Mbit/s (or 1.5Mbit/s) digital systems. When the bundles of subscriber lines originate at one point, e.g., an office or large multi-tenant building, the remote multiplexing equipment (necessary to terminate subscriber lines on 2 Mbit/s bearers; see Chapter 9) can be located on the subscribers' premises. When only one or two lines are required at individual subscribers' premises, the remote multiplexers may be housed in a street cabinet. In this case, the optical fibre terminates at the remote multiplexer in the street cabinet, from which copper-pair cable is used to serve each of the subscribers' premises (see Fig. 21.10*b*).

The use of digital transmission, particularly over optical fibre, enables the length of subscriber lines to be increased significantly whilst giving improved quality of transmission. This facilitates increasing the size of local-exchange catchment areas to an optimum for the digital SPC switching unit, thus providing overall savings compared to analogue networks serving the same number of subscribers. Figs. 21.10*a* and *b* illustrate the effect of local-network restructuring following the replacement of an analogue network by digital SPC exchanges and the digitalisation of the subscriber lines.

21.3.4 Network resilience

The result of planning an IDN, based on cost optimisation and the exploitation of SPC and digital technology, is a network design comprising few, but large, switching centres, and few, but large, traffic routes between them. In addition, many of the large local exchanges act as parents (or 'hosts') to several dependent remote subscriber units (RSU), as described in Chapter 9. The security of an exchange and transmission equipment in an IDN is thus even more important than in an analogue network. The effect of route congestion and exchange-processor overload will clearly cause loss of traffic at the exchange. However, the use of automatic alternative routeing at an exchange experiencing route congestion will throw additional traffic on to other routes, possibly causing them to become congested. Thus, there is a danger that route congestion may propagate through the network. The term 'network resilience' is used to describe the ability of a network to cope with transmission and switching failure, route congestion and processor overload.

The planning of an IDN must ensure an adequate level of network resilience. Clearly, the fundamental requirement for a resilient network is for the transmission and switching elements to have known, and preferably high, reliabilities. This is achieved using high-grade components and systems with appropriate degrees of redundancy in the hardware (see Chapters 6 and 7),

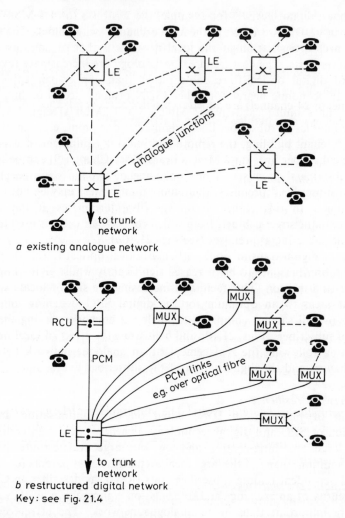

a existing analogue network

b restructured digital network
Key: see Fig. 21.4

Fig. 21.10   *Local-network restructuring*

together with resilient software (see Chapter 18). In addition, the capabilities of digital SPC exchanges facilitate the use of network traffic management to monitor the network performance and centrally control the initiation of remedial action, as described in Chapter 19. However, despite such features, the resilience of an IDN is also very much influenced by the network structure adopted.

The various methods of improving IDN resilience by use of suitable network structures are briefly described below:

(i) Mesh-routeing strategy: Such a strategy uses a mesh structure superimposed on the strictly hierarchical traffic routeing patterns in both the trunk and local

switching networks. This involves the provision of routes from local exchanges to non-parent trunk exchanges and to other local exchanges ('sideways routes'), even though not justified by traffic levels.

(ii) Network redundancy: This relates to a suitable degree of over provision of transmission and switching capacity. For example, subsequent PCM modules (30-channel or 24-channel) are added to traffic routes in advance of additional capacity being warranted by forecast traffic growth.

(iii) Provision of a service-protection network: A service-protection network is a special network, superimposed on the IDN to provide reserve capacity. It comprises transmission capacity, e.g., 140 Mbit/s blocks, extending over the major links of the transmission-bearer network. These are provided on the basis of one service-protecting system per N parallel traffic-carrying systems ('1-in-N protection'). More ellaborate methods include automatic re-routeing of transmission links, thus achieving 'm-in-n protection' levels. The principal use of a service-protection network is the substitution of its capacity for transmission links that have failed. This requires the provision of broadband switches[10] (e.g. at 140 Mbit/s) at all points of interconnection between the service-protection network and the IDN. Such switching may be automatic, initiated on detection of transmission-system failure, or manual, under local or remote control. A service-protection network may also be used to offload transmission capacity during periods of planned outages to enable work on the network to be undertaken, e.g., during transmission-link augmentation or rerouteing.

(iv) Transmission-routeing diversity: The vulnerability of large traffic routes to line-plant failures can be minimised by the use of route diversity. This involves the spreading of traffic routes over two or more separate transmission paths, as described in Section 21.1.

(v) Automatic alternative routeing (AAR): The managed use of automatic alternative routeing allows exchanges to avoid congestion and circumvent link failures within the network. As mentioned earlier, the use of this technique must be bounded to ensure that it improves rather than degrades performance. One method of constraining AAR is that of partitioning traffic routes, reserving certain channels for incoming, outgoing and both-way use. Thus, large surges of traffic on to a route in one direction, resulting from AAR, will not totally block other traffic from using that route.

The implementation of any of the above means of increasing network resilience will increase the network cost. Clearly, an administration must balance the cost of achieving a certain level of resilience against the cost of failure. The latter covers not only the operational costs but also the loss of call revenue (perhaps to a competitor) and the consequences of not meeting service obligations to subscribers (e.g., financial damages) and to the Government.

21.4 The planning of auxiliary networks

As with all networks, each of the auxiliary networks (see Fig. 21.1) requires careful planning to ensure a coherent structure of nodes and links. The deployment of the auxiliary networks is dictated by the need to provide adequate support to the IDN. However, subject to this constraint, these networks may be considered as separate entities with their own planning rules.

21.4.1 Common-channel signalling network

This network comprises the links and nodes of the common-channel signalling (CSS) systems (e.g., CCITT No.7) within the IDN. The nodes, which are normally formed by digital SPC exchanges, may be terminal (originating and terminating CCS messages) or transit [known as signal-transfer points, (STP)], as described in Chapter 20. The transmission path for a CCS link is usually provided by one or more of the PCM systems carrying traffic between two exchanges (e.g. in TS16 of 2Mbit/s systems). When digital paths are not available, the signalling link may be provided at the lower rate of 4.8 kbit/s via voice-frequency modems over analogue circuits. A CCS-network plan is needed to ensure that an adequate network of signalling links is available to support its dependent IDN.

The plan for the CCS network is based on the practical constraints of avoiding the need to provide additional line plant while meeting reliability and cost criteria. The network structure is based on the following routeing rules:

(i)   Quasi-associated routeing may be used if a CCS link is required between two exchanges which are not interconnected by a traffic route.

(ii)   A single associated-signalling link is used with a traffic route of one PCM-module capacity.

(iii)   A single duplicated-signalling link is used with traffic routes of two, but less than N, PCM modules. (The value of N is set by reliability considerations, but is typically 20.) Where possible, the duplicate CCS links should be diversely routed over the transmission-bearer network.

(iv)   Traffic routes consisting of N or more PCM modules should have two duplicated, diversely routed signalling links.

(v)   Quasi-associated-signalling links may be used in the event of failure of associated-signalling links.

21.4.2 Synchronisation network

A vital element of the planning for an IDN is the establishment of a set of synchronisation links and nodes to control the timing of the whole network. A typical synchronisation network is described in Chapter 10. Fig. 10.6 shows a four-level synchronisation hierarchy with unilateral links between levels (.i.e., effective at the lower level) and bilateral links within levels (i.e., effective at both ends). A plan is required to ensure that an IDN is adequately synchronised as it develops and expands.

Typically, the objectives of a network synchronisation plan are:

(i)   Providing line plant specifically for synchronisation purposes should be avoided; synchronisation links should preferably be derived from normal traffic-carrying PCM modules.

(ii)   The second level of the synchronisation network hierarchy (i.e., the level below the reference node) should be well interconnected. Thus, in the event of failure of the reference node, the rest of the network is synchronised to the common frequency of the level-2 nodes.

(iii)   The number of effective links at any node should be sufficient to ensure that there is a presribed probability that at least one link is operative despite line-plant failures on the other links.

(iv)   Nodes can have their hierarchical status changed by appropriate changes to the number, and control direction, of their synchronisation links.

(v)   Synchronisation links should be organised so that each node can 'see' the reference node via at least two different paths via different parent nodes.

(vi)   For satisfactory operation, the number of links between a synchronisation node and the reference node should not exceed at set limit.

Based on the above objectives, a set of planning rules must be established so that an adequate synchronisation network can be installed in line with the introduction of digital switching and transmission into the national network. These planning rules are also called 'synchronisation-network topology rules' because they provide the network planners with guidelines for the structure and interconnection of the synchronisation network. The topology rules need to specify the minimum number of effective links at the various levels of the synchronisation hierarchy and the interconnection between the nodes. The rules should take into consideration the predicted outage times (a function of failure rate and mean time to repair) of line plant and nodal equipment, compared to the targets for each level of the synchronisation-network hierarchy.

An objective synchronisation network is then defined in the form of a synchronisation-network master plan. As the plans for the introduction of digital links and exchanges become known, a corresponding rolling synchronisation network implementation plan is produced to ensure that the emerging IDN is synchronised.

21.4.3 Administration network

The administration network has a number of different components, each fulfilling a specific aspect of operations and maintenance support to the exchanges and transmission links of the IDN. These components include operations-and-maintenance centres, network-management centres and billing centres. The links between these various centres and the IDN may be direct or indirect. Where the level of data flow between an IDN node (i.e., exchange or transmission station) and the administration-network centre is sporadic and of low volume, modem links achieved via the PSTN maybe suitable. Alternatively, access via a packet-switched service may be used. Leased-line connections are necessary where the level of data is high. Alternatively, bulk data, such as the periodic exchange output for billing, may be transported manually (off-line) using magnetic tapes or cartridges and so many not involve a telecommunications link.

21.4.4 Centralised-intelligence network

The centralised-intelligence databases (see Chapter 9) are located at a number of centres, either at key exchanges or at network-management centres. Local and trunk exchanges are connected to their databases by a variety of means, according to the level of traffic involved. Where the flow of data between an exchange and a database is low, access may be provided via a public packet-switched network. Very high levels of data flow warrant the use of leased-line connections to the database. Alternatively, the common-channel signalling network may be used for all links between exchanges and the database.

21.5 Network-conversion strategies

A national IDN takes a number of years to create. Where an existing analogue network is well established, the IDN needs to be introduced in a controlled way to ensure continuity of service to subscribers on both networks. The rate of introduction of the IDN is constrained by the high levels of capital investment involved and the extent to which existing plant can be discarded, prematurely or otherwise, in favour of digital equipment. The rate of introduction is influenced by the expected savings from operating digital SPC equipment. Careful planning is required during the transitional period, when the emerging IDN and the established analogue network coexist, to ensure that the penalties of operating two networks are minimised. These comprise the financial burden of interworking equipment, together with the additional operating costs resulting from the need for two sets of spares, retraining two different sets of maintenance and provisioning procedures, etc. The planning of an IDN must, therefore, be made in the context of a network-conversion strategy.

A network-conversion strategy is a coherent set of rules that encompass all aspects of the conversion of an existing network into an IDN. The rules cover the criteria for deploying digital SPC exchanges and digital transmission systems and the recovery of analogue equipment. Criteria are required for premature replacement of analogue equipment; alternatively, the digital equipment might be provided to take network growth only, with the existing analogue plant being retained until the end of its economic life. The digital equipment might initially be confined to specific geographical areas, or to exchange areas serving specific market sectors, e.g., major business subscribers with digital PABX requirements.

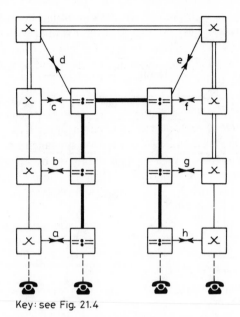

Key: see Fig. 21.4

Fig. 21.11   *Routeing between analogue network and the IDN*

The network-conversion strategy also needs to specify the new network structure of the IDN and the points of interconnection between the old and new networks. Traffic-routeing rules are needed to control the degree of shuttling between old and new networks that calls may undergo. Excessive shuttling will incur both increased transmission degradation (due to quantisation noise introduced by the A/D conversions) and an unnecessarily high provision of interworking equipment. Fig. 21.11 illustrates two separate network structures, the existing analogue network and the emerging IDN, and their possible points of interconnection (labelled *a* to *h*). Traffic which switches between networks at the originating local or trunk exchanges (*a* or *b*) is known as 'near-end' interconnected traffic; traffic switching between networks at the destination exchanges (*g* or *h*) is known as 'far-end' interconnected traffic. Where possible, near-end interconnection is used because this loads the digital network with as much traffic as possible and allows the analogue network to be offloaded and withdrawn.

Full flexibility in traffic routeing requires the existence of a complete trunk-distribution network associated with all analogue and digital trunk exchanges. Where full distribution networks for both exchanges are not available, a proportion of the trunk terminating traffic on either exchange needs to be double-switched and routed via the co-sited trunk unit. This is illustrated in Fig. 21.12, where the digital trunk exchange has a route to only one of the three dependent local exchanges. Therefore, traffic arriving over the digital

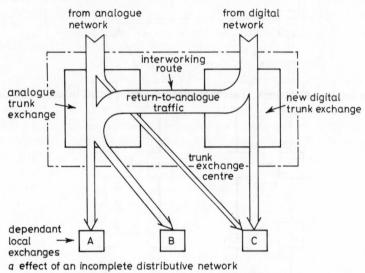

Fig. 21.12 *Incoming trunk-distribution network for analogue-digital dual-unit trunk exchange*

network destined for exchanges A and B must be routed via the inter-connection route to the analogue trunk exchange, which has a full distribution network. The arrangement, therefore, relies on the interconnection link between the IDN and analogue network, known as the 'return-to-analogue' route,[11] which must be adequately dimensioned to ensure the prescribed grade of service. However, this requires careful planning because the level of return-to-analogue traffic will fluctuate as new exchanges are added to the IDN. A similar need for double switching applies, in the reverse direction, to trunk traffic originating from exchanges A, B and C (Fig. 21.12).

The establishment of an overlay IDN may prove an attractive component of a network-conversion strategy. Fig. 21.13 shows the concept of an overlay IDN in which a relatively small volume of plant is installed alongside the existing analogue network. This IDN veneer, being small in capacity, can be installed relatively quickly. It also has the benefit of creating a coherent IDN across a country. Thus, any new services which are dependent on the IDN can be offered on a national basis rapidly, compared to the alternative strategy of piecemeal replacement. Although the overlay is normally provided to take traffic growth from the network, it may also replace some analogue capacity.

The choice of network-conversion strategy is based on an evaluation of many factors, including:

- programme of capital expenditure,
- installation and circuit-provision manpower requirements,
- marketing strategy for new services,
- degree of development and age of existing analogue network,
- maintenance strategy,
- exchange accommodation constraints,
- IDN planning and management resources.

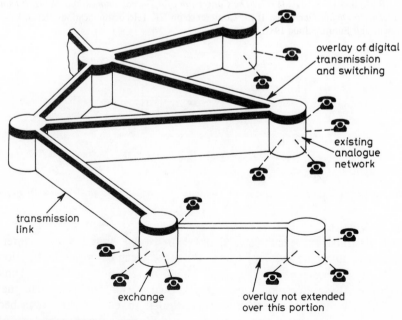

Fig. 21.13 *IDN overlay concept*

The reader is referred to the range of material covering the complex subject of digital exchange and IDN planning. A sample is given in the list of references which follows.

21.6 References

1 ITTLS (1973): *Telecommunications Planning,* ITT Laboratories of Spain
2 HARRISON, K.R. (1980): *Telephony Transmission Standards in the Evolving Digital Network*, Post Office Electrical Engineers' Journal, **73**, pp.74–81.
3 BACK, R.E. (1975): 'Network Planning', in *Telecommunication Networks,* ed. FLOOD, J.E. (Peter Peregrinus), Chap. 14
4 SMART, J.R. and TEESDALE, R.R. (1975): 'Case Studies': Introduction of digital switching and transmission, Ibid., Chap. 16
5 BREARY, D. (1974): A Long-term Study of the United Kingdom Trunk Network, Post Office Electrical Engineers' Journal, **66, 67**
6 MCDONALD, J.C. (1983): 'Digital Networks', in *Fundamentals of Digital Switching,* ed. MCDONALD, J.C., (Plenum Press). Chap. 10
7 FRY, R.A. (1975): 'Transmission Standards and Planning,' in *Telecommunications Networks* Op. cit.
8 WHORWOOD, B. (1983): 'Voice Communication Requirements', in *Local Communications,* ed. GRIFFITHS,J.M. (Peter Peregrinus), Chap. 2
9 GARBUTT, B.N. (1985): *Digital Restructuring of the British Telecom Network'*, British Telecommunications Engineering, **3**, pp.300–303
10 SUTTON, P.J. (1986): *Service Protection Network,* Paper presented to IEEE Global Telecommunications Conference (Globecom), Houston, Texas, USA, December 1986
11 MUIR, A. and HART, A. (1987): *The Conversion of a Telecommunications Network from Analogue to Digital Operation,* IEEE Conference on UK Telecommunications Networks — 'Present and Future', June 1987, London

Chapter 22

ISDN and the introduction of new services

22.1 Introduction

The early 1980s witnessed a flurry of activity at national and international forums concerning the specification of the integrated-services digital network (ISDN) concept. Several administrations have developed pilot ISDNs, the first of which was opened for public service by British Telecom in 1985.[1] The CCITT published the first set of recommendations on ISDNs in their Red Book in 1985. Thus, the concept of ISDN has been defined and the technology of its implementation has been developed. Subsequent work has been directed to determining the full implications on customer services, the relationship between ISDN and other networks, and refining the recommendations.[2] This chapter describes the ISDN concept, and the way that it may enable new services to be introduced, and it concludes with a look ahead.

The CCITT defines an ISDN as 'as integrated services network that provides digital connections between user-network interfaces'.[3] Thus, at its simplest level, as ISDN may be considered as a provider of end-to-end digital connections between two subscribers' terminals. This results in the boundary of an IDN (see Chapter 2) being shifted from the local exchange to the subscriber's premises. Also, the CCITT defines an integrated-services network as 'a network that provides or supports a range of different telecommunication services'.[3] An ISDN is therefore not just an extension of digital transmission to the subscriber's terminal, but it is also a provider of a spectrum of services. This, in fact, is the main advantage of an ISDN, namely the ability to use just one network to provide a wide range of services, rather than to use different networks for each service. The capability to support both telephony and non-voice (e.g., data) services on a common network in an efficient and economic way results from the use of digital transmission, digital SPC switching and common-channel signalling, which are the key elements of modern telephony-based networks.

One of the main advantages of digital transmission, either over network links or via an exchange switch block is that, once in digital format, traffic of all types of service can be easily mixed over the same bearer (Chapter 5). The ISDN exploits this characteristic by mixing traffic that originates in digital format (e.g., data) with digitally encoded analogue traffic (e.g., voice). Each channel is conveyed as a 64 kbit/s entity through the exchanges and transmission links (multiplexed into 2 Mbit/s or 1.5 Mbit/s frames as appropriate), irrespective of the type of service.

The first generation of ISDN exchanges are based on digital SPC telephone exchange systems, and all calls are connected on a 64 kbit/s circuit-switched basis. An ISDN exchange is not able to provide data-packet switching within its digital circuit-switched switch blocks. However, an ISDN exchange and its subscriber lines can still provide a convenient form of communication for packet services between subscribers and packet exchanges within the network.

Common-channel signalling, with its large capacity and the flexibility to handle different forms of signal format, enables the ISDN to establish connections across the network for a wide range of services. A special version of this signalling is extended from the exchange to the subscriber's terminal to enable maximum benefit to be gained. In addition to providing fast multiservice signalling between subscriber and network, common-channel signalling can pass further signals during and after call set-up, between the two subscribers' terminals via the established network connection (known as 'user's end-to-end signalling'). This gives a powerful additional communication facility to the users of an ISDN, which enables terminal-based facilities to be operated, e.g., the passing of encryption keys. The exchanges do not interpret or process such end-to-end signalling.

The most fundamental feature of an ISDN is that all the various services can be connected to the network via a small set of defined interfaces. These 'user-network' interfaces have been defined by the CCITT and are designed to be universally applicable. (The term 'user', which is favoured in the CCITT literature on ISDN, is synonymous with the term 'subscriber' used throughout this book). The interface definitions include not only the physical and electrical characteristics of an ISDN plug and socket, but also the ISDN subscriber signalling for both the basic call set-up protocols and a wide range of supplementary services.

The fundamental premise is that an ISDN will be based on a telephone IDN, which is progressively developed to incorporate the additional functions required to support the extra services. This means that services on an ISDN must be compatible with 64 kbit/s connections, although switched connections through ISDN exchanges at other rates may be introduced in the future.[4] It also means that all ISDN lines have numbers within the national telephone numbering scheme.

In order for an ISDN to be established on a foundation of a telephone IDN and still be capable of providing a wide range of existing and new services, the network needs various additional connection capabilities. An approximate mix of these capabilities will enable the various services to be provided. Each of the services requires different forms of network connection: some require the conveyance of speech, some data, some are unidirectional, some very high speed, some packet mode, etc. However, the wide range of specified services can be provided by an ISDN using just a limited range of network connections, known as 'connection types' by the CCITT.[5] The most important connection type is that necessary for standard telephone calls, which must be provided by ISDNs in all countries. Other connection types, e.g., those necessary for data packet switching, may not be provided by all national ISDNs initially and may never be provided in some countries.

Two forms of access are possible: 'basic rate' (at 144 kbit/s) and 'primary rate' (at 2 Mbit/s or 1.5 Mbit/s). The basic access has been designed for use over the standard copper-cable pairs of the local telephone network. It is suitable for users requiring simultaneous communication from only two terminals. The primary access is suitable for higher-capacity requirements and, in particular, for connection to integrated service PABXs. Transmission of the primary rate requires the provision of special cable or modification to the existing telephone copper-pair network. ISDN exchanges need to terminate both forms of access. In addition, basic rate access may be provided by copper-cable pairs connected to remotely located multiplexers which are linked to an ISDN exchange via a 2 Mbit/s (or 1.5 Mbit/s) digital line system. This

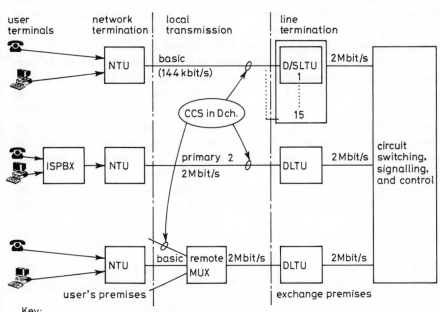

Key:
DLTU: digital line-termination unit
NTU: network-terminating unit
D/SLTU: digital subscriber line-termination unit

Fig. 22.1 *Components of an ISDN*

arrangement enables a centrally located ISDN exchange to serve users in distant non-ISDN exchange areas.

Fig. 22.1 shows the components of an ISDN for all three forms of access. The components comprise:

(i)   User terminals.
(ii) Subscriber common-channel signalling.
(iii) Network termination.
(iv) Digital local transmission.
(v)   Subscriber line termination at the exchange.
(vi) Circuit switching and control.

These components are briefly described in the following Sections.

22.2 Reference configurations

It is useful to use the CCITT recommendation on reference configurations to describe the various components of an ISDN.[6] The recommendation, like others in the I-series, relies on a functional model. This enables the ISDNs of various countries to be implemented in an evolutionary way to suit national conditions and to take account of the variety of regulatory conditions around the world. A simplified version is shown in Fig. 22.2, in which a series of reference points and functional groups have been identified.

Reference points R, S and T have been defined as points at which physical interfaces may be specified and between which functions have been defined. Potential reference points U and V have also been identified, but these are not currently subject to CCITT recommendations. However, national-standard interfaces may be produced for the U reference point in order for a 'wires-only' interface to be offered to users' terminal equipment. Similarly, national or proprietary interfaces at the V reference point may be defined in order to allow administrations a choice of digital local-line systems to terminate at ISDN exchanges.

Reference points R, S, T (and U) are located at the subscriber's premises. The three functional units, NT1, NT2 and TA, are provided by equipment on

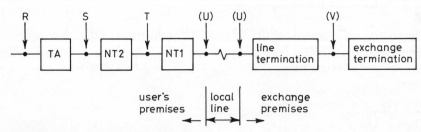

Fig. 22.2 *ISDN reference configuration*

the subscriber's premises; they may or may not form part of the administration's network, according to the regulatory environment. Briefly, the functions included in each of the groups are:

(i) The network termination 1 (NT1) comprises network-termination functions that must be provided at the subscriber's premises. These include functions broadly equivalent to those defined at layer 1 (physical) of the OSI reference model, providing physical and electrical termination. Functions include digital line-transmission termination, power feeding, timing and primary maintenance.

(ii) The network termination 2 (NT2) comprises the higher-level network termination functions of channel multiplexing, traffic concentration (i.e., channel switching) and secondary maintenance functions (e.g., functional testing). Examples of NT2 devices include PABXs and call-connection systems.

(iii) The terminal adaptor (TA) performs the functional conversion between non-ISDN-type subscriber apparatus and the ISDN user-network interfaces at the S or T reference points.

(Although not defined by the CCITT, it is generally accepted that the line-termination functional group provides all the necessary functions involved in terminating a local digital line system. Similarly, the exchange termination functional group provides an appropriate interface to the exchange switch block. In practice, both functional groups may be provided by a single piece of equipment, e.g., the DLTU.)

The same set of ISDN user-network interfaces has been defined for both the S and T reference points. This gives full flexibility in the way that ISDN-type users' apparatus may be deployed. For example, the NT2 function may not be required in some cases (i.e., only an NT1 device exists); in other cases NT1 and NT2 functions may be physically provided by one device. The R reference point covers any existing non-ISDN-type user-network interfaces. Typical examples of terminal interfaces that would apply at R, and hence require terminal adaptors, are CCITT X21 and X21 bis (V24 and V35 compatible).

22.3 Digital local transmission

The digital local transmission for an ISDN is provided between a network terminating unit (NTU) on the subscriber's premises and a line-termination unit at the ISDN exchange (Fig. 22.1).

22.3.1 Basic-rate transmission systems

The 144 kbit/s basic ISDN access (192 kbit/s gross line rate) may be provided over the existing copper-pair telephone cables. There are two considerations:

(i) Provision of full duplex (i.e., simultaneous transmit and receive) transmission

(ii) Efficient and economic digital transmission over practical distances from the exchange, without causing interference to other subscribers' lines.

Duplex transmission may be achieved by using separate pairs for transmit and receive channels. Alternatively, electrical separation of transmit and receive channels can enable a single pair of wires to be used. Two methods of electrical separation are employed.[1,7] The first method time-shares the path by interleaving bursts of bits alternately from each direction. This is known as burst-mode (or time-compression duplex) transmission. The second method, echo cancellation, relies on the subtraction, at the hybrid transformers at each end of the line, of a simulated replica of the unwanted component of the transmit signal that would otherwise interfere with the receive signal. With improvements in technology, the echo cancellation method appears increasingly, to be the favoured choice.

Efficient digital transmission over the local copper-cable network requires the use of a suitable line code. The choice of line code for digital baseband (i.e., not superimposed on a carrier) transmission needs to meet two objectives, namely: conveyance of timing information and the production of a suitable energy spectrum. Adequate timing information in the signal transmitted to line is essential because the receiving NTU must use it to derive its timing. In turn, the NTU will synchronise the attached user terminals. The energy spectrum must be shaped so as to minimise interference with, and from, adjacent cable pairs, while giving adequately sharp pulse shapes to ease detection at the receiver and ensure good transmission quality over the required length of cable. Suitable codes include WAL2[8] and the new standard of 2B1Q.[9]

22.3.2 Primary-rate transmission systems
There is a wide range of digital transmission systems suitable for providing primary-rate access (2 Mbit/s or 1.5 Mbit/s) to the ISDN local exchange. The transmission may be over copper-cable pairs, where suitable equalisation and conditioning can be applied. Alternatively, special plant may be provided, e.g. small-bore coaxial cable, transverse-screened pair cable or optical-fibre cable. For all these solutions, transmit and receive paths are generally provided by spatial separation, i.e., by the use of separate bores, pairs or fibres. Digital microwave radio, using different carrier frequencies for transmit and receive directions, may also be used to provide primary-rate ISDN access.

22.3.3 ISDN subscriber line termination at the exchange
The line-termination arrangements at the ISDN local exchange for primary, basic and multiplexer access are shown in Fig. 22.1. The basic (144 kbit/s) access on copper-pair cable terminates on digital subscriber line-termination

units (D/SLTU), as described in Chapter 7 and shown in Fig. 7.7. The B channels from 15 D/SLTUs are multiplexed into 30 channels of one 2 Mbit/s bus into the subscriber-concentrator switch block. Each B channel may then be routed independently through the network. The signalling messages from the D channel are either taken directly, or indirectly via one of the digital switch blocks, to a signalling receiver, as described in Chapter 8.

The 2 Mbit/s inputs from primary accesses and remote multiplexer sites terminate on DTLUs, as described in Chapter 7 and shown in Fig. 7.9. ISDN exchanges need to support both analogue and digital (ISDN) subscriber lines. Therefore, the design of the line terminations must support various mixes of analogue and digital lines. The grouping of lines into 2 Mbit/s modules, as described above, represents a convenient method of mixing the various types of subscriber lines at an exchange.

22.3.4 Network termination unit

The ISDN terminates on the user's premises at a network termination unit (NTU). As already described, the CCITT identifies two functional network termination entities: 'NT1' and 'NT2' (Fig. 22.2). The NT1 set of functions is the minimum that can be provided in an NTU. These comprise:

(i)   Termination of the local digital line-transmission system.
(ii) Power feeding from line or local mains or battery-powering facility.
(iii) Basic maintenance functions (e.g., application of test loop).
(iv) Provision of standard ISDN user-network interface.
(v)   Access contention on subscriber's CCS channel.

Additional functions fall into the NT2 category, for example:

(vi) Channel multiplexing.
(vii) Data-protocol conversion (e.g., data character code or speed conversion).
(viii) Call routeing (e.g., PABX).
(ix) Supplementary service features (e.g., dialled-digit storage, short-code dialling, directory storage).
(x)   Data detection and correction.
(xi) Network management for users.
(xii) Display of supervisory information.

An NTU may provide both NT1 and NT2 functions. In addition, the NTU may provide the terminal adaptor (TA) functions. The range of functions included in the NTU is determined by the regulatory regime within the country. Where the NTU is considered part of the network, and is thus provided by the telecommunications administration, the inclusion in the NTU of functions such as NT2 and TA, which are related to user terminals, may not be appropriate. Fig. 22.3 illustrates an example of the split of functions between the NTU (NT1 only) and an NT2 and TA-type terminal.

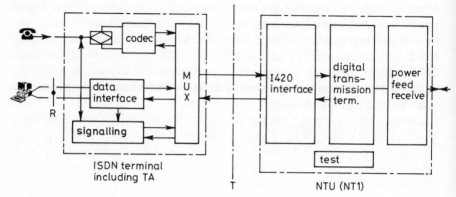

Fig. 22.3 *Functional split between NTU and terminal equipment (example)*

22.4 Network access

The term 'network access' is used generally to describe the link between a subscriber and the network. The access to an ISDN has been defined by the CCITT[10] in terms of the types of channel that may be carried and the frame structures of groupings of such channels.

22.4.1 Types of channel

Four types of information (or traffic) channel have been defined, namely:

(i)   B channel: 64 kbit/s with byte timing. This channel may be used to carry 64 kbit/s encoded voice (standard PCM or other forms of coding), data at 64 kbit/s or a 64 kbit/s multiplex of lower-speed data channels, a 64 kbit/s multiplex of low-bit-rate encoded speech or a 64 kbit/s multiplex of data and voice, etc. The B channel can carry traffic that may be 'non-switched' (i.e., a semi-permanent connection through the exchanges, as described in Chapter 8), circuit-switched or packet-switched.

(ii) H_0 channel: 384 kbit/s. This is intended to be the standard broadband channel. It is suitable for a number of broadband applications, e.g., video communications.

(iii) H_{11} channel: 1536 kbit/s (for 1.5 Mbit/s-based countries).

(iv) H_{12} channel: 1920 kbit/s (for 2 Mbit/s-based countries). Both H_{11} and H_{12} may be used as single broadband channels carrying any form of traffic at 1.5 Mbit/s or 1.92 Mbit/s, respectively. Alternatively, they may carry a multiplex of H_0 channels.

The B channel can be used with either primary or basic access. However, the H_0, H_{11} and H_{12} channels can only be used with primary access.

Two signalling channels have been defined:

(v)   D channel: 16 kbit/s or 64 kbit/s. This channel is primarily for user-network signalling on circuit-switched connections. It may also be used for the conveyance of telemetry signals and slow-speed packeted data.

(vi)   E channel: 64 kbit/s. The E channel is an alternative signalling channel at 64 kbit/s for an ISDN access which uses the message-transfer part of CCITT No.7 signalling system. This mode maybe appropriate where the signalling information for a number of different access links is to be conveyed over one E signalling channel.

Connection from subscribers to the ISDN is in the form of two standard frame structures, known as 'basic' and 'primary' access, in which the above channels are carried. Both forms of access use standard interfaces at S and T reference points.

22.4.2 Frame structures

The basic-access channel structure comprises two B channels and a D channel at 16 kbit/s and is known as '2B + D(16)', giving a total user bit rate of 144 kbit/s. The addition of a 16 kbit/s echo channel and 32 kbit/s for frame alignment, etc., gives a total bit rate of 192 kbit/s, as described below.

The basic-rate channel-access structure (2B + D(16)) is transmitted in a TDM frame. The effect of the frame overhead (i.e., frame alignment bits, etc.) is minimised by spreading the frame over 250 μs rather than 125 μs, as shown in Fig. 22.4. Each 48-bit frame consists of two consecutive octets from the first B channel, two consecutive octets from the second B channel, two 2-bit samples of the D channel, plus frame-alignment and balancing bits. Thus, the 144 kbit/s of duplex user information is transmitted at the rate of 192 kbit/s (48 bit every 250 μs). The 12 extra bits represent a modest overhead of 25% with the double-length frame, compared to the 66.7% overhead that would have been incurred if a single-frame arrangement had been employed.

The frame structure is shown in Fig. 22.4 for transmission in the direction of the user terminal to the network; a different structure applies in the reverse direction. A high-timing component is ensured by a pseudo-ternary line code, in which binary '1' is represented by zero volts and binary '0' is represented alternately by negative and positive pulses. A 'balance bit' is injected into each

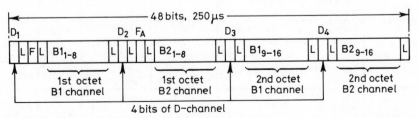

Key: L = balance bit   F and F_A = framing bits

Fig. 22.4   *Basic rate frame structure[13]*

frame in order to maintain a zero net voltage. The user's ISDN terminal derives its bit, octet and frame timing from the 192 kbit/s digital signal leaving the network-termination unit.

Two versions of primary access are defined: one for 2 Mbit/s-based countries and the other for 1.5 Mbit/s-based countries, as follows:

(i)   For 2 Mbit/s countries: 30 B channels plus a D channel at 64 kbit/s (30B + D(64)). This, together with the 64 kbit/s for frame alignment, etc, gives a total bit rate of 2048 kbit/s.

(ii)   For 1.5 Mbit/s countries: 23 B-channels plus a D channel at 64 kbit/s (23B + D(64)). This gives a total bit rate of 1544 kbit/s, including an alignment bit.

There are also standard multiplexed channel structures for broadband accesses. Two examples of 2 Mbit/s access are $5H_0 + D(64)$ and $H_{12} + D(64)$. Other access structures will be defined as the ISDN concept evolves.

22.4.3 User-network interfaces

As mentioned earlier, an important feature of an ISDN is that a small set of user-to-network interfaces are defined to cope with a wide range of services, users' applications and the configurations of users' terminals. This feature will ensure that, in the future, subscribers should be able to buy data and telephone equipment confident that they will work on any ISDN (irrespective of the network administration) and provide the required service. To this end, ISDN interfaces have been defined by the CCITT at the basic and primary rates.

The ISDN user-network interfaces are defined in terms of the three functional layers of the ISDN-protocol reference model.[11] This model is derived from the open system interconnection (OSI) 7-layer model,[12] although there are some important differences. These are due to the wider

Table 22.1 *CCITT recommendations on ISDN user-network interface*

Layer	Coverage	CCITT recommendations		
		Basic	Primary (D)	Primary (E)
1	Electrical characteristics total frame structure	I430	I431	I431
2	Link access procedure (LAP) on the D (or E) channel	I440 (also Q290)	I440 (also Q920)	Q170
	Frame structure for LAP D (or E)	I441 (also Q921)	I441 (also Q921)	Q710
3	Call connection control protocols	I450 & (also Q930) I451 (also Q931)	I450 & (also Q930) I451 (also Q931)	I450 & (also Q930) I451 (also Q931)

range of network-service protocols that an ISDN must support, compared to the OSI model of open communication between two computers. The ISDN model also has seven layers. However, although these are numbered was in the OSI model, use of the layer names is considered misleading in an ISDN context. This subject is beyond the scope of this book. Suffice to say that the two interfaces are defined according to Table 22.1. They are briefly described in the following Section.

22.4.3.1 Basic-rate interface There are three possible configurations for the basic-rate interface identified by the CCITT, namely point-to-point, passive bus and extended passive bus. Fig. 22.5a illustrates how an ISDN basic-rate interface may be used with a passive-bus configuration. The network termination, NT1 or NT2 according to whether the interface is at the S or T reference point, can support up to eight user terminals. The terminals may be ISDN-type (TE1), offering the I-interface, or non-ISDN-type (TE2), attached to the bus via a terminal adaptor. When more than one terminal attempts to transmit at one time, the ISDN network termination resolves the contention, as described below.

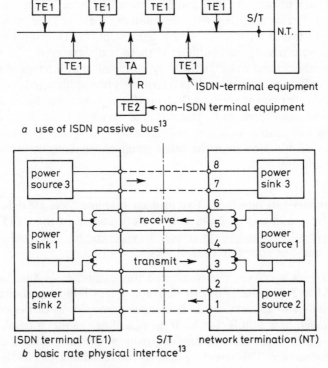

a use of ISDN passive bus[13]

b basic rate physical interface[13]

Fig. 22.5 *ISDN basic rate interface*

Physically, the interface is provided by a plug-and-socket arrangement with up to eight connectors (Fig. 22.5*b*). The user's traffic is conveyed over the transmit and receive connectors. Normally, power is fed to the user terminal over a phantom circuit derived from the four-wire combination of the transmit and receive connectors. However, there is also provision for up to two additional sets of power-feeding connectors. This arrangement can cope with a wide range of user terminals.

The 16 kbit/s transmission capability of the D-channel is operated as a packet-transport mechanism. This is capable of carrying the common-channel-signalling packets from up to eight terminals attached to the NT1. In addition, the D-channel can interleave user's data packets with the signalling packets. The interleaving process is controlled by the NT1, which gives priority to signalling packets to ensure that call-control messages are not delayed. This means that user's data packets will be delayed or lost whenever there is contention for the D-channel between data and signalling traffic. There will also be contention between the signalling packets from the various user terminals on a passive-bus (or extended passive-bus) termination. Both forms of contention are resolved by the NT1.

The mechanism used for contention resolution is based on the use of D-channel echos. Each D-channel digit received by an NT1 from a user terminal is repeated back towards the bus, and hence to all the attached terminals (TE1 or TA/TE2). The sending terminal then compares the echo D bit with a stored version of the last D bit sent. If there is agreement, the D-channel signalling and B-channel transmission continue; if there is disagreement, the terminal ceases transmission and waits. In this way, terminals may attempt to transmit whenever the D-channel is clear. If two terminals transmit simultaneously, the echo will indicate that the received D-channel signals were mutilated and both terminals will cease and wait. The terminal which was in the process of transmitting call-control signalling will have a higher-priority setting within the network termination than the other terminal, enabling it to attempt a retransmission sooner.[13]

22.4.3.2 Primary-rate interface. Two primary-rate interfaces are defined; one at 1.544 kbit/s and one at 2.048 kbit/s. Both adhere to the CCITT G703 digital-transmission standard interface; this covers bit rate, pulse shape, impedance and line code (see chapter 5):

(i)   1.544 kbit/s interface: The frame structure comprises 24 8-bit time slots (numbered 1 to 24) with a leading single-framing bit ('F-bit'), as shown in Fig. 5.8. Time slots 1 to 23 carry the 23 B-channels; time slot 24 carries either the D or E channel, or a 24th B-channel if no signalling channel is required. The F-bit is used to provide the 6-bit framing pattern '001011' within a 24-frame multiframe structure. This structure enables the remaining 18 bits to be used for maintenance messages (12 bits) and for error checking (6 bits), if required

by an administration. In non-ISDN applications, the 24-frame multiframe structure is used to convey the 6-bit framing pattern and common-channel signalling messages (Chapter 5).

(ii) 2.048 kbit/s interface: The frame structure comprises 32 8-bit time slots (numbered 0 to 31), as shown in Fig. 5.7. Time slot 0 ('TS0') is used to convey the full 7-bit frame alignment pattern '0011011' in every alternate frame. This leaves TS0 available for network and maintenance signalling during the other alternate frames (Chapters 5 and 10). Time slot 16 ('TS16') carries either the D- or E-channel. The 30 B-channels are conveyed in time slots 1 to 15 and 17 to 31.

22.4.4 *ISDN access (User-network) signalling*

A special common-channel signalling system has been specified for use between ISDN terminals and an ISDN exchange. This signalling is carried in the D-channel of primary and basic accesses. Hence, it is often referred to as 'D-channel signalling'. The use of the signalling is defined in the CCITT recommendations on the 'line access procedure on the D-channel', known as 'LAPD',[14] for layer 2 and in the recommendations on message structure and sequences for layer 3.[15]

This ISDN-access signalling enables a wide range of facilities to be provided and allows the exchange to give more information on the progress of calls. The following functions are facilitated:

(i)   Setting up of independent simultaneous connections (over the B-channels).

(ii)   Support of a variety of ISDN supplementary services.

(iii)   Signalling during the call.

(iv)   Transmission of low-volume data when signalling is absent.

An important feature of the signalling system is its indication of the service capability for each call. This information is generated by the sending end to indicate the type of service, e.g., telephony, digital data or mixed digital data/telephony. The indication is examined by the exchange in the connection to determine the type of routeing through the network for that call. At the distant end, the receiving terminal, or network-termination unit, can examine the compatibility elements to determine which of the terminals connected to the ISDN line should receive the call. This procedure ensures that calls are not established between incompatible terminals, e.g., between a telephone and a facsimile machine.

For data-call connections, call-progress information is conveyed in the D-channel signalling and is displayed by the ISDN user terminal. With telephony connections, the call-progress information (e.g., ringing tone) is transmitted in the speech channel (i.e., the appropriate B-channel) as well as the D-channel. Chapter 20 describes the ISDN-access signalling system in more detail.

22.5 Service aspects of ISDNs

ISDNs are networks capable of supporting a range of services; they are not new services in themselves. However, the existence of an ISDN should enable new services to be more readily provided. The CCITT[16] has defined two categories of telecommunications services that an ISDN can support, namely teleservices and bearer services (see Fig. 22.6). Teleservices comprise both the transport of the information and its encoding and formatting, and thus depend on the users' terminals, such as telephones or data terminals. Examples of teleservices are speech, high-quality sound, facsimile, videotex, video, etc.[17] Bearer services comprise only the transport of information across the ISDN between two user-network interfaces (at R, S or T reference points). Two examples of bearer services are:

(i)   64 kbit/s, 3.1 kHz audio, point-to-point, bidirectional, circuit-switched connection.

(ii)   64 kbit/s, unrestricted-digital information, multipoint connection[18].

In describing services that can be provided by an ISDN, it is useful to note that teleservices encompass all seven layers of the OSI model, whereas bearer services only cover layers 1 to 3. The CCITT has described a technique of using a set of attributes to define teleservices and bearer services.[16] Thirteen attribute parameters have been identified for bearer services,[17] and these include:

(i)   Information-transfer mode (e.g., circuit switched connection).
(ii)   Information-transfer rate (e.g., 64 kbit/s).
(iii)   Information-transfer capability (e.g., 3.1 kHz audio).
(iv)   Structure (e.g., 8 kHz integrity).
(v)   Establishment of communication (e.g., on demand).
(vi)   Communication configuration (e.g., multipoint).
(vii)   Symmetry (e.g., unidirectional).

By an appropriate choice of attribute for each of the 13 parameters, a unique bearer service can be defined. Similarly, a teleservice can be uniquely defined by a further eight parameters in addition to the 13 of the bearer service.[18]

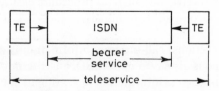

Key: TE = terminal equipment

Fig. 22.6 *ISDN service categories*

22.5.1 *ISDN user terminals*

ISDN user terminals offer either a basic (2B + D(16)) or primary ((30B + D(64)) or 23B + D(64)) interface to the network. These terminals, known as 'I-series terminals', have interfaces in accordance with CCITT Recommendation I420 for basic access and Recommendation I421 for primary access. The I420 terminals may be of the following types: data only (e.g., facsimile machines, personal computers, videotex or telemetry terminals), telephony-only, or combined telephony and data types. In general, I421 terminals undertake the multiplexing or switching of a number of channels on to a 2 Mbit/s or 1.5 Mbit/s digital system. An example of an I421 terminal is that of an integrated services PBX (ISPBX) which combines the traffic from telephone and data terminals within a common switch block.

Some of the I-series terminals will support other terminals. For example, ISPBXs will support a range of terminals on their extension lines, including standard telephones, I420 data terminals and non-ISN-type data terminals (see Fig. 22.7*b*). The support of non-ISDN-type terminals requires the provision of a terminal adaptor (TA), as shown in Fig. 22.2. The TA may be incorporated

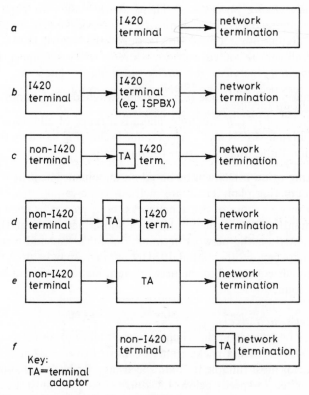

Fig. 22.7 *Connection of terminals to an ISDN*

in the I-series terminal (Fig. 22.7c). Alternatively, it could be provided as a self-contained terminal, offering non-ISDN interfaces (e.g., X21 and X21 bis) for non-ISDN terminal attachments and the I420 or I421 interface for attachment to the network termination (Figs. 22.7d and e). Otherwise, the TA may be incorporated in the network termination (Fig 22.7f).

It is important to note that the terminal-adaptor unit and the I420 terminal device must generate the basic-interface electrical signal. This not only includes production of the access-channel structure, e.g., '2B + D(16)', but also the termination (send and receive) of the ISDN-access common-channel signalling.

22.5.2 Circuit switching and control
The handling of D-channel signalling from ISDN subscriber lines is described in Chapter 8 (see Fig. 8.3).

Most recent digital SPC telephony switching systems are capable of being enhanced to provide ISDN connections. The appropriate CCITT recommendations are in the Q500 series. The enhancements required to give an ISDN capability include:

(i)   New call-routeing procedures: These are necessary to cope with the routeing of a 'digital data call' (as indicated by a service indicator code), which requires a digital-only connection through the network. However, a 'telephony' call can be routed over a mixture of analogue and digital links.

(ii)   Supplementary services for ISDN subscribers: These are additional service features specifically for ISDN users. An example is the provision of closed-user-group facilities.

(iii)   Special forms of call charging: There is a range of additional charging facilities required by ISDN users. An example is charging based on throughput of information rather than duration for data calls.

The numbering of an ISDN will be based on an enhancement of the existing telephone numbering plan. There are methods of 'escaping' from the ISDN numbering range to those of other networks, e.g., packet or telex networks. An ISDN number will uniquely address a subscriber's line at the S or T reference point (i.e., the network termination unit). 'Subaddressing', using additional digits beyond the national ISDN number, may be used to indicate specific terminals connected to the network termination unit, but beyond the T-reference point.[19]

22.5.3 Packet-switch access
Reference has been made in this chapter to the ability of an ISDN to support packet data services. Data packets may be carried over a B- or D-channel.[20] The B-channel is most suitable for data speeds at 4.8 kbit/s and above; the D-channel is limited to speeds below 4.8 kbit/s. Fig. 22.8 illustrates the three routeing possibilities for a packet-switched connection between two

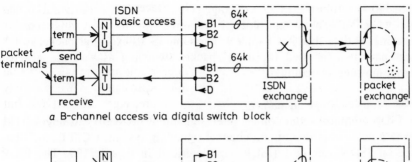

a B-channel access via digital switch block

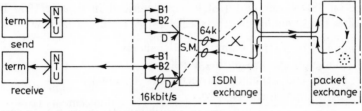

b D-channel access via digital switch block

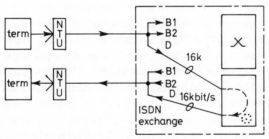

c D-channel access to internal packet switch block

Key:

✗ : digital (circuit-switch) switch block giving 64kbit/s interconnectivity

S.M. : statistical multiplexer

⋰ : packet switch block

note: routeings are shown conceptually at the 64kbit/s or 16kbit/s time-slot lev

Fig. 22.8 *Packet-switch access in an ISDN*

subscribers' lines (basic access) on an ISDN exchange. Packet-data traffic, carried in a B-channel, is transported via a 64 kbit/s circuit-switched connection through the ISDN exchange on to a trunk or junction route to a separate dedicated package exchange (Fig. 22.8*a*). Data packets carried over D-channels are extracted from the D/SLTUs and statistically multiplexed with the packets from a number of other D-channels on to a 64 kbit/s bearer, as described in Chapter 8. This multiplexed circuit may then be routed to the packet exchange directly or via the ISDN exchange switch block (Fig. 22.8*b*). Where the total packet traffic is expected to be large and a high proportion is intra-exchange, system economies may be gained by including a packet switch block within the ISDN exchange (Fig. 22.8*c*). This configuration has merit when such a switch block is required for intra-exchange control messaging, as described in Chapter 9. The same ISDN exchange packet switch block may then be used for users' packet switching as well as packeted internal-message switching.

22.5.4 Inter-exchange signalling

An ISDN requires common-channel signalling between exchanges. The international standard common-channel signalling system, CCITT No.7, has a special ISDN-user part (ISUP), as described in Chapter 20. The ISUP provides the signalling for the special ISDN features, namely:

(i)   Support of a large number of B-channels.
(ii)   Change of services during a call.
(iii) End-to-end signalling between users' terminals.
(iv)   ISDN supplementary services.
(v)   Conveyance of end-to-end compatibility information for teleservices.

22.5.5 Interfaces between ISDN and other networks

An ISDN will need to interwork with other networks in order to provide the necessary interconnectivity for its users. Such interworking may be with existing dedicated networks in the same country, e.g., data-packet switching, data-circuit switching or private-circuit networks. Fig. 22.9 shows the range of possible interconnection points with an ISDN. In addition to the reference points R, S and T (U and V are internal to the ISDN), reference points K, L, M, N and P have been identified by the CCITT.[21] However, in practice there may not be a need for defined interfaces at all these reference points.

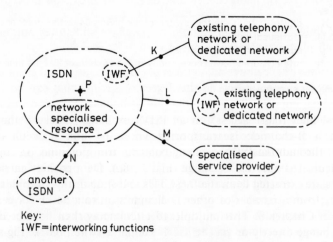

Key:
I WF = interworking functions

Fig. 22.9   *Reference points between an ISDN and other networks*[21]

22.6 The way ahead

22.6.1 The spread of ISDN

The degree to which the ISDN approach is adopted within a digital network depends on its advantages as perceived by users and the administration. The advantages of an ISDN as seen by the users may be summarised as follows:

(i)   Single access and interface for many services with consequent reduction in office wiring.
(ii)   Two independent channels (basic access) over a single 'connection'.
(iii)   D-channel signalling enables the users to signal during the call.
(iv)   D-channel signalling may be used as a low-speed channel for users' data.
(v)   Up to eight terminals may be supported on the basic access.
(vi)   Simultaneous use of the ISDN access for different services.
(vii)   One number for all services.
(viii)   Many supplementary services.

Clearly, the relative merits of the above depend on the tariff relationship between services provided over an ISDN and those provided by separate dedicated networks.

The advantages of an ISDN to an administration stem from having only one common-service network to plan and provide, rather than several dedicated-service networks. This has the following effects:

(i)   The plans for the network are less sensitive to errors in the forecasts for individual services.
(ii)   New services may be made available nationwide quickly without the delay incurred in establishing a new network infrastructure (i.e. the network and its support systems).
(iii)   Only one set of maintenance and one set of provision procedures are required.
(iv)   Economies are gained in equipment purchase and stores holding.
(v)   Overall planning effort is reduced.

The importance of the above advantages depends on the actual costs involved in the ISDN implementation and the degree of savings that can be achieved in practice. An important factor in the rate of implementation of an ISDN is the user demand for non-telephony services. It is expected that data-communication and video services will continue to increase in importance. However, the administration must balance the potential savings and improvements that an ISDN might make with the penalties involved in abandoning any existing dedicated networks. Most countries have established telex and private-circuit (leased-line) networks; many also have circuit-switched data and video distribution (cable TV) networks.[10,20] Since both the dedicated existing networks and an ISDN can be used to provide the same or similar services, there is a need for a planned evolution towards the ISDN, which must take account of network economics and resources, as well as service needs.

The range of telecommunications (and bearer) services has extended vastly over the last 50 years or so from just telegraphy and simply telephony to that shown in Fig. 22.10. This trend will continue at an ever-increasing rate due to the spread of digital networks, SPC switching and common-channel signalling. In addition, the availability of intelligent terminals and a growing need for data communications will increase the number of nonvoice services required. Users' applications will require ever-greater data throughputs and digital capacities. It is likely that the future digital networks, whether based on an ISDN approach or not, will need to support switched services at rates greater than the ubiquitous 64 kbit/s. (These are known as 'broadband services'.)

Furthermore, the non-voice services, both data and video, are expected to grow very much faster than simple telephony. This growth is a result of the

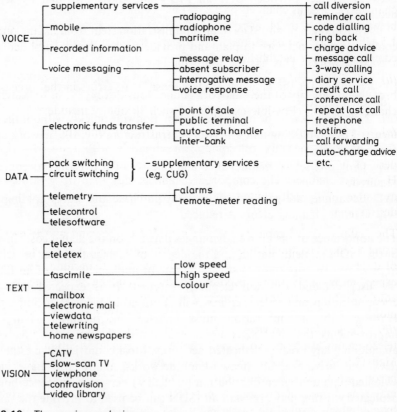

Fig. 22.10 *The service explosion*

increasing importance of data communications to the running of businesses and industry. The applications of software-supported business systems, in shops as well as offices and factories, are also growing. Thus, telecommunications services, which historically have been little more than communication bearers (bit-transporters in an ISDN), are now beginning to have additional capabilities associated with them. These so-called 'value-added services' perform some enhancement to the information content of the communication, e.g., protocol conversion and store-and-forward of data and spoken messages.

22.6.2 Exchange systems of the future

The growth in the demand for new telecommunications (and bearer) services indicates that future exchange systems will need to provide a wider set of more complex services. The characteristics of future requirements are summarised below:

(i)   Broadband services: In addition to the standard 64 kbit/s circuit-switched connection, future digital services will include connections at a variety of higher-bit-rate capacities. These include connections of H_0, H_{11} and H_{12} channels, as well as channels of $n \times 64$ kbit/s capacities, where n is any integer between 2 and 30 or 24, as required. Connections at greater than 2 Mbit/s or 1.5 Mbit/s may also be required, particularly if switched high-resolution video communication is established.

(ii)   Flexibility and responsiveness: Users will want the ability to allocate the broadband capacity of their network connections flexibly, so as to match their changing needs.[22] Possible examples of such flexibility include:[23]

(a)   changing the value of n for $n \times 64$ kbit/s call connections,

(b)   changing the mix of component channels within an H_{12} channel, perhaps by including non-switched (i.e., leased-line) connections within the H_{12} access channel. The component channels may then be separated at an exchange and each routed individually through the network to their destinations.

The reallocation of capacity may be provided on demand by the user (e.g., using a control terminal) or on a pre-assigned basis (e.g., according to the time of day) as previously agreed with the administration. In addition, users will expect increasingly rapid responses to their changing service needs. Thus, administrations will have to implement new forms of services in a matter of months rather than the 2 to 3 years curently incurred.

(iii)   Value-added services: The value-added services associated with telecommunications networks will continue to expand in volume and range. Many of these services, such as signalling and protocol conversion, may become accepted enhancements to standard exchange-based services. Others may remain more appropriately provided outside an exchange environment, by equipment located on service suppliers' premises.

As part of the process of digitalising the subscribers' access network, optical fibre will be used increasingly in place of copper cable. Optical fibre provides the capability of transmitting wide-bandwidth communication to subscribers' premises. The use of optical fibre also enables the local network to be restructured, as described in Chapter 21. The ISDN architecture, described in this chapter as a means of combining the switching of voice and data connections within one circuit-switched 64 kbit/s-based digital SPC exchange, will be extended to cover other forms of connection.[24,25] Thus, the trend is for future ISDNs to combine the following capabilities:

(i)   Narrow and broadband connections.
(ii)   Circuit-switched and packet-switched connections.
(iii) Switched and non-switched (leased-line) connections.

The network trends described above are likely to influence the way that exchange functions are implemented. A number of possibilities have been identified, [26,27] and these are summarised below:

(i) Dispersed control: Much of the control software, both special-feature programs and subscriber-specific data, will be located at remote centres (SCPs) in accordance with the intelligent network architecture (Chapter 9). In addition, the remaining software, which covers the standard telephone services and is currently resident in exchange-control processors, may be dispersed to terminal devices on the subscriber's premises. This becomes possible when optical fibre is used to link subscribers' premises in the form of a widespread local-area network (LAN) architecture. Communication between subscribers' terminals on the LAN is then broadcast, with the recipient terminal applying selection techniques to extract the appropriate channel.

(ii) Optical switching: As the use of optical-fibre transmission for trunk and local circuits increases, the switching of channels in optical rather than electrical mode will enable significant savings in electro-optical interworking equipment at exchanges. In addition, optical switching should offer an efficient method of providing broadband connections, with both fixed and variable capacities.

(iii) ATD: The asynchronous time-division (ATD) technique,[28] applies the principle of packet switching, hitherto restricted to low-speed data, to a wide range of services, including telephony. The technique relies on the conversion of all input channels (telephony, broadband data or video, etc.) to packets, each with appropriate headers to assist routeing, and of an appropriate length according to the type of service. The various packets* are statistically multiplexed within the system and switched at high speed so that real-time services, such as telephony, are not delayed perceptibly. ATD is expected to

---

* also called 'cells' to avoid confusion with CCITT × 25-type packets

offer an efficient method of variable-bandwidth switching in a multiservice network.

It is evident that the design of future generations of exchanges will be radically different from the digital SPC exchange systems described in this book.

22.7 References

1 PRICE, C.D. and BOULTER, R.A. (1985): *Integrated Services Digital Network*, British Telecommunications Engineering, **3** pp.311–317

2 IRMER, T. (1987): *ISDN In Europe — How, Why and When*,, British Telecommunications Engineering, **5**, pp.296–298

3 CCITT: Blue Book (IXth Plenary Assembly, Melbourne, November 1988), Rec I112

4 CCITT Rec. I120, Ibid.

5 CCITT Rec. I340, Ibid.

6 CCITT Rec. I411, Ibid.

7 TAYLOR, G (1983): 'Electronics in the Local Network, in *Local Communications*, ed. GRIFFITHS, J.M. (Peter Peregrinus Ltd) Chap. 13

8 BREWSTER, R. (1983): 'Data Transmission', Ibid.,Chap. 7

9 RONAYNE, J. (1987): *The Integrated Services Digital Network: from concept to application'* (Pitman), pp.65–66

10 CCITT Rec. I420, Op. cit.

11 CCITT Rec. I320, Op. cit.

12 CCITT X200 series recommendations, Op. cit.

13 CCITT Rec. I430, Op. cit.

14 CCITT Recs. I440, (also Rec. Q.920), I441 (also Rec. Q.921), I450 (also Rec. Q.930), Op. cit.

15 CCITT Rec. I450 (also Rec. Q.930) and I451 (also Rec. Q.931), Op. cit.

16 CCITT Rec. I210, Op. cit.

17 CCITT Rec. I212, Op. cit.

18 CCITT Rec. I211, Op. cit.

19 CCITT Rec. E164, Op. cit.

20 LISLE, P.H. and WEDLAKE, J.O. (1984): *Data Services and the ISDN'*, British Telecommunications Engineering, **3**, p.79

21 CCITT Rec. I310, Op. cit.

22 EBERT, I, ABDOU, E. and RICHARDS P.S. (1987): *Network Flexibility Through Dynamic Transport Control*, Fifth World Telcom. Forum, Technical Symposium: 'Telecommunications Services For a World of Nations', ITU, Geneva, October 1987, Vol. II, pp.5–7

23 OLIVER G. (1983): 'The Future of the Local Network', in *Local Communications*, ed. GRIFFITHS, J.M. (Peter Peregrinus Ltd.), Chap. 16, pp. 256–262.

24 CHAGNON, P.J., DUCHESNE, M.A. and VENIER, D.J. (1987): *Bell Canada — Network Evolution for the 1990s*, Ibid., pp.39–43

25 PLEHIERS, P. and BAUWENS, J. (1987): *Evolution of IDN Towards ISDN in the Belgian Network*, Ibid., pp.55–59

26 OHTSUKI, I, *et al* (1987): *A New Switching System Architecture for the 1990s*, Ibid., pp.15–19

27 LEAKEY, D.M. (1988): *The Future of the Public Telecommunication Network'*, Telecommunications, pp. 35–47.

28 LITTLEWOOD, M., GALLAGHER, I.D. and ADAMS, J.L. (1987): *Network Evolution Using Asynchronous Time-Division Techniques*, British Telecommunications Engineering, **6**, pp.95–104.

Glossary

Addressing, direct Accessing an address which contains the required information.

Addressing, indirect Accessing an address which contains the address of the required information.

ADPCM (Adaptive differential pulse-code modulation) A form of PCM in which the changes in consecutive samples are encoded. The CCITT have standardised a 32 kbit/s version.

A-law The PCM coding law used by the CCITT (and CEPT) 30-channel PCM system.

Alternative routeing The carrying of traffic on a 2nd- or Nth-choice route when the 1st-choice route is found to be fully loaded.

AMI (Alternative mark inversion) A technique for converting a binary signal to a pseudo-ternary signal in which successive marks have alternately positive and negative polarity.

Associated signalling The mode of common-channel signalling in which the CCS channel is routed physically separately from the traffic channel it serves, although the originating and terminating points are common.

Asynchronous network A digital network in which the timing of the exchange clocks is not linked.

Availability A measure of the delivery of the specified service as a fraction of a given time interval. Thus, it is the probability of delivery of the specified service.

Bit map One or more words in computer memory used so that each binary bit is a representation of some entity or condition. For example, when a particular bit is set to 1, a task represented by that bit is activated. An individual bit used in the way is referred to as a 'flag'.

BORSCHT This acronym describes the functions of terminating a subscriber's exchange line: battery, overvoltage, ringing, supervision, coding, hybrid and test.

Broadband Description of a telecommunications system that can support services with bandwidths greater than those used for telephony. In a digital network, broadband services are those operating at rates greater than 64 kbit/s.

Bus See *highway*.

Call-duration time The time for which the calling subscriber is charged, i.e., the time between the called subscriber answering and the first subscriber's *clear-down*.

Call-holding time The total time that equipment is held, including the time before the called subscriber answers.

Call record A record set up in memory for each call handled by an SPC exchange. Every event is recorded, so the call record may provide all types of information for many purposes.

CAS (Channel-associated signalling) A signalling method in which the signalling related to a traffic channel is carried in the channel itself or in a separate signalling channel permanently associated with it.

Centrex The provision, by the local exchange, of extension-to-extension communication within a business subscriber's private system. In addition to PABX-type facilities, centrex subscribers have normal telephone service for calls leaving or entering their private system.

CEPT (Conference Europeane de Postes et Telecommunications) The European standards body.

CCITT (Comite Consultatif International Telegraphique et Telephonique). The ITU body which sets standards for telephone and telegraph communications.

CCS (Common-channel signalling) A form of signalling system in which message-based signals are sent for a number of traffic channels via a separate common channel.

CHILL (CCITT high-level language) A computer language developed by the CCITT for programming SPC systems.

Class-of-service record A record stored for each line termination on an SPC exchange. In it is stored the information which defines the termination (its 'class-of-service' information).

Clear-down The process of disconnecting the temporarily established communication path between two users on completion of a call over a circuit-switched network.

Cluster A tightly coupled group of processors, working in the multiprocessor mode, each having an allocation of dedicated random-access memory, as well as access to common memory available to all the processors in the cluster.

Configuration management A formal engineering discipline which provides developers and users with the methods and tools to identify the software developed, establish criteria, control documentation or software changes against these criteria, record status and audit the product. It is the means by which the integrity and continuity of the product are recorded, communicated and controlled.

CM (Connection memory) The element within a digital time switch which stores an indication of the time slots that each speech sample should be read out of the speech memory.

Codec Equipment which provides both encoding and decoding at the termination of a digital system.

Companding A process of compression of the amplitude of signals according to frequency or magnitude, before transmission, and corresponding expansion after receipt, in order to reduce the effective noise.

CPE Customers' -premises equipment)

CVSD (Continuously variable-slope delta modulation). This digital encoding technique is a derivative of delta modulation in which the size of sampling is varied (i.e., variable-slope) according to the input signal magnitude.

DASS (Digital access signalling system) The UK interim standard for primary and basic rate ISDN access.

Database A collection of files stored as an integrated system. It is designed to achieve efficient updating and retrieval of data in specific applications, as well as data consistency, security and minimised redundancy.

Database, relational A database, modelled on a tabular file structure, in which the files are structured and linked together by pointers in accordance with the actual relationships between them.

Data structure The configuration, in computer storage, of data of a given type. The data structure is chosen to be appropriate for a given purpose: for example, speed of access, economy of storage or efficiency of processing.

dBm Power level in decibels relative to 1 mW.

dB (RE) The subjective assessment of loss of a telephone system relative to the CCITT NOSFER reference system, known as the reference equivalent (RE) value.

Delta modulation A digital encoding technique in which a single bit is used to code whether each PAM sample is bigger or smaller than the previous one.

Despotic (synchronised network) A network synchronising arrangement in which a unique master clock controls all other clocks in the network.

Dimensioning Determining and planning the appropriate amount of equipment.

DMS (Database-management system) A set of computer programs designed to carry out the operations on a database (e.g., updating and retrieving data) efficiently.

DPNSS (Digital private network signalling system) The UK standard for *common-channel signalling* between digital PABXs over leased lines.

Duplex Bidirectional transmission. The term 'half-duplex' is used to indicate that transmission may operate in only one direction at a time. 'Full duplex' therefore describes simultaneous both-way transmission.

EFS Error-free one-second intervals. A unit used in the quantification of error performance of data transmission.

Equipment, long-holding time Equipment which is connected into the circuit for the duration of a call.

Equipment, short-holding time Equipment which is connected into the circuit only while the call is being set up. It is then released and used for setting up another call.

Erlang The unit of telephone traffic. If a circuit is held continuously for the duration of a chosen monitoring period, that circuit has carried 1 erlang of traffic.

Fault avoidance Eliminating faults during design and production (by strict procedures, appropriate tools and techniques, testing, quality assurance, etc.) to try to ensure trouble-free operation.

Fault tolerance The ability to continue operation, with no effect on service, in the presence of one (or more) faults. It must be designed into the system and may be achieved by redundancy and automatic recovery procedures.

FDM (Frequency-division multiplex(ing)). In which channels are multiplexed by virtue of discreet allocations of frequency bandwidth.

GOS (Grade of service). The probability that a call will be successful.

HDB3 (High-density binary modulus 3) A line code used for PCM systems and as a standard CCITT interface for digital transmission systems.

Highway (or bus) A common path within an exchange or computer used to distribute signals from several channels, usually operating in TDM mode.

IDF (Intermediate distribution frame) A frame, on which cables are terminated, which provides a flexibility point between the cable and exchange equipment. It is used, when necessary, in addition to the main distribution frame.

IDN (Integrated digital network) A network in which all transmission and switching is in digital format.

In-band signals Where the *CAS* for a traffic circuit is carried within the 300–3400Hz speech bandwidth.

Indexing See *Key transformation*

Interpreter A program which translates commands, instruction by instruction, into the machine language of a computer, forming tasks which are then executed.

Interrupt A signal received by a computer's control unit which causes it to suspend its current operations and commence processing another task.

ISDN (Integrated services digital network) An integrated digital network, which supports a range of services, using a standard set of interfaces and protocols at the subscriber's premises using digital access to the local exchange.

ISUP ISDN user part. The part of the CCITT Signalling System No.7 that formats and conveys signalling messages for services on an ISDN.

ITU (International Telecommunications Union) The telecommunications arm of the United Nations and the parent body of the CCITT.

Jitter Short-term variations of the elements of a digital signal from their ideal positions in time.

Key A known item of information which is used as the basis of a search for other information stored in a table in computer memory.

Key transformation The translation of a search key into an address in computer memory in which the required data are stored. The key itself is not stored, but a relationship exists between the key and the location (see *scanning*).

LAN Local-area network A homogeneous network for communication between computers, usually within a radius of about 5 km and often within a single building. Transmission is by packets of data, or 'messages'.

Language, assembly A computer language of which the syntax comprises mnemonic symbols, and is thus at a higher level than the computer's machine language.

Language, high-level A computer language whose syntax consists of words of normal human language. It allows a programmer to function with a knowledge of the problem and the language, but not of the computer.

Language, machine A language, composed of binary-coded instructions, which is specific to a given model or range of models of computer. For a computer to function, all commands must be translated into its machine language.

Language, man-machine A language designed for communication between terminal users and computers. It is usually translated, instruction by instruction, into the computer's machine language by an interpreter.

Language, query A very high-level language by which a user may gain access to data, software or a process within a computer. For example, a database-management system includes a query language to allow users to use the database.

LAP (Link-access procedure) The data link-level protocol specified in the CCITT X25 recommendation. Additional LAPs have been specified. LAP-balanced (known at LAPB) and LAPD, the link-level protocol specified for ISDN connections.

Line signals The signals associated with the control of a circuit. This category includes circuit 'seize', 'backward busy' and 'clear' signals.

List, linked A data structure, consisting of a number of storage nodes, in which the nodes are linked in the form of a chain or list by pointers. The nodes, therefore, do not need to be physically adjacent in storage. The algorithms for entering data into nodes and extracting data from them treat the nodes as a list and follow the chain of pointers until the required node is found.

List, sequential A data structure, consisting of a number of storage nodes, in which the nodes are sequential in storage. The nodes are treated as a list by the algorithms which enter data into them and extract it from them.

LSI (Large-scale integration) Used to describe semiconductor chips with very high densities of components.

Maintenance centre A centre, remote from the equipment area, in which alarms, error messages and the results of diagnoses are displayed, and from which maintenance actions are initiated.

MDF (Main distribution frame) A terminating point, for cables entering an exchange, which provides flexibility between the circuits in the cables and the exchange equipment.

Memory, bubble Computer memory fabricated on to the surface of a wafer of garnet, in which bits are stored as tiny magnetic bubbles. It offers high-density storage for random access.

Memory, random-access See *RAM*

Memory, virtual The organisation of secondary storage by the operating system so that a programmer need not consider the limitations of the computer's primary storage, nor even be aware that secondary storage is necessary.

Mesochronous network A synchronous network which has the frequency of its exchange clocks controlled to operate at the same average frequency (but allowing short-term deviations within limits).

Modem Modulator/demodulator. A device that enables digital data signals to be sent over analogue transmission and switching systems.

MTP (Message transfer part) The part of the CCITT Signalling System No.7 that transfers signalling messages over the network and performs subsidiary functions, such as error control.

Mu-law The coding law used by the CCITT standard 24-channel PCM systems.

Multiplex A combination of many traffic channels on to a single transmission bearer.

Multiplexer The equipment which multiplexes input channels on to a single transmission bearer. It is usually used to describe terminal equipment of multiplexed systems, incorporating both the multiplexing of channels in the transmit direction and the demultiplexing of channels in the receive direction.

Multiplexing The process of combining several tributary channels into a single composite (i.e., 'multiplexed') channel for transmission over a common bearer.

Network management Monitoring network performance and taking action to maximise the utilisation of the network at all times and under all conditions. Actions may be 'expansive' (for example, providing temporary alternative routes) to cope with traffic overload, or 'restrictive' (for example, barring certain calls) to reduce the amount of traffic.

NT (Network termination) CCITT terminology for functional units on the customer's premises used for ISDN access. Two functional units are defined: NT1 and NT2

OMC (Operations and maintenance centre) A centre, remote from the exchange equipment, which is a maintenance centre and from which all changes to exchange data are made over data links from terminals.

Operating system A set of programs which carries out the internal functions and housekeeping of a computer. It forms the interface between the application programs and the computer hardware, and is part of the system software.

Out-band signals Where the channel-associated signalling for a traffic circuit is carried within the channel bandwidth but outside the speech band, i.e., between 3.4 kHz and 4kHz.

PABX (Private branch automatic exchange) Privately owned switching equipment used by a company of its premises to provide extension-to-extension connections and access to the PSTN.

PAM (Pulse-amplitude modulation) The process of representing an analogue signal by a series of discrete samples. The PAM process is analogue because the sample values are continuous, directly equating to the value of the input signal at the sampling instants.

PCM (Pulse-code modulation) A process that coverts continuous analogue signals into a series of binary words. It is the standard form of voice encoding for telecommunications networks.

Performance monitoring Analysing relevant data at regular intervals and comparing the derived statistics with expected values, in order to assess whether performance has remained within certain limits.

Phantom circuit A new circuit that is formed from existing circuits by the use of transformers at each end of the line. Each leg of the phantom circuit is derived from one physical pair.

Plesiochronous network As asynchronous network in which the exchange clocks operate at nominally the same frequency, but within closely defined limits.

Processing, parallel The concurrent processing of two or more tasks.

PSTN (Public switched telephone network) The dial-up telephone network.

Queue A means of storage in computer memory such that items are input at one end and extracted at the other. It is a first-in-first-out device.

RAM (Random-access memory) Digital memory which can be accessed directly and within which any location is accessed as easily and quickly as any other, it forms a computer's main memory or primary storage.

RE See *dB (RE)*.

Redundancy Providing more equipment of a given type than is necessary purely for functionality, in order to have a means of maintaining service in the event of a failure.

Reliability The probability of continuity of operation. It may be measured by the 'mean time between failures (MTBF)'.

Roll-back In an attempt to recover from an exchange fault, the system is reloaded with data which define the configuration of the system a short time before. Thus, the system is 'rolled back'. Repeated failure results in successive roll-backs, each going further back in time, until complete reinitialisation of the system is achieved.

ROM (Read-only memory) A store into which contents may be written once only and thereafter may only be read.

Scanning   A means of searching for information stored in computer storage, such that the search key, which is also stored, is compared with successive entries until a match is found.

SDL (Specification and description language)   A pictorial language developed by the CCITT for the specification and design of SPC software.

SDM (Space-division multiplex(ing)   A multiplexed system in which channels are allocated discrete paths in space. The term SDM or 'space division' is used to describe the types of switch block that provide separate physical paths or storage locations exclusively to each connection for the duration of its call.

Selection signals   The signals associated with the routeing of a call through one or more telephone exchanges. This category includes address signals as well as class-of-service signals.

Simplex   Unidirectional transmission.

SM (Speech memory)   The element within a digital time switch which stores the speech samples for the durations necessary to achieve the required time-slot changes.

Software, applications   Software produced to perform tasks which are specific to a given application.

Software build   The assembly of all the software (including data) necessary for a specific exchange to function correctly. This consists of the exchange-independent software and the exchange-dependent software.

Software engineering   The application of engineering and quality-control procedures, techniques and disciplines to software development, control and maintenance.

Software, exchange-dependent   The data which describe an exchange; for example, its equipment and connections and the class of service of each line termination.

Software, exchange-independent   The programs which provide the exchange functions and facilities. The programs are stored in a library and each is applicable to any exchange (of the same model) which requires the function or facility which it provides.

Software, system   Software produced to provide the facilities necessary to make a computer system operational and efficient.

Space switch   A switch that provides connections between different physical inputs and outputs (i.e., different points in space). A space switch may be analogue or digital and it may operate in SDM or TDM mode. A crossbar exchange is an example of an analogue SDM space switch. Digital exchanges have switch blocks which include digital TDM space switches.

Stack   A means of storage in computer memory such that items are inserted and removed from the same end. It is a last-in-first-out device (see *queue*).

Statistical multiplexing   The multiplexing of channels by the process of interleaving variable-sized packets from each active tributary. Unlike TDM, there is no permanent allocation of time slots to the tributaries; instead, the capacity of the system is allocated on a dynamic basis according to the amount of traffic at the time.

Storage, primary   The storage provided within the computer itself, consisting of directly-addressable, random-access memory.

Storage, secondary   Storage provided external to the computer, usually on magnetic disc. Also known as ancillary storage, bulk storage or external memory.

Synchronous or synchronised network   A network which has the frequency of its exchange clocks controlled to operate at the same rate.

TA (Terminal adaptor)   CCITT terminology for the unit that converts between a non-I series terminal and an I-series interface to an ISDN.

TDM (Time-division multiplex(ing)   A multiplexed system in which channels are allocated exclusive periodic elements of time (i.e., time slots).

TE (Terminal equipment)   CCITT terminology for users' terminals attached to an ISDN.

Telephone traffic   The occupancy of a circuit is the traffic carried by that circuit. The traffic intensity is the occupancy as a proportion of the total time, and the unit of traffic intensity is the erlang.

Time switch   A switch that provides connections between different time slots (i.e., different points in time). A time switch is always digital, but it may operate in TDM or SDM mode.

Translation table   A means of storage in computer memory, such that known information is used as a key in the search for other stored information. (The known information is, thus, translated into the required information.) The key may either be stored (see *scanning*) or not (see *key transformation*).

Transmulitplexer   A piece of equipment that transforms an FDM signal into a corresponding TDM signal that has the same structure as if it had been derived from PCM multiplexing equipment, and vice versa.

TUP (Telephone user part)   The part of the CCITT Signalling System No.7 that formats and conveys the signalling messages for telephone services.

Wander   Long-term variations of the elements of a digital signal from their ideal positions in time.

2B1Q   A line code in which pairs of bits are converted to one quaternary signal element.

4B3T   A line code in which four bits are converted to three ternary signal elements.

6B4T   A line code in which six bits are converted to four ternary signal elements.

Index